my **revision** notes

AQA Level 2

Certificate in Further
MATHEMATICS

Michael Ling

Series editor:
Roger Porkess

HODDER
EDUCATION
AN HACHETTE UK COMPANY

Hachette UK's policy is to use papers that are natural, renewable and recyclable products and made from wood grown in well-managed forests and other controlled sources. The logging and manufacturing processes are expected to conform to the environmental regulations of the country of origin.

Orders:
please contact Bookpoint Ltd, 130 Park Drive, Milton Park, Abingdon, Oxon OX14 4SE. Telephone: +44 (0)1235 827827.
Fax: +44 (0)1235 400401.
Email education@bookpoint.co.uk
Lines are open from 9 a.m. to 5 p.m., Monday to Saturday, with a 24-hour message answering service. You can also order through our website: www.hoddereducation.co.uk

ISBN: 978 1 5104 6078 2

First published in 2019 by
Hodder Education,
An Hachette UK Company
Carmelite House
50 Victoria Embankment
London EC4Y 0DZ
www.hoddereducation.co.uk
Impression number 10 9 8 7 6 5 4 3 2 1
Year   2023 2022 2021 2020 2019

Cover photo
Typeset in Bembo Std Regular 11/13 by Integra Software Services Pvt. Ltd., Pondicherry, India
Printed in India

A catalogue record for this title is available from the British Library.

Get the most from this book

Welcome to your Revision Guide for My Revision Notes: AQA Level 2 Certificate in Further Mathematics. This book will provide you with reminders of the knowledge and skills you will be expected to demonstrate in the exam with opportunities to check and practice those skills on exam-style questions. Additional hints and notes throughout help you to avoid common errors and provide a better understanding of what is needed in the exam.

Included with the purchase of this book is valuable online material that provides full worked solutions to all the 'Target your revision questions', 'Exam-style questions' and 'Review questions'. The online material is available at www.hoddereducation.co.uk/myrevisionnotesdownloads

Features to help you succeed

Target your revision

Use these questions at the start of each section to focus your revision on the topics you find tricky. Short answers are at the back of the book, but use the full worked solutions online to check each step in your solution.

About this topic

At the start of each chapter, this provides a concise overview of its content.

Before you start, remember ...

A summary of the key things you need to know before you start the chapter.

Key facts

Check you understand all the key facts in each subsection. These provide a useful checklist if you get stuck on a question.

Worked examples

Full worked examples show you what the examiner expects to see in order to ensure full marks in the exam. The examples cover a sample of the type of questions you can expect.

Hint

Expert tips are given throughout the book to help you do well in the exam.

Common mistakes

Your attention is drawn to typical mistakes students make, so you can avoid them.

Exam-style questions

For each topic, these provide typical questions you should expect to meet in the exam. Short answers are at the back of the book, and you can check your working using the online full worked solutions.

Review questions

After you have completed each section in the book, answer these questions for more practice. Short answers are at the back of the book, but the full worked solutions online allow you to check every line in your solution.

At the end of the book, you will find the following useful information:

Exam preparation

Includes hints and tips on revising for the AQA Level 2 Certificate in Further Mathematics exam, and information about the structure of the exam papers.

Formulae you need to learn

Provides a succinct list of all the formulae you need to remember and the formulae that will be given to you in the exam.

Please note that the formula sheet as provided by the exam board for the exam may be subject to change.

During your exam

Includes key words to watch out for, common mistakes to avoid and tips if you get stuck on a question.

My revision planner

REVISED   TESTED   EXAM READY

SECTION 1 ALGEBRA

Target your revision (Chapters 1–4)

Try answering each question below. If you get stuck, follow the page reference underneath to revise that topic.

1 Ratios and percentages in arithmetic and algebra
(i) Split 900 in the ratio 13 : 5.
(ii) Increase £15 by 20%.

(see page 4)

2 Manipulating fractions
Without using a calculator, work out $\frac{1}{2} + \frac{3}{4} \times \frac{5}{6}$.

(see page 4)

3 Manipulating algebraic terms
Simplify $2(m - 3n) - 4(m - 5n)$.

(see page 5)

4 Manipulating fractions in algebra
Work out $\frac{b}{3} - \frac{a^2}{2ab} \times \frac{b^2}{2a}$.

(see page 5)

5 Linear equations
Solve the equation $2x - 3 = 1 - 4x$.

(see page 7)

6 Ratios
$a : b = 3 : 2$.
Work out the ratio $a + 2b : b$.

(see page 8)

7 Percentages
John's salary of £p is increased by 5%. Jean's salary of £q is decreased by 6%.
The amount that John's salary is increased is the same as the amount that Jean's salary is decreased.
Find p as a percentage of q.

(see page 8)

8 Factorial notation
Work out $\frac{6!}{(3!)^2}$.

(see page 10)

9 Permutations
A group of 10 students take part in a race. In how many ways can 1st, 2nd and 3rd places be achieved?

(see page 10)

10 Combinations
How many ways are there of choosing 7 people from a group of 11?

(see page 10)

11 Expanding a bracket
Expand $2(x + y) + 3(2x - 3y)$.

(see page 12)

12 Multiplying out two brackets
Expand $(x + 2y)(2x - y)$.

(see page 12)

13 Multiplying out three brackets
Expand $(x + 2)(2x - 1)(x - 3)$.

(see page 12)

14 Pascal's triangle
Write down the 8th row of Pascal's triangle.

(see page 14)

15 Terms of the binomial expansion
Find the term in x^3 in the expansion of $(2 - x)^5$.

(see page 14)

16 Work out the term that is independent of x in the expansion of $\left(2x - \frac{3}{x}\right)^8$.

(see page 14)

17 Square roots
Simplify $\frac{\sqrt{8} \times \sqrt{96}}{\sqrt{6}}$.

(see page 15)

18 Manipulating algebraic expressions involving square roots
Simplify $\sqrt{50} + \sqrt{18}$.

(see page 15)

19 Rationalising the denominator of a fraction
Simplify $\frac{2 + \sqrt{2}}{4 + \sqrt{2}}$.

(see page 15)

20 Prime factors of a number
Write 140 as a product of its prime factors.

(see page 17)

21 Algebraic factors of a number
Factorise $6x^2 + 9xy$.

(see page 17)

22 Quadratic factorisation
Factorise the following.
(i) $x^2 - 8x + 15$ (ii) $2x^2 + x - 15$

(see page 17)

23 Difference of two squares
Factorise $4y^2 - 25$.

(see page 17)

24 Rearranging formulae

Make x the subject of the formula $y = \dfrac{x+1}{x-2}$.

(see page 19)

25 Make r the subject of the formula $V = \dfrac{4}{3}\pi r^3$.

(see page 19)

26 Simplifying algebraic fractions

Simplify $\dfrac{2}{x+1} - \dfrac{x}{x-1}$.

(see page 20)

27 Solving equations involving fractions

Solve the equation $\dfrac{x-3}{6} = \dfrac{2x+3}{5} - 6$.

(see page 20)

28 Completing the square

Write $x^2 - 12x + 31$ in the form $(x+a)^2 + b$ where a and b are to be determined.

(see page 22)

29 Maximum and minimum values

By completing the square, find the minimum value of $x^2 - 4x + 7$ and the value of x at which it occurs.

(see page 22)

30 The domain and range

You are given the function $f(x) = 2 - x^2$ over the domain $0 \leqslant x \leqslant 2$.

Work out the range of $f(x)$.

(see page 25)

31 Composite functions

You are given that $f(x) = 2 - x$ and $g(x) = x^2$. Work out $fg(x)$ and $gf(x)$.

(see page 25)

32 Inverse functions

You are given that $f(x) = \dfrac{2-x}{3}$. Work out $f^{-1}(4)$.

(see page 28)

33 Graphs of linear functions

(i) Work out the equation of the line that joins the points $(-1, 2)$ and $(3, 7)$.

(ii) Draw the line on a graph.

(see page 29)

34 Quadratic functions

A quadratic curve has its vertex at $(3, 4)$ and the curve passes through the point $(0, -5)$.

(i) Work out the equation of the curve.

(ii) Sketch the curve.

(see page 30)

35 Exponential functions

Find, by sketching their graphs, the points of intersection of the curve $y = 3 \times 2^x$ and the line $4y - 7x = 17$.

(see page 32)

36 A curve has equation $y = 5 \times 2^{-x}$.

(i) Sketch the curve.

(ii) Find, correct to 1 significant figure, the smallest value of x for which $5 \times 2^{-x} < 1$.

(see page 30)

37 Functions with more than one domain

Sketch the graph of $y = f(x)$ where

$f(x) = x + 2$ for $-2 \leqslant x < 0$

$f(x) = 2 + x - x^2$ for $0 \leqslant x < 2$

$f(x) = 0$ for $x \geqslant 2$.

(see page 34)

38 Sketching graphs

Sketch the graph of the curve $y = x^2 - x + 1$.

(see page 36)

39 Solving quadratic equations that can be factorised

Solve the equation $y^2 - 2y - 8 = 0$.

(see page 37)

40 Solving quadratic equations that cannot be factorised: completing the square

By first completing the square solve the equation $x^2 + 6x - 9 = 0$.

(see page 37)

41 Solving quadratic equations that cannot be factorised: using the formula

Solve the equation $x^2 - x - 8 = 0$.

(see page 37)

42 Checking for roots of a quadratic equation

Determine whether the equation $y^2 - 2y + 9 = 0$ has any real roots.

(see page 37)

43 Finding the maximum or minimum value of a quadratic function

By first completing the square, find the maximum value of the function $7 + 4x - x^2$.

(see page 36)

44 Disguised quadratic equations

By making a suitable substitution, solve the equation $x^4 - 5x^2 + 6 = 0$.

(see page 46)

45 Solving two simultaneous equations when both are linear

Solve the following equations simultaneously.

$x - y = 7$

$7x + 2y = 4$

(see page 40)

46 Solving two simultaneous equations when one is linear and the other quadratic

Solve the following equations simultaneously.

$y + 2x = 4$

$y = x^2 - 3x + 4$

(see page 40)

47 Solving three simultaneous equations in three unknowns

Solve the following equations simultaneously.

$3x - y + z = 4$

$2x + 3y - z = 4$

$4x + 2y + 3z = -3$

(see page 40)

48 Finding the quotient when a cubic function is divided by a linear function

Find the quotient and the remainder when $x^3 + 2x^2 - x + 5$ is divided by $x - 2$.

(see page 42)

49 Roots of cubic equations and the factor theorem

(i) Show that $(x - 1)$ is a factor of $f(x) = x^3 - 2x^2 - 5x + 6$.

(ii) Hence solve the equation $f(x) = 0$.

(see page 42)

50 Linear inequalities

(i) Solve the inequality $-3 \leqslant 2x - 1 < 5$.

(ii) Illustrate the solution on a number line.

(see page 44)

51 Quadratic inequalities

(i) Solve the inequality $x^2 - 7x + 12 < 0$.

(ii) Illustrate the solution on a graph.

(see page 44)

52 The index laws

Write $\sqrt{\dfrac{x^3 \times x^{-5}}{x^2}}$ as a single power of x.

(see page 46)

53 Proof

Prove that, for all integers $n > 1$, $(n + 2)^2 - (n - 2)^2$ is divisible by 8.

(see page 48)

54 Position to term sequences

The first four terms in a linear sequence are 2, 5, 8, 11. Find the first term in the sequence which exceeds 100.

(see page 49)

55 Term to term sequences

A sequence is such that $a_{n+1} = a_n + 2$ with $a_1 = 1$. Find a_{23}.

(see page 49)

56 Finding the rule for a sequence given its terms

The first three terms of a sequence are 2, 8, 16. Given that its position to term rule is $a_n = pn^2 + qn + r$, find the values of p, q and r.

(see page 49)

57 The limit of a sequence

In a sequence the nth term is given by $u_n = \dfrac{n}{n+1}$. Find the limiting value of this sequence as $n \to \infty$.

(see page 49)

Short answers on page 117–118

Full worked solutions online

CHECKED ANSWERS

Chapter 1A: GCSE-level revision for number and algebra

About this topic

This section reminds you of the basic skills in arithmetic and algebra needed for this course.

Before you start, remember ...

- ratios and percentages
- rounding techniques
- manipulation of algebraic terms.

1.1 Numbers and the number system

REVISED

Key facts

1 Ratios

Like fractions, ratios should be expressed in their simplest terms, e.g. $4 : 6 = 2 : 3$

A ratio has no units.

2 Percentages

A percentage is a hundredth part of the whole,

e.g. 32% of 500 is $\frac{32}{100} \times 500 = 160$.

3 Manipulating fractions

When adding or subtracting, find the LCM of the denominators.

When multiplying or dividing, look for common factors.

Common mistake: The LCM is often confused with the HCF.

The LCM of two numbers is the Lowest Common Multiple which means the smallest number into which both the numbers can be divided.

The Highest Common Factor (HCF) of two numbers is the largest number which divides into both the numbers.

For example, consider 12 and 18. The LCM is 36. The HCF is 6.

Worked examples

Ratios

1 Express the ratio 35 : 49 in its simplest form.

Solution

$35 : 49$

$= 5 : 7$

The two numbers, 35 and 49, have a common factor of 7. So, divide through by 7.

2 Split 350 into two parts in the ratio 3 : 7.

Common mistake: Take care to add the two parts together to get the total number of parts.

Solution

Smaller part is $\frac{3}{3+7}$ of total.

i.e. $\frac{3}{10} \times 350 = 105$

Larger part is $\frac{7}{3+7}$ of total.

i.e. $\frac{7}{10} \times 350 = 245$

So, the two parts are 105 and 245.

The smaller number is 3 parts and the larger number is 7 parts. So, there are 3 + 7 = 10 parts.

Check that the two parts sum to 350.

Worked example

Manipulating fractions

3 Work out $\frac{3}{7} \times \frac{5}{9} + \frac{2}{3}$.

Solution

$$\frac{3}{7} \times \frac{5}{9} + \frac{2}{3}$$

$$= \frac{5}{21} + \frac{2}{3}$$

$$= \frac{5}{21} + \frac{14}{21}$$

$$= \frac{19}{21}$$

> The top and bottom lines of $\frac{3}{7} \times \frac{5}{9}$ have a common factor of 3 so this is cancelled to give $\frac{5}{21}$.

> The LCM of 21 and 3 is 21.

Practice question (GCSE level)

TESTED

Adam, Beryl and Chloe have been given a gift of £31 500 which they are to split in proportion to their ages. Adam is 14, Beryl 15 and Chloe 16.

Work out how much each person receives.

Short answer on page 118

Full worked solution online

CHECKED ANSWER

1.2 Simplifying algebraic expressions

REVISED

Key facts

1 Collecting like terms
Like terms should be collected together,
e.g. $a + 2b + 3a - b = (a + 3a) + (2b - b) = 4a + b$

> 'Like terms' are those that can be added together or subtracted from each other.

2 Expanding brackets
Every term inside the bracket must be multiplied by the number or expression outside,
e.g. $2(x + 3) - 4(1 - 2x) = 2x + 6 - 4 + 8x$
$$= 10x + 2$$

> Take care with signs.

3 Multiplying fractions
Look for common factors which cancel. They may be numbers or letters or even expressions,
e.g. $\frac{2x^2}{y} \times \frac{y}{4x} = \frac{2x^2}{4x} = \frac{x}{2}$

> y can be cancelled.

> Then $2x$.

4 Adding fractions
As with arithmetic, you need to find a common denominator which is the lowest common multiple of the denominators (LCM).

> **Common mistake:** When removing brackets, consider the negative signs in order to avoid making an error. When subtracting a bracket in which there is a negative sign, apply the general principle 'two negatives make a positive'.

Worked example

Collecting terms and factorisation

1   Simplify $3xy - 2y^2 + xy - 4y^2$.

Solution

$3xy - 2y^2 + xy - 4y^2$

$= 4xy - 6y^2 = 2y(2x - 3y)$

> The terms in xy are 'like' terms, as are the terms in y^2.

Worked example

Multiplying out and collecting

2   Simplify $5(x-2) + 3(2-x)$.

Solution

$5(x - 2) + 3(2 - x)$

$= 5x - 10 + 6 - 3x$

$= 2x - 4$

> Expand brackets.

> Collect like terms.

Worked example

Multiplying fractions

3   Simplify $\dfrac{4x^2 y}{3z^3} \times \dfrac{6z^4}{5xy}$.

Solution

$\dfrac{4x^2 y}{3z^3} \times \dfrac{6z^4}{5 \cancel{x} y} = \dfrac{4x \cancel{y}}{3z^3} \times \dfrac{6z^4}{5 \cancel{y}}$

$= \dfrac{4x}{3z^{\cancel{3}}} \times \dfrac{6z^{\cancel{4}}}{5}$

$= \dfrac{4x}{\cancel{3}} \times \dfrac{\cancel{6}z}{5}$

$= \dfrac{4x}{1} \times \dfrac{2z}{5}$

$= \dfrac{8xz}{5}$

> Cancel y top and bottom.

> Cancel x top and bottom.

> Cancel z^3 top and bottom.

> Cancel 3 top and bottom.

> **Hint:** Cancelling can be in any order or all at once!

Worked example

Adding fractions

4 Simplify $\dfrac{1}{x-1}+\dfrac{2}{x-2}$.

Solution

$= \dfrac{x-2}{(x-1)(x-2)}+\dfrac{2(x-1)}{(x-1)(x-2)}$

$= \dfrac{x-2+2(x-1)}{(x-1)(x-2)}$

$= \dfrac{3x-4}{(x-1)(x-2)}$

> The LCM is the product of the denominators i.e. $(x-1)(x-2)$.

Practice question (GCSE level)

TESTED

Simplify fully the following.

$$\dfrac{2x^2-4xy}{2y^3-xy^2}$$

Short answer on page 118

Full worked solution online

CHECKED ANSWER

1.3 Solving linear equations

REVISED

Key facts

1 **A linear equation is an equation in an unknown (often denoted x) whose power is one**

There is only one value of x that satisfies this equation.

2 **Steps for solving linear equations:**
  - clear fractions
  - multiply out brackets
  - collect all terms in the unknown on one side and numbers on the other.

> It may be that not all these operations will be necessary.

Worked examples

Solving linear equations

1 Solve the equation $2(x+1)+3(2x+1)=21$.

Solution

$2(x+1)+3(2x+1)=21$

$\Rightarrow \quad 2x+2+6x+3=21$

$\Rightarrow \qquad\qquad 8x=16$

$\Rightarrow \qquad\qquad\quad x=2$

> Multiply out the brackets.

> Collect terms in x on the left and numbers on the right.

2 Solve the equation $\frac{x}{2} - 3 = 1 - \frac{x}{3}$.

Solution

$$\frac{x}{2} - 3 = 1 - \frac{x}{3}$$

$$\Rightarrow \frac{x}{2} \times 6 - 3 \times 6 = 1 \times 6 - \frac{x}{3} \times 6$$

$$\Rightarrow \frac{6x}{2} - 18 = 6 - \frac{6x}{3}$$

$$\Rightarrow 3x - 18 = 6 - 2x$$

$$\Rightarrow 3x + 2x = 18 + 6$$

$$\Rightarrow 5x = 24$$

$$\Rightarrow x = \frac{24}{5}$$

> Clear fractions by multiplying throughout by the LCM of denominators, in this case 6.

> **Common mistake:** Every term (in this example there are four terms) must be multiplied by the same number, in this case 6.

> Collect terms in x on the left and numbers on the right.

Practice question (GCSE level)

TESTED

Solve the equation $\frac{2(x-1)}{3} - x = \frac{1}{4}$.

Short answer on page 118

Full worked solution online

CHECKED ANSWER

1.4 Algebra and number

REVISED

Key facts

Use of number skills

Working with algebra involves essentially the same skills as you would use with numbers but using letters.

1 **Percentages**

A percentage is a fraction with denominator 100.

A quantity, known or unknown, can be increased or decreased by a given percentage.

2 **Ratios**

When a number is split into two parts in the ratio $a : b$ they can be found by multiplying it by the fractions $\frac{a}{a+b}$ and $\frac{b}{a+b}$.

Worked example

Percentages

1 John weighed x kg when he went on holiday. While on holiday his weight increased by 10%.
Write down an expression for his weight on return from his holiday.

Solution

Weight at the end of the holiday $= x + \dfrac{10}{100}x$

$$= \dfrac{110x}{100}$$

$$= 1.1x$$

So John's weight on return from his holiday was $1.1x$ kg.

> After the holiday John's total weight is the original, x kg, plus the increase which is 10% of x.

Worked example

Ratios

2 Given that $a : b = 2 : 3$ and $b : c = 4 : 7$, work out the ratio $a : c$.

Solution

$a : b = 2 : 3 = 8 : 12$

$b : c = 4 : 7 = 12 : 21$

So $a : b : c = 8 : 12 : 21$

$\Rightarrow \quad a : c = 8 : 21$

> Multiply both by a number that makes the b value the same, i.e. 4 and 3 respectively.

Practice question (GCSE level)

TESTED ☐

Paul has a bag of marbles which are coloured red or blue.

Initially the ratio red : blue is 4 : 3.

Paul withdraws 5 marbles; 2 of these are red and 3 are blue.

The ratio red : blue is now 3 : 2.

Work out how many blue and how many red marbles there were initially.

Short answer on page 118

Full worked solution online

CHECKED ANSWER ☐

Chapter 1B: Further revision for number and algebra

About this topic

When you are arranging a number of objects in a certain way there are two cases to consider: when order matters and when order does not matter.

These are called respectively permutations and combinations.

Before you start, remember ...

- expanding brackets
- manipulation of algebraic terms.

1.5 The product rule for counting

Key facts

1 Factorial notation

$1 \times 2 \times 3 \times ... \times n = n!$

This is called 'n factorial'.

$(m+1) \times (m+2) \times ... \times n = \dfrac{1 \times 2 \times 3 \times ... \times m \times (m+1) \times ... \times n}{1 \times 2 \times 3 \times ... \times m}$

$= \dfrac{n!}{m!}$

> Multiply and divide by all numbers up to and including m.

2 Permutations

A selection of objects in which the order is important is called a permutation.

> For example, the number of ways that 1st, 2nd and 3rd place medals can be awarded to the first 3 runners in a race with 7 competitors is $7 \times 6 \times 5 = 210$.

3 Combinations

A selection of objects in which the order is not important is called a combination.

> For example, 3 out of 7 runners will qualify for the next round of a competition. The number of ways that this can happen is $\dfrac{7 \times 6 \times 5}{3!} = 35$ since there are 3! ways in which each combination of 3 runners can appear in the list of permutations.

Worked example

Factorials

1 Work out $\dfrac{8!}{3! \times 5!}$.

Solution

$\dfrac{8!}{3! \times 5!} = \dfrac{8 \times 7 \times 6 \times \cancel{5 \times 4 \times 3 \times 2 \times 1}}{3 \times 2 \times 1 \times \cancel{5 \times 4 \times 3 \times 2 \times 1}}$

$= \dfrac{8 \times 7 \times \cancel{6}}{\cancel{3 \times 2} \times 1}$

$= 8 \times 7$

$= 56$

Worked example

Permutations

2   From a group of 40 people, the roles of chairman, secretary and treasurer are to be selected.
In how many ways can this be done?

Solution

The chairman can be chosen in 40 ways. There are then 39 choices for secretary and 38 for treasurer.

So the number of ways is $40 \times 39 \times 38 = 59280$.

> The three roles being filled are all different, so this is a permutation question.

Worked example

Combinations

3   From a group of 40 people in a club, 3 are to be selected to represent the group at a meeting.
In how many ways can this be done?

Solution

There are $40 \times 39 \times 38 = 59280$ ways of selecting 3 people in order.

However, in this case order does not matter.

So the 59280 ways will include sets of the same 3 people repeated.

For every 3 people there will be 3! ways.

So the number of selections is

$$\frac{40 \times 39 \times 38}{1 \times 2 \times 3} = \frac{59280}{6} = 9880.$$

> In this example order does not matter so it is a combination.

> **Common mistake:** It is important that you distinguish between a combination and a permutation. You should ask yourself 'Does order matter?'.

Exam-style question

TESTED

A group of six friends go to the theatre.

(i)   How many ways are there for them to sit in a row of six seats in the theatre with no restrictions?
(ii) Mia sits in the seat at the left-hand end of the row. How many ways are there for the remaining five to seat themselves?
(iii) Additionally, Jane and Julian must sit next to each other. Now how many ways are there for the group to sit?

Short answers on page 118

Full worked solution online

CHECKED ANSWERS

1.6 Expanding brackets

REVISED ☐

Key facts

1 Expanding a bracket

Every term inside the bracket is multiplied by the term or expression outside it,

e.g. $2(x + 1) = 2x + 2$

e.g. $5a(2a + 3b) = 10a^2 + 15ab$

2 Multiplying out two brackets

Every term in the second bracket is multiplied by every term in the first bracket. 'Like terms' should then be collected.

3 Multiplying out three brackets

Multiply out two brackets first and then multiply out the third bracket.

> You have met this topic in Section 1.2.

> The principle is the same for more than three brackets, but you will not be asked to multiply out more than three brackets in this course.

Worked example

Expanding a bracket

1 Work out $2(2x - 3) - 3(x - 1)$.

Solution

$2(2x - 3) - 3(x - 1)$

$= (4x - 6) - (3x - 3)$

$= 4x - 6 - 3x + 3$

$= x - 3$

> The two terms in the first bracket are multiplied by 2 and the two terms in the second bracket are multiplied by 3.

> Remove brackets. Take care of the negative signs. Collect like terms together.

Worked example

Multiplying out two brackets

2 Work out $(2x + 3)(x - 2)$.

Solution

$(2x + 3)(x - 2) = 2x(x - 2) + 3(x - 2)$

$= 2x^2 - 4x + 3x - 6$

$= 2x^2 - x - 6$

> Multiply the second bracket by $2x$ and then by 3. Multiply out the brackets. Take care over signs.

> Collect like terms.

Worked example

Multiplying out three brackets

3 Work out $(2x+1)(x-2)(3x-2)$.

Solution

$(2x+1)(x-2)(3x-2)$

$= (2x+1)(x(3x-2)-2(3x-2))$

$= (2x+1)((3x^2-2x)-(6x-4))$

$= (2x+1)(3x^2-2x-6x+4)$

$= (2x+1)(3x^2-8x+4)$

$= 2x(3x^2-8x+4)+1(3x^2-8x+4)$

$= 6x^3-16x^2+8x+3x^2-8x+4$

$= 6x^3-13x^2+4$

> Multiply out the second and third brackets in the usual way to give a quadratic.

> Collect like terms.

> Now multiply out the resulting two brackets. The principle is the same even though there are now three terms in the second bracket.

> Collect like terms as before. The terms in x cancel in this example.

Exam-style question

TESTED ☐

A step ABCDEFGHIJKL has a shape as shown with the base ABCD a horizontal rectangle. All other faces other than the side faces ABEHIL and DCFGJK are rectangular.

The lengths of the edges are given in the diagram and units are cm.

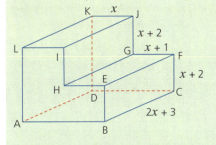

(i) Find expressions for
  (a) the volume of the shape
  (b) the total surface area of the shape.
(ii) Find the volume and the surface area when $x = 1$.

Short answers on page 118

Full worked solution online

CHECKED ANSWERS ☐

1.7 Binomial expansions – using Pascal's triangle

Key facts

1 Definition

The binomial expansion concerns the expansion of $(a + b)^n$ where n is a positive integer.

2 Terms of the binomial expansion

There are $n + 1$ terms in the expansion of $(a + b)^n$.

Each term is of the form $Ca^{n-r}b^r$ where $r = 0$ to n.

The number C is called the coefficient of the term. The value of C is found in Pascal's triangle.

3 Pascal's triangle

$$
\begin{array}{ll}
(a+b)^0 & \qquad\qquad\quad 1 \\
(a+b)^1 & \qquad\qquad 1 \quad 1 \\
(a+b)^2 & \qquad\quad 1 \quad 2 \quad 1 \\
(a+b)^3 & \qquad 1 \quad 3 \quad 3 \quad 1 \\
(a+b)^4 & \quad 1 \quad 4 \quad 6 \quad 4 \quad 1 \\
(a+b)^5 & 1 \quad 5 \quad 10 \quad 10 \quad 5 \quad 1
\end{array}
$$

> Note how this array is constructed: it can be extended as far as required.

> For larger values of n using Pascal's triangle becomes cumbersome. There is also a formula for finding binomial coefficients in such cases, but it is beyond the scope of this course.

4 Calculation of each row

Each row can be calculated from the one before it as shown.

i.e. $5 = 1 + 4$, $10 = 4 + 6$, etc.

$$
\begin{array}{cccccc}
1 & 4 & 6 & 4 & 1 \\
1 & 5 & 10 & 10 & 5 & 1
\end{array}
$$

> By this means the triangle can be extended as far as required. However, since each line requires the one before it, to find any given line the whole of the triangle before it is required.

Worked examples

Binomial expansion

1 The binomial coefficients for power 4 are: 1, 4, 6, 4, 1.
Expand $(2 + x)^4$, simplifying all the terms.

Solution

Applying the binomial expansion and including the coefficients gives

$(2 + x)^4 = 1 \times 2^4 + 4 \times 2^3 \times x + 6 \times 2^2 \times x^2 + 4 \times 2^1 \times x^3 + 1 \times x^4$

$\qquad\qquad = 16 + 32x + 24x^2 + 8x^3 + x^4$

> There are 5 terms with increasing powers of x and decreasing powers of 2.
>
> The coefficients are from Pascal's triangle.

2 Expand $(2 - 3x)^3$, simplifying all the terms.

Solution

The binomial coefficients for power 3 are: 1, 3, 3, 1.

Applying the binomial expansion gives

$(2 - 3x)^3 = 1 \times 2^3 + 3 \times 2^2 \times (-3x)^1 + 3 \times 2^1 \times (-3x)^2 + 1 \times (-3x)^3$

$\qquad\qquad = 8 - 36x + 54x^2 - 27x^3$

> There are 4 terms with increasing powers of $(-3x)$ and decreasing powers of 2.
>
> The coefficients are from Pascal's triangle.

> Note that the second term in the expression to be expanded is $(-3x)$ and so it is this that is raised to a power. Particular care must be taken over the negative signs.

Common mistake: The whole of the 2nd term must be raised to a power. Errors in failing to deal with the negative sign and the power of $3x$ can be eliminated by using brackets to write it as $(-3x)$, as shown in this example.

Exam-style question

TESTED

Expand $(3x - 4y)^5$, simplifying all the terms.

Short answer on page 118

Full worked solution online

CHECKED ANSWER

1.8 Manipulating surds

REVISED

Key facts

1 Irrational numbers

A number which cannot be expressed as a fraction (or integer) is described as irrational.

For example, $5\sqrt{2}$, 2π and $\dfrac{3+\sqrt{5}}{2}$ are irrational numbers.

2 Square roots

A square root of an integer is either itself an integer (e.g. $\sqrt{16} = 4$) or it is irrational (e.g. $\sqrt{5}$).

Square roots can sometimes be simplified.

For example, $\sqrt{20} = \sqrt{4 \times 5} = \sqrt{4}\sqrt{5} = 2\sqrt{5}$.

> If this can be done then you should usually do so.

3 Rationalising the denominator when it is a square root

In the fraction $\dfrac{2}{\sqrt{3}}$ the denominator is irrational. By multiplying top and bottom of the fraction by $\sqrt{3}$ the denominator becomes rational since $\dfrac{2}{\sqrt{3}} = \dfrac{2 \times \sqrt{3}}{\sqrt{3} \times \sqrt{3}} = \dfrac{2\sqrt{3}}{3}$.

4 Rationalising the denominator when it contains a square root

In the fraction $\dfrac{3}{2+\sqrt{3}}$ the denominator is irrational. By multiplying top and bottom of the fraction by $2 - \sqrt{3}$ the denominator becomes rational since $(2 - \sqrt{3})(2 + \sqrt{3}) = 4 + 2\sqrt{3} - 2\sqrt{3} - \sqrt{3}\sqrt{3}$.
$$= 4 - 3 = 1$$

> This process is called 'rationalising the denominator'.

> **Hint:** Note the connection with the difference of squares.

Worked examples

Manipulating numeric expressions involving square roots

Square roots are treated in exactly the same way as ordinary numbers or as letters in algebra.

1 Find $(\sqrt{3} - \sqrt{2})(\sqrt{3} + \sqrt{2})$.

Solution

$(\sqrt{3} - \sqrt{2})(\sqrt{3} + \sqrt{2})$
$= \sqrt{3} \times \sqrt{3} + \sqrt{3} \times \sqrt{2} - \sqrt{2} \times \sqrt{3} - \sqrt{2} \times \sqrt{2}$
$= 3 + \sqrt{6} - \sqrt{6} - 2 = 1$

> Multiply out in the usual way $\sqrt{3} \times \sqrt{3} = 3$.

2 Simplify $\sqrt{54} - \sqrt{24}$.

Solution

$\sqrt{54} - \sqrt{24} = \sqrt{9 \times 6} - \sqrt{4 \times 6}$
$= 3\sqrt{6} - 2\sqrt{6} = \sqrt{6}$

> **Hint:** Sometimes you will be asked for an approximate answer (usually to 3 significant figures) and at other times you will be asked for an exact value. If you are asked for an exact answer leave any square root signs in and do not use your calculator to find their values.

> Note that $\sqrt{54}$ and $\sqrt{24}$ are not in their simplest form. When you extract the factors that are perfect squares they become $3\sqrt{6}$ and $2\sqrt{6}$ and so are like terms.

Worked example

Manipulating algebraic expressions involving square roots

Remember that algebra is simply 'generalised arithmetic'.

3 Simplify $\sqrt{8a^3} + \sqrt{32a^5}$.

Solution

$$\sqrt{8a^3} + \sqrt{32a^5}$$
$$= \sqrt{4a^2 \times 2a} + \sqrt{16a^4 \times 2a}$$
$$= 2a\sqrt{2a} + 4a^2\sqrt{2a}$$
$$= 2a(1 + 2a)\sqrt{2a}$$

Extract from the square root factors which are perfect squares. These are, in this example, 4 and 16 and also a^2 and a^4.

Worked example

Rationalising the denominator of a fraction

Simple fractions will terminate (e.g. $\frac{3}{5}$) or recur (e.g. $\frac{3}{7}$) when turned into decimals but this is not true for irrational numbers such as π and some square roots. Their decimals go on for ever and have no recurring pattern.

'Rationalising' means making the denominator a rational number; it may be an integer.

4 Simplify $\frac{5}{5 - \sqrt{2}}$.

Solution

$$\frac{5}{5 - \sqrt{2}} = \frac{5}{5 - \sqrt{2}} \times \frac{5 + \sqrt{2}}{5 + \sqrt{2}}$$
$$= \frac{25 + 5\sqrt{2}}{25 - 2}$$
$$= \frac{5}{23}(5 + \sqrt{2})$$

When leaving an arithmetic or algebraic term as an exact answer it is usual to express the term in the form $a\sqrt{b}$ or $a + b\sqrt{c}$ where a, b and c are rational numbers. Terms such as $\frac{a}{\sqrt{2}}$ or $\frac{1}{1 + \sqrt{a}}$ are not in this form so they need to be manipulated so that they are. The process usually involves multiplying the top and bottom of such a fraction by a number that makes the denominator rational.

Hint: Note the form of the expression you use to rationalise the denominator.

Exam-style question

TESTED ☐

Simplify the following.

$$\frac{2}{5 + \sqrt{3}} + \frac{3}{5 - \sqrt{3}}$$

Short answer on page 118

Full worked solution online

CHECKED ANSWER ☐

Chapter 2: Algebra II

About this topic

This chapter extends the basic algebra you met in Chapter 1.

Before you start, remember ...

- prime factors
- LCM.

2.1 Factorising

REVISED

Key facts

1 Prime factors of a number

When the prime factors of a number are multiplied together you get the number,

e.g. 2, 3 and 5 are prime factors of 30. $2 \times 3 \times 5 = 30$.

> Sometimes a prime factor is repeated, e.g. $2 \times 2 \times 5 = 20$.

2 Algebraic factors

In algebra the factors multiply together to give an algebraic expression,

e.g. $x(y + 2) = xy + 2x$

3 Factorisation

Factorisation is the process of splitting a number or expression into its (usually prime) factors,

e.g. $42 = 2 \times 3 \times 7$

e.g. $3x + 6y = 3(x + 2y)$

e.g. $3x + xy = x(3 + y)$

4 Quadratic factorisation

Quadratic factorisation is the process of splitting a quadratic expression into its factors,

e.g. $x^2 + 5x + 6 = (x + 3)(x + 2)$

e.g. $2x^2 + 7x + 6 = (2x + 3)(x + 2)$

5 Difference of two squares

For all a, b $a^2 - b^2 = (a - b)(a + b)$.

Worked examples

Factorising expressions with a number of terms

1 Factorise fully $2axy + 2by - abx - b^2$.

Solution

$2axy + 2by - abx - b^2$

$= 2axy - abx + 2by - b^2$

$= ax(2y - b) + b(2y - b)$

$= (ax + b)(2y - b)$

> Pair the terms with common factors.

> Extract the common factors. The expressions within the brackets are the same.

> So, extract this as a common factor.

Worked examples

Factorising quadratic expressions of the form $x^2 + bx + c$

2 Factorise $x^2 + 12x + 35$.

Solution

$x^2 + 12x + 35$
$= x^2 + 7x + 5x + 35$
$= x(x + 7) + 5(x + 7)$
$= (x + 5)(x + 7)$

> The two numbers that add to give 12 and multiply to give 35 are 5 and 7.

> Factorisation can often be completed in one step by inspection.

3 Factorise $x^2 + x - 12$.

Solution

$x^2 + x - 12$
$= x^2 + 4x - 3x - 12$
$= x(x + 4) - 3(x + 4)$
$= (x - 3)(x + 4)$

> In this case, since the constant term is negative you need to find two numbers whose difference is 1 and product is −12. One of the numbers is negative. For this expression the numbers are 4 and −3.

4 Factorise $x^2 - 10x + 21$.

Solution

$x^2 - 10x + 21$
$= x^2 - 7x - 3x + 21$
$= x(x - 7) - 3(x - 7)$
$= (x - 7)(x - 3)$

> In this case the x coefficient is negative and the constant term is positive. So, both numbers are negative.
>
> For this expression the numbers are −3 and −7.

5 Factorise $x^2 - 9xy - 10y^2$.

Solution

$x^2 - 9xy - 10y^2$
$= x^2 - 10xy + xy - 10y^2$
$= x(x - 10y) + y(x - 10y)$
$= (x - 10y)(x + y)$

> In this expression the sum of powers in each term is 2. The technique for factorisation is the same as above but with y included.

> The two numbers in this expression are −10 and 1.

Worked example

Factorising quadratic expressions of the form $ax^2 + bx + c$

6 Factorise $2x^2 + 11x + 12$.

Solution

$2x^2 + 11x + 12$
$= 2x^2 + 8x + 3x + 12$
$= 2x(x + 4) + 3(x + 4)$
$= (2x + 3)(x + 4)$

> The two numbers must add to give 11 and multiply together to give $2 \times 12 = 24$.
>
> The two numbers are 8 and 3.

> Dealing with negative signs is the same as if $a = 1$.

> **Worked example**
>
> **Difference of squares**
>
> **7** Factorise $4a^2 - 9b^2$.
>
> **Solution**
>
> $$4a^2 - 9b^2 = (2a)^2 - (3b)^2$$
> $$= (2a - 3b)(2a + 3b)$$

Each term needs to be a perfect square.

Exam-style question

TESTED

Factorise $6x^2 + xy - 15y^2$.

Short answer on page 119

Full worked solution online

CHECKED ANSWER

2.2 Rearranging formulae

REVISED

> **Key fact**
>
> **1** A formula usually gives one variable (the subject) in terms of others, for example $p = 2(x + y)$. Rearranging a formula involves manipulating it so that a different variable is the subject.

> **Worked examples**
>
> **Rearranging formulae**
>
> **1** Make n the subject of the formula $m = 2n - 7p$.
>
> **Solution**
>
> $$m = 2n - 7p$$
> $$\Rightarrow 2n = m + 7p$$
> $$\Rightarrow \quad n = \frac{m + 7p}{2}$$

Isolate the term involving n on one side and the other terms on the other side.

> **2** Make t the subject of the formula $v = u + at$.
>
> **Solution**
>
> $$v = u + at$$
> $$\Rightarrow at = v - u$$
> $$\Rightarrow \quad t = \frac{v - u}{a}$$

Isolate the term involving t on one side and the other terms on the other side.

3 Make R the subject of the formula $A = P\left(1 + \dfrac{Rn}{100}\right)$.

Solution

$A = P\left(1 + \dfrac{Rn}{100}\right)$

$\Rightarrow \quad \dfrac{A}{P} = 1 + \dfrac{Rn}{100}$ Divide by P.

$\Rightarrow \quad \dfrac{Rn}{100} = \dfrac{A}{P} - 1$ Isolate the term involving R on one side and the other terms on the other side.

$\Rightarrow \quad R = \dfrac{100}{n}\left(\dfrac{A}{P} - 1\right)$ Multiply by 100 and divide by n.

4 Make r the subject of the formula $V = \pi r^2 h$.

Solution

$V = \pi r^2 h$

$\Rightarrow r^2 = \dfrac{V}{\pi h}$ Divide by πh.

$\Rightarrow \quad r = \pm\sqrt{\dfrac{V}{\pi h}}$ Take the square root.

Hint: Note that this formula is for the volume of a cylinder with circular cross section, radius r and height h. If the formula is set in context then you could not have a negative volume so the ± would be omitted.

Exam-style question

TESTED ☐

Temperature can be measured using two different scales: Celsius, °C, and Fahrenheit, °F.

The formula that connects them is

$F = \dfrac{9}{5}C + 32$

Rearrange this formula so that C is the subject.

Short answer on page 119

Full worked solution online

CHECKED ANSWER ☐

2.3 Simplifying algebraic fractions and solving equations

REVISED ☐

Key facts

1 Simplifying algebraic fractions

Fractions in algebra obey the same rules as fractions in arithmetic.

Common factors in the top line (numerator) and the bottom line (denominator) should be cancelled.

Two fractions to be added should be adjusted so that they have a common denominator.

2 Solving equations involving fractions

Sometimes you will need to solve equations involving fractions. This requires the methods you have met for solving equations and those for simplifying fractions.

Worked examples

Simplifying algebraic fractions

1 Simplify $\dfrac{2x}{3y} \times \dfrac{9y^2}{4x}$.

Solution

$$\dfrac{\cancel{2}x}{\cancel{3}y} \times \dfrac{\overset{3}{\cancel{9}}y^2}{\underset{2}{\cancel{4}}x}$$

> As in arithmetic there are common factors of 2 and 3 top and bottom which are cancelled.

$$= \dfrac{\cancel{x}}{\cancel{y}} \times \dfrac{3y^{\cancel{\kern0.4em}}}{2\cancel{x}}$$

> There are also common factors of x and y top and bottom which should be cancelled.

$$= \dfrac{1}{1} \times \dfrac{3y}{2} = \dfrac{3y}{2}$$

2 Simplify $\dfrac{1}{x-1} - \dfrac{1}{x+1}$.

Solution

$$\dfrac{1}{x-1} - \dfrac{1}{x+1}$$

> The denominators have no common factors so the LCM is the product.

$$= \dfrac{x+1}{(x-1)(x+1)} - \dfrac{x-1}{(x-1)(x+1)}$$

> Combine the fractions.
> Note that the
> LCM $= (x-1)(x+1) = x^2 - 1$.
> Take care with signs.

$$= \dfrac{(x+1) - (x-1)}{x^2 - 1}$$

$$= \dfrac{2}{x^2 - 1}$$

Worked example

Solving equations involving fractions

3 Solve the equation $\dfrac{2x+1}{3} + \dfrac{4x-3}{4} = 1$.

Solution

$$\dfrac{2x+1}{3} + \dfrac{4x-3}{4} = 1$$

> The LCM is 12, so multiply throughout by 12. (Don't forget to also multiply the 1 on the right-hand side!)

$$\Rightarrow \dfrac{12(2x+1)}{3} + \dfrac{12(4x-3)}{4} = 12$$

> Expand the brackets.

$$\Rightarrow 4(2x+1) + 3(4x-3) = 12$$
$$\Rightarrow 8x + 4 + 12x - 9 = 12$$
$$\Rightarrow 20x - 5 = 12$$

> Collect like terms and complete the solution.

$$\Rightarrow 20x = 17$$
$$\Rightarrow x = \dfrac{17}{20}$$

Exam-style question

TESTED ☐

Solve the equation $2x + \dfrac{1-x}{3} = 1 + \dfrac{2(x+2)}{5}$.

Short answer on page 119

Full worked solution online

CHECKED ANSWER ☐

2.4 Completing the square

Key facts

1 Completing the square

A quadratic expression $x^2 + bx + c$ can always be rewritten in the form $(x + p)^2 + q$. ◄───

> This is a special case of the next fact, where $a = 1$.

A quadratic expression $ax^2 + bx + c$ where $a \neq 1$ can always be rewritten in the form $a(x + p)^2 + q$.

2 Maximum and minimum values ◄

> This is one example where this rearrangement is useful – other examples will be met later in the course.

If $f(x) = a(x + p)^2 + q$ then $f(x)$ has a turning point when $x = -p$.

If $a > 0$ then $f(-p) = q$ is a minimum value.

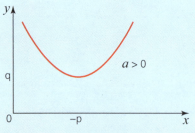

If $a < 0$ then $f(-p) = q$ is a maximum value.

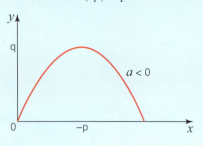

Worked examples

Completing the square

1 Express the quadratic function $x^2 + 6x + 7$ in the form $(x + p)^2 + q$ where p and q are numbers to be determined.

> **Hint:** 'Numbers to be determined' means that you are required to calculate the values and state what they are.

Solution

When $(x + p)^2 + q$ is expanded, the x term is $2px$.

$2p = 6$ so $p = 3$

$x^2 + 6x + 7$

$= (x + 3)^2 - 9 + 7$ ◄───

> Write $x^2 + 6x$ as $(x + 3)^2 - 9$.

$= (x + 3)^2 - 2$

$p = 3$ and $q = -2$ ◄───

> The actual values should be given at the end.

2 Express the quadratic function $x^2 - 7x + 4$ in the form $(x - p)^2 + q$ where p and q are numbers to be determined.

Solution

$x^2 - 7x + 4$

$= \left(x - \dfrac{7}{2}\right)^2 - \left(\dfrac{7}{2}\right)^2 + 4$ ← Write $x^2 - 7x$ as $\left(x - \dfrac{7}{2}\right)^2 - \left(\dfrac{7}{2}\right)^2$.

$= \left(x - \dfrac{7}{2}\right)^2 + 4 - \dfrac{49}{4}$ ← Simplify the constant term.

$= \left(x - \dfrac{7}{2}\right)^2 - \dfrac{33}{4}$

$p = \dfrac{7}{2}, q = -\dfrac{33}{4}$ ← Give the values you have found. Note the signs.

3 Express the quadratic function $2x^2 - 8x + 7$ in the form $a(x - p)^2 + q$ where a, p and q are numbers to be determined.

Solution

$2x^2 - 8x + 7$

$= 2\left(x^2 - 4x + \dfrac{7}{2}\right)$ ← Take out 2 as a factor which means that $a = 2$.

$= 2\left((x - 2)^2 - 4 + \dfrac{7}{2}\right)$ ← Inside the bracket write $x^2 - 4x$ as $(x - 2)^2 - 4$.

$= 2\left((x - 2)^2 - \dfrac{1}{2}\right)$ ← Expand the bracket to find the value of q.

$= 2(x - 2)^2 - 1$

So $a = 2, p = 2, q = -1$.

4 Express the quadratic function $2 + 4x - x^2$ in the form $q - (x - p)^2$ where p and q are numbers to be determined.

Solution

$2 + 4x - x^2$ ← You must be careful of the negative signs in such an example.

$= 2 - (x^2 - 4x)$ ← Rewrite using brackets, inside which the x^2 term is positive.

$= 2 - (x^2 - 4x + 4 - 4)$ ← Add 4 inside the bracket to make a squared expression, but remember to take it away as well!

$= 6 - (x^2 - 4x + 4)$ ← Take the extra 4 outside the bracket and add to the 2. Remember the signs!

$= 6 - (x - 2)^2$

So $q = 6$ and $p = 2$. ← Then finish it off. Be careful writing down the answer. Look at the way the question was worded.

Worked example

Finding the maximum or minimum value of a quadratic function

5 **(i)** Work out the minimum value of $x^2 - 8x + 9$ and the value of x when it occurs by completing the square.

(ii) Explain why it is a minimum.

Solution

(i) $x^2 - 8x + 9$

$= (x - 4)^2 - 16 + 9$

$= (x - 4)^2 - 7$

When $x = 4$, $x - 4 = 0$ and the expression $= -7$.

(ii) For any other value of x, $(x - 4)^2 > 0$ and so the value of the expression is greater than -7.

So the minimum value is -7 when $x = 4$.

> Complete the square.

> $(x - 4)^2$ is a squared term and so is always positive except when it is 0.
> This occurs when $x = 4$.

Exam-style question

TESTED ☐

By completing the square, work out the minimum value of $2x^2 - 6x + 5$.

Short answer on page 119

Full worked solution online

CHECKED ANSWER ☐

Chapter 3: Algebra III

About this chapter

This chapter is about functions: their meaning, the notation used to write them and some of their graphs. The language of functions is really important in more advanced mathematics, allowing you to express mathematical ideas in words and symbols.

Before you start, remember ...

- algebraic manipulation of expressions
- the gradient of a line
- plotting graphs
- speed–time graphs
- distance–time graphs.

3.1 Function notation and composite functions

REVISED

Key facts

1 The function notation

A relationship like $y = x^2 - 4$ may be written using function notation as $y = f(x)$ where $f(x) = x^2 - 4$. In a case like this, y is said to be a function of x.

$f(x)$ may contain a number of terms in x.

> Other letters may be used, not just x, y and f. For example $s = g(t)$ means that the quantity s is a function of t.

2 Input and output values

When a value for x is input, the function produces an output number. The output value is unique for the input number.

input value $\xrightarrow[\text{function}]{}$ output value,

e.g. $f(x) = x^2 - 3$ gives $f(2) = 4 - 3 = 1$.

3 The domain and range

The domain is the set of possible input values for x.

The range is the set of possible output values.

4 Combining functions

Adding, multiplying and dividing functions,

e.g. If $f(x) = x^2$ and $g(x) = x + 2$.

then $f(x) + g(x) = x^2 + x + 2$

then $f(x) \times g(x) = x^2(x + 2)$

then $\dfrac{f(x)}{g(x)} = \dfrac{x^2}{x + 2}$

5 Composite functions

$fg(x)$ means use x as the input for the function g to obtain an output $g(x)$. Then use $g(x)$ as an input for the function f to obtain an output $f[g(x)]$. This is usually written as $fg(x)$.

If $f(x) = x^2$ and $g(x) = x + 2$ then $fg(x) = f(g(x)) = f(x + 2) = (x + 2)^2$.

So for $x = 3$, $g(x) = 5$ and $fg(x) = 5^2 = 25$.

> Note that if $f(x) = x^2$ and $g(x) = x + 2$ then $gf(x) = g(f(x)) = g(x^2) = x^2 + 2$. In general, $gf(x) \neq fg(x)$.

Worked examples

Values of the function

1 You are given that $f(x) = x^2 + 1$ for all values of x.
    (i) Work out $f(3)$.
    (ii) Given that $f(x) = 17$, find the possible values of x.

Solution

(i) $f(3) = 3^2 + 1 = 10$

(ii) $f(x) = x^2 + 1 = 17$

$\Rightarrow x^2 = 16$

$\Rightarrow \quad x = \pm 4$

> Equate and solve. Remember that the value of $f(x)$ is unique for a given value of x, but that more than one value of x can give a particular value of $f(x)$.

2 You are given that $f(x) = 2x^2 - 3$.
Write down expressions for

    (i) $f\left(\dfrac{x}{2}\right)$

    (ii) $f(x + 1)$

Solution

(i) $f\left(\dfrac{x}{2}\right) = 2\left(\dfrac{x}{2}\right)^2 - 3$

$\qquad = \dfrac{2x^2}{4} - 3$

$\qquad = \dfrac{x^2}{2} - 3$

(ii) $f(x + 1) = 2(x + 1)^2 - 3$

$\qquad = 2x^2 + 4x + 2 - 3$

$\qquad = 2x^2 + 4x - 1$

> Replace x by $\dfrac{x}{2}$.

> Simplify your answer.

> Replace x by $x + 1$.

> Simplify your answer.

Worked example

Domain and range

3 You are given that $f(x) = 2x^2 - 1$ for all values of x.
Write down the range of $f(x)$.

Solution

For all real values of x, $x^2 \geqslant 0 \Rightarrow 2x^2 \geqslant 0 \Rightarrow 2x^2 - 1 \geqslant -1$.

So the range is $f(x) \geqslant -1$.

Worked example

Combining functions

4 You are given that $f(x) = x + 3$ and $g(x) = x^2$.
   (i) Write down the expressions represented by
      (a) $f(x) + g(x)$
      (b) $\dfrac{g(x)}{f(x)}$
   (ii) Work out the values of
      (a) $f(2) + g(2)$
      (b) $\dfrac{g(2)}{f(2)}$

Solution

(i) (a) $f(x) + g(x) = x + 3 + x^2$

   (b) $\dfrac{g(x)}{f(x)} = \dfrac{x^2}{x+3}$

(ii) (a) $f(2) + g(2) = 2 + 3 + 4 = 9$

   (b) $\dfrac{g(2)}{f(2)} = \dfrac{2^2}{2+3} = \dfrac{4}{5}$

Note that you could have worked this out given that $f(2) = 5$ and $g(2) = 4$ so $f(2) + g(2) = 4 + 5 = 9$.

Worked example

Composite functions

5 You are given that $f(x) = 2x + 1$ and $g(x) = x^2 + 2$.
   By finding $fg(x)$ and $gf(x)$, show that $fg(x) \neq gf(x)$.

Solution

$f(x) = 2x + 1$ and $g(x) = x^2 + 2$
$\Rightarrow fg(x) = f(g(x)) = f(x^2 + 2)$
$\qquad = 2(x^2 + 2) + 1 = 2x^2 + 5$
$gf(x) = g(f(x)) = g(2x + 1)$
$\qquad = (2x + 1)^2 + 2 = 4x^2 + 4x + 3$
These functions are not the same.
So $fg(x) \neq gf(x)$.

In general $fg(x) \neq gf(x)$ but they may be numerically the same for particular values of x.

Exam-style question

TESTED

You are given that $f(x) = x - 1$ and $g(x) = x^2 + 1$.

Work out values of x for which $fg(x) = gf(x)$.

Short answer on page 119

Full worked solution online

CHECKED ANSWER

3.2 Inverse functions

Key facts

1 Inverse functions

The relationship $y = f(x)$ is called a function if, for a given input, there is a unique output.

The reverse mapping is called the inverse function, denoted $f^{-1}(x)$.

The mapping $y = f(x)$ is called a 'many-to-one' function if a set of input numbers gives the same output number. If the output number is unique to the input number then the function is called 'one-to-one'.

An inverse only exists if the function is one-to-one.

2 Finding an inverse function

There are two methods to find an inverse function given a function.

Method 1: reversing the flow diagram

The inverse function, $f^{-1}(x)$, reverses the rules of a function, $f(x)$.

e.g. If $f(x) = 4x + 3$ then the rules are, to get from x to $f(x)$:

multiply by 4

add 3.

The inverse function reverses the operation and the order:

e.g. subtract 3

divide by 4.

i.e. $f^{-1}(x) = \dfrac{x-3}{4}$

Method 2: interchange x and y

First interchange y and x in the function formula $y = f(x)$.

Then make y the subject of the new formula.

So for $f(x) = 4x + 3$ first write $y = 4x + 3$.

Then interchange x and y giving $x = 4y + 3$.

Then make y the subject:

$$x = 4y + 3 \Rightarrow 4y = x - 3 \Rightarrow y = \frac{x-3}{4}$$

> You can show this in a flow diagram, like this:
>
> $f(x) \quad x \xrightarrow[\times 4]{} 4x \xrightarrow[+3]{} 4x+3$
>
> $f^{-1}(x) \quad \dfrac{x-3}{4} \xleftarrow[\div 4]{} x-3 \xleftarrow[-3]{} x$

Worked example

Finding the inverse function

1 You are given that $f(x) = 2x - 5$.
   (i) Find the inverse function, $f^{-1}(x)$.
   (ii) Show that $ff^{-1}(4) = 4$.

Solution

(i) $y = 2x - 5$

$\Rightarrow x = 2y - 5$

$\Rightarrow y = \dfrac{x+5}{2}$

i.e. $f^{-1}(x) = \dfrac{x+5}{2}$

(ii) $f(4) = 2 \times 4 - 5 = 3$

$f^{-1}(3) = \dfrac{3+5}{2} = 4$

$\Rightarrow f^{-1}f(4) = 4$

> This is equivalent to:
> add 5
> divide by 2

> Note that $f^{-1}f(4) = f^{-1}(3) = 4$
> But also $f^{-1}(4) = \dfrac{4+5}{2} = \dfrac{9}{2}$
> and $f\left(\dfrac{9}{2}\right) = 9 - 5 = 4$
> i.e. $f^{-1}f(x) = ff^{-1}(x) = x$.

Worked example

One-to-one functions

2 Determine the domain for which the function $f(x) = x^2 + 3$ is one-to-one and, over this domain, find the inverse function.

Solution

The rules for the function are:

square x

add 3

The rules for the inverse, if it exists, are:

subtract 3

take the square root.

i.e $f^{-1}(x) = \sqrt{x - 3}$

However, it has to be possible to take the square root and so $x \geqslant 3$.

Exam-style question

TESTED ☐

You are given the two functions $f(x) = x + 3$ and $g(x) = x^2 + 1$.
Work out the inverse of $fg(x)$ and determine the domain for which $fg(x)$ is one-to-one.

Short answers on page 119

Full worked solution online

CHECKED ANSWERS ☐

3.3 Drawing graphs

REVISED ☐

Key facts

You will also meet the equation of a straight line in Chapter 5.

1 **The gradient of a line**
   The gradient of the sloping line in this right-angled triangle is given by $\dfrac{\text{height}}{\text{base}}$.

   height

   base

2 **The equation of a line.**
   The general form of the equation of a straight line is $y = mx + c$ where the gradient is m and the intercept on the y-axis is $(0, c)$.

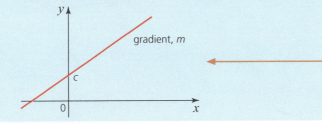

gradient, m

The line $y = mx + c$

3 Polynomial functions

Linear, quadratic and cubic functions are all examples of polynomials.

> In this course these are the only polynomials you will meet.

4 Drawing curves

To plot a curve:
- make out a table of values
- mark the points on your graph
- join them with a smooth curve.

> You can plot any function providing you know its expression.

To sketch a curve:
- use your knowledge of the functions to draw its approximate shape
- mark on important points, such as where it crosses the axes and any turning points
- you have to know how the function behaves.

> A turning point is an example of a stationary point.

To draw a curve:
- you are expected to make decisions about how accurate it should be; this will depend on the context
- usually the instruction 'draw' is taken to suggest an accuracy somewhere between plot and sketch.

Worked example

The equation of a line

1 The graph of a line is shown in the diagram below.

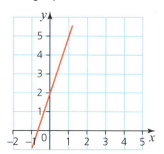

Find the equation of this line.

Solution

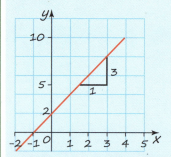

From the triangle you can see that the gradient is $\frac{3}{1} = 3$.

The intercept on the y-axis is at $(0, 2)$.

So the equation of the line is $y = 3x + 2$.

Worked example

Plotting a quadratic function

2 Plot the curve $y = x^2 - 5x + 7$ in the range $-1 \leqslant x \leqslant 5$.

Solution

Points on the curve:

x	−1	0	1	2	3	4	5
y	13	7	3	1	1	3	7

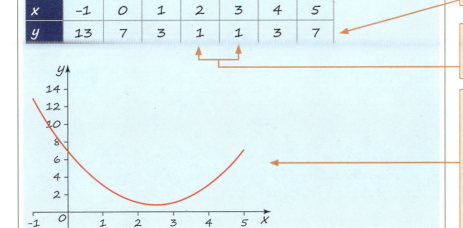

Calculate the values for y given the value for x on your calculator.

Note that a quadratic curve is symmetric in the vertical line through the turning point.

Note that a quadratic curve can be this way up or 'upside down' as in the graph below. The curve will be 'upside down' if the coefficient of the term in x^2 is negative.

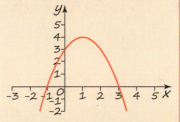

Worked example

Plotting an unknown curve

3 Plot the curve with points given in the table below, for $0 \leqslant x \leqslant 5$.

x	0	1	2	3	4	5
y	0	−2.2	−0.5	3.5	9.7	18

Solution

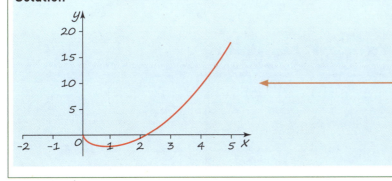

The curve will be U-shaped if the coefficient of the term in x^2 is positive.

Plot the given points and join them with a smooth curve.

Exam-style question

(i) On the same graph plot
    (a) the line $y = 3x - 1$
    (b) the curve $y = x^2 + 3x - 2$.
(ii) Write down the coordinates of the points where they intersect.

Short answers on page 119

Full worked solution online

3.4 Graphs of exponential functions

Key facts

1 **Exponential functions**

In an exponential function the variable is the power, e.g. $y = 3^x$.

2 **Properties of the exponential functions of the form $y = ka^x$ where $a > 1$ and k is any number.**

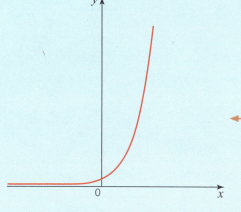

- y is always positive.
- The curve passes through the point $(0, k)$.
- The gradient is always positive and increasing for increasing x.
- The curve approaches the negative x-axis.

This is an example of exponential growth.

3 **An equivalent exponential function is of the form $y = ka^{-x}$ where $a > 1$.**

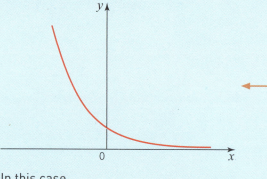

In this case
- y is always positive
- the curve passes through the point $(0, k)$
- the gradient is always negative and increasing for increasing x
- the curve approaches the positive x-axis.

This is an example of exponential decay.

Worked examples

Exponential graphs

1 Plot the graph of $y = 3 \times 2^x$ in the range $-2 \leqslant x \leqslant 2$.

Solution

x	-2	-1	0	1	2
y	0.75	1.5	3	6	12

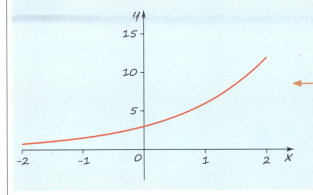

This is an example of exponential growth.

2 Plot the graph of $y = 3^{-x}$ in the range $-2 \leqslant x \leqslant 2$.

Solution

x	-2	-1	0	1	2
3^{-x}	9	3	1	0.33	0.11

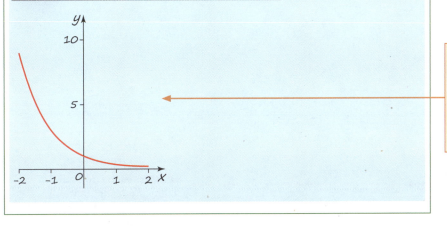

This is an example of exponential decay.
Notice that $3^{-x} = \left(\frac{1}{3}\right)^x$ and so an alternative way of writing the equation of this curve is $y = \left(\frac{1}{3}\right)^x$.

Exam-style question

TESTED

(i) Plot on the same grid the curves $y = 2^x + 1.5$ and $y = 3^x$.
(ii) From your graph estimate the root of the equation $2^x + 1.5 = 3^x$.
(iii) Substitute the value of x you obtained in part **(ii)** in each of the equations $y = 2^x + 1.5$ and $y = 3^x$. Hence check your answer to part **(ii)**.

Short answers on page 119

Full worked solution online

CHECKED ANSWERS

3.5 Graphs of functions with more than one part to their domains

REVISED ☐

> ## Key facts
>
> **1 Functions with more than one part to the domain**
>
> In polynomial and exponential functions the domain of x is often assumed to be infinite.
>
> However, some functions are defined with different formulae for different parts of their domains.
>
> e.g. $f(x) = x^3$ for $x \geqslant 0$
>
> $= 0$ for $x < 0$
>
> **2 Kinematics graphs**
>
> Speed–time and distance–time graphs often have more than one domain.
>
> When speed is plotted against time
> - a straight line indicates constant acceleration; it is the gradient
> - the area under the function represents the distance travelled.
>
> When distance is plotted against time
> - a straight line indicates constant speed; it is the gradient.

Worked examples

Functions with more than one part to the domain

1 Draw the graph of $y = f(x)$ where

$f(x) = x + 1$ for $0 \leqslant x < 3$

$= 4$ for $3 \leqslant x < 7$

$= 18 - 2x$ for $7 \leqslant x \leqslant 9$.

Solution

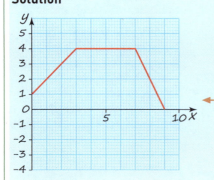

> The range for y in the first domain is from 1 to 4.
>
> For the second domain the value of y is constantly 4.
>
> For the third domain the range for y is from 4 to 0.

2 The graph of $y = f(x)$ has two domains. The first domain is the line AB which is parallel to the x-axis, and the second domain has the curve $y = 2(x - 3)^2$ passing through B, C and D. The point C lies on the x-axis.

The points A and C have coordinates (0, 2) and (3, 0).

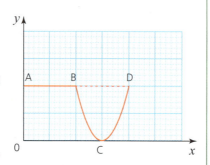

(i) Write down the coordinates of the points B and D.

(ii) Define the function $y = f(x)$.

Solution

(i) Since A, B and D lie on the same line then they all have a y-coordinates of 2.
Solving $2(x - 3)^2 = 2$
$\Rightarrow \qquad (x - 3)^2 = 1$
$\Rightarrow \qquad (x - 3) = \pm 1$
$\Rightarrow \qquad x = 2 \text{ or } 4$
So coordinates of B and D are (2, 2) and (4, 2).

(ii) The first domain is therefore from 0 to 2 and the second from 2 to 4.
This gives
$y = 2$ for $0 \leqslant x \leqslant 2$
$y = 2(x - 3)^2$ for $2 \leqslant x \leqslant 4$.

Worked example

Distance–time graph

3 The graph below shows the distance travelled by a car from its starting point.

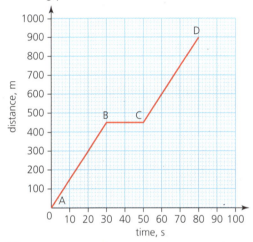

(i)   State what is happening between the points B and C on the graph.
(ii) How far has the car travelled in 80 seconds?
(iii) How realistic is this as a model for the motion of the car?

Solution

(i)   There is no change in distance so the car is stationary.
(ii) The end of the graph is at (80, 900) so in 80 seconds the car travels 900 metres.
(iii) The model assumes that the car instantaneously starts and stops which is not realistic.

> The straight line implies that the car is immediately travelling at 15 m s⁻¹ until the instant that it stops. A car would accelerate with speeds that would increase from 0 to 15 m/s.

Exam-style question

TESTED ☐

Draw the graph of the function f(x) which is defined as follows.
$y = x^2$ for $-3 \leqslant x < 3$
$y = 18 - 3x$ for $3 \leqslant x < 6$
$y = 0$ for $x \geqslant 6$

Short answers on page 119

Full worked solution online

CHECKED ANSWER ☐

Chapter 4: Algebra IV

About this topic

In this chapter you revise and extend your knowledge of a number of techniques in algebra: quadratic and simultaneous equations, index laws, inequalities and sequences. You will need to be fluent in the algebra you have learnt throughout this course.

Before you start, remember ...

- factorising quadratic expressions
- simultaneous equations
- inequalities
- index laws
- sequences.

4.1 Quadratic equations

> **Key facts**
>
> **1   Quadratic graphs**
> The graph of the curve $y = ax^2 + bx + c$ is a parabola.
> If $a > 0$ then the curve is the U-shaped and is symmetric about the line $x = -\dfrac{b}{2a}$ with a minimum point on this line.
>
>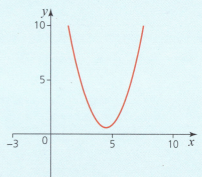
>
> If $a < 0$ then the curve is $\cap$-shaped and is symmetric about the line $x = -\dfrac{b}{2a}$ with a maximum point on this line.
>
>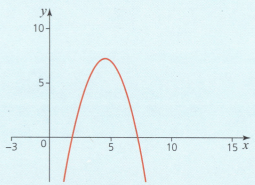

2 Solving quadratic equations that can be factorised

The solution of the factorised equation $(x - a)(x - b) = 0$ is $x = a$ and $x = b$.

3 Solving quadratic equations that cannot be factorised: completing the square

If a quadratic equation $ax^2 + bx + c = 0$ is rewritten $a(x + p)^2 - q = 0$ then $x = -p \pm \sqrt{\dfrac{q}{a}}$.

4 Solving quadratic equations that cannot be factorised: using the formula

The formula for solving the quadratic equation $ax^2 + bx + c = 0$ is

$$x = \frac{-b \pm \sqrt{b^2 - 4ac}}{2a}.$$

5 Checking for roots of a quadratic equation

The expression $b^2 - 4ac$ in the formula is called the discriminant.
- If $b^2 - 4ac > 0$ there are two distinct roots.
- If $b^2 - 4ac = 0$ the roots are coincident.
- If $b^2 - 4ac < 0$ there are no real roots.

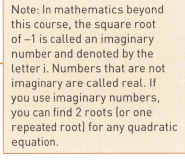

Note: In mathematics beyond this course, the square root of −1 is called an imaginary number and denoted by the letter i. Numbers that are not imaginary are called real. If you use imaginary numbers, you can find 2 roots (or one repeated root) for any quadratic equation.

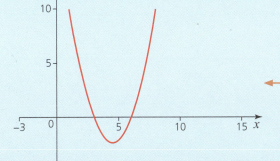

$b^2 - 4ac > 0$; 2 distinct roots.

$b^2 - 4ac = 0$; coincident roots.

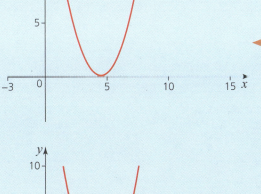

$b^2 - 4ac < 0$; no roots.

Worked example

Sketching quadratic graphs

1 Sketch the graph of the curve $y = x^2 - 11x + 28$.

Solution

The coefficient of the x^2 term is positive so the graph is U-shaped.

It crosses the y-axis at $(0, 28)$.

$x^2 - 11x + 28 = (x - 4)(x - 7)$

So the curve crosses the x-axis at $(4, 0)$ and $(7, 0)$.

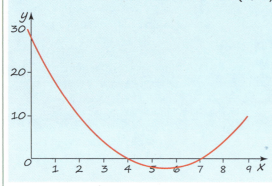

Worked examples

Solving quadratic equations

In this example the equation can be factorised.

2 Solve the equation $2x^2 - 7x - 4 = 0$.

Solution

$$2x^2 - 7x - 4 = 0$$
$$\Rightarrow \quad 2x^2 - 8x + x - 4 = 0$$
$$\Rightarrow 2x(x - 4) + 1(x - 4) = 0$$
$$\Rightarrow \quad (2x + 1)(x - 4) = 0$$
$$\Rightarrow \quad x = -\frac{1}{2} \text{ or } x = 4$$

> Factorise the first two terms and factorise the last two terms, giving a common factor.

> Factorise completely.

> Don't forget to give your answers!

In this example the equation cannot be factorised and the method of completing the square is used.

3 Express $x^2 + 2x - 5 = 0$ in the form $(x + p)^2 = q$ and hence solve the equation.

Solution

$$x^2 + 2x - 5 = 0$$
$$\Rightarrow \quad x^2 + 2x = 5$$
$$\Rightarrow \quad x^2 + 2x + 1 = 6$$
$$\Rightarrow \quad (x + 1)^2 = 6$$
$$\Rightarrow \quad x + 1 = \pm\sqrt{6}$$
$$\Rightarrow \quad x = -1 \pm \sqrt{6}$$
$$x = 1.45 \text{ or } -3.45$$

> Add a number to both sides so that the left-hand side is a perfect square

> Take square roots of both sides. This is the answer you should give if an exact answer is required

> Or give it to 3 significant figures.

In this example the equation cannot be factorised and the formula is used.

4 Solve the equation $2x^2 - 4x - 7 = 0$.

Solution

$2x^2 - 4x - 7$ cannot be factorised.

Comparing $2x^2 - 4x - 7 = 0$ with $ax^2 + bx + c = 0$

gives $a = 2$, $b = -4$, $c = -7$.

Substituting in $x = \dfrac{-b \pm \sqrt{b^2 - 4ac}}{2a}$ gives

$$x = \frac{4 \pm \sqrt{16 + 56}}{4} = \frac{4 \pm \sqrt{72}}{4}$$

$$\Rightarrow x = 1 \pm \frac{3}{2}\sqrt{2}$$

$$\Rightarrow x = 3.12 \quad \text{or} \quad x = -1.12$$

> There are no numbers that multiply to give $2 \times -7 = -14$ and add to give -4 so the formula must be used.

> If you are asked to give the answer exactly then this is the answer you should give.

> If not, then use your calculator. The answers will then need to be rounded; unless you are told otherwise give them to 3 significant figures.

Worked example

Checking for roots of a quadratic equation

5 Determine whether the equation $3x^2 - 4x + 5 = 0$ has 0, 1 or 2 roots.

Solution

For $3x^2 - 4x + 5 = 0$, $a = 3$, $b = -4$, $c = 5$.

Substitution gives $b^2 - 4ac = 4^2 - 4 \times 3 \times 5 = -44 < 0$.

So there are no roots.

> The numbers are the coefficients of the terms of the equation.

Worked example

Finding the maximum or minimum value of a quadratic function

6 Find the minimum value of $f(x) = x^2 - 4x + 7$.

Solution

$f(x) = x^2 - 4x + 7$

$\quad\ = x^2 - 4x + 4 + 3$

$\quad\ = (x - 2)^2 + 3$

Since the minimum value of the squared term is 0 when $x = 2$, the minimum value of $f(x)$ is 3 when $x = 2$.

> This is completing the square.

Exam-style question

TESTED ☐

(i) Express the quadratic expression $x^2 - 3x - 1$ in the form $(x - p)^2 - q$ where p and q are to be determined.
(ii) Find the coordinates of the turning point of the curve $y = x^2 - 3x - 1$.
(iii) Sketch the graph of $y = x^2 - 3x - 1$.
(iv) Hence solve the equation $x^2 - 3x - 1 = 0$.

Short answers on page 119

Full worked solution online

CHECKED ANSWERS ☐

4.2 Simultaneous equations

Key facts

1 Solving two simultaneous equations when both are linear

The two equations can be solved by elimination or by substitution.
- Elimination means manipulating the equations so that the coefficients of one variable are the same. Then the two equations are added or subtracted to leave a linear equation in one unknown.
- Substitution means making one of the variables the subject of one of the equations and substituting in the second equation.

2 Solving two simultaneous equations when one is linear and the other quadratic

The equations need to be solved by substitution. Make one variable the subject of the linear equation and substitute in the quadratic equation.

3 Solving three simultaneous equations in three unknowns
- Take two pairs of equations and eliminate one variable.
- Solve the resulting pair of equations in two unknowns using one of the methods described above.
- Using the two values substitute into one of the equations to give the third unknown.

In this course the three equations will be linear.

Worked example

Solving simultaneous equations

In this example there are two linear equations.
1 Solve the following simultaneous equations
   (i) by elimination and
   (ii) by substitution.
$2x + 5y = 19$, $3x - y = 3$

Solution

(i) Method 1 – elimination

$2x + 5y = 19$ ①
$3x - y = 3$ ②

② × 5: $15x - 5y = 15$ ③
$2x + 5y = 19$ ①

Add $17x = 34$ ① + ③
$\Rightarrow x = 2$

Substitute for x into ①
$4 + 5y = 19$
$\Rightarrow 5y = 15$
$\Rightarrow y = 3$

So the solution is $x = 2$, $y = 3$.

Make one coefficient the same or the negative of the other.

Add the two equations since in this case one of the coefficients is negative. Then solve for x.

Then substitute into one of the equations to find y.

(ii) Method 2 – substitution

$2x + 5y = 19$ ①
$3x - y = 3$ ②

② $y = 3x - 3$
In ①: $2x + 5(3x - 3) = 19$
$\Rightarrow 17x - 15 = 19$
$\Rightarrow 17x = 34$
$\Rightarrow x = 2$ and so $y = 3$

Make one variable the subject of one equation and then substitute.

In the next example one equation is linear and the other is quadratic

2 Find the points of intersection of the line $2x - y = 11$ and the curve $y = x^2 - 5x - 5$.

Solution

$$2x - y = 11$$
$$\Rightarrow \qquad y = 2x - 11$$
$$\Rightarrow \qquad 2x - 11 = x^2 - 5x - 5$$
$$\Rightarrow \qquad x^2 - 7x + 6 = 0$$
$$\Rightarrow \qquad (x - 6)(x - 1) = 0$$

Either $x = 6$ giving $12 - y = 11 \Rightarrow y = 1$.

or $x = 1$ giving $2 - y = 11 \Rightarrow y = -9$.

The coordinates of the points of intersection are (6, 1) and (1, –9).

> Make y the subject of the linear equation and substitute to give a quadratic equation in x.

> Solve for x. Remember that there will be two values.

> Substitute each value into the linear equation to find the associated value of y.

> Remember to pair off the values correctly.

Worked example

Solving three equations in three unknowns

3 Solve the following simultaneous equations.
$$x + 2y - z = 0$$
$$2x - y + z = 5$$
$$3x + 4y - 2z = 1$$

Solution

$$x + 2y - z = 0 \qquad ①$$
$$2x - y + z = 5 \qquad ②$$
$$3x + 4y - 2z = 1 \qquad ③$$
$$① + ② \Rightarrow \quad 3x + y = 5 \qquad ④$$
$$③ + 2② \Rightarrow \quad 7x + 2y = 11 \qquad ⑤$$
$$⑤ - 2④ \Rightarrow x = 1$$

Substitute into ⑤ $\Rightarrow 7 + 2y = 11 \Rightarrow y = 2$

Substitute into ① $\Rightarrow 1 + 4 - z = 0 \Rightarrow z = 5$

> **Hint:** It does not matter which variable you eliminate first – choose what looks like the easiest!

> In this solution z is eliminated, reducing the question to two simultaneous equations in two variables which are solved in the usual way.

Exam-style question

TESTED

(i) Draw the graphs of $x + y = 4$ and $y = x^2 - x$.
(ii) Use your graph to solve the simultaneous equations $x + y = 4$ and $y = x^2 - x$.
(iii) Solve algebraically the equations $x + y = 4$ and $y = x^2 - x$.

Short answers on page 120

Full worked solution online

CHECKED ANSWERS

4.3 Solving cubic equations

Key facts

1 Roots of cubic equations

A cubic equation will have three roots, two of which may be repeated, or one root.

2 The factor theorem

$(x - a)$ is a factor of $f(x)$ if $f(a) = 0$.

So $x = a$ is a root of the equation $f(x) = 0$,

e.g. for $f(x) = x^3 - 2x + 1$ $f(1) = 1 - 2 + 1 = 0$

So $x = 1$ is a root of the equation $f(x) = 0$.

3 Finding all the roots of a polynomial equation

If a cubic equation has three integer roots then the factor theorem can be applied repeatedly to find roots.

If the equation has only one integer root, $x = a$, with the other two roots irrational then the factor $(x - a)$ must be divided into the cubic function to find the quadratic function. The quadratic equation will have two roots, repeated roots or no real roots.

4 Finding roots of a higher order equation

In general a polynomial equation can have as many roots as the order of the equation.

> In this course you will be expected to be able to find roots of cubic or equations which are integers for which the factor theorem is appropriate.

> Note, that if a cubic equation has three roots, a, b, c, then the equation $f(x) = 0$ can be written $(x - a)(x - b)(x - c) = 0$. When multiplied out the constant term $= -abc$.

> So when looking for possible roots, identify all factors of the constant term.

Worked example

Dividing a cubic expression by a factor to find the quotient

1 Divide $x^3 + 2x^2 - 3x + 2$ by $x - 3$.

Solution

Lay out your work as shown.

Bring down the next term so that you have $5x^2 - 3x$.

Multiply $x - 3$ by $5x$ and place under as shown. Place $+5x$ on the top line.

$$\begin{array}{r} x^2 + 5x + 12 \\ (x-3)\overline{)x^3 + 2x^2 - 3x + 2} \\ \underline{x^3 - 3x^2} \\ 5x^2 - 3x \\ \underline{5x^2 - 15x} \\ 12x + 2 \\ \underline{12x - 36} \\ 38 \end{array}$$

> Multiply $(x - 3)$ by x^2 and place under the first two terms. Subtract and put x^2 on the top line.

> Repeat the process, leaving a remainder of 38. The quotient, the number of times that $x - 3$ goes into the cubic function, is shown on the top line.
> In this example it is $x^2 + 5x + 12$; the remainder is 38.

Worked examples

Using the factor theorem to solve a cubic equation

In the next example, the equation has three roots, all of which are integers.

2 Solve the equation $f(x) = x^3 + 4x^2 + x - 6 = 0$.

Solution

The factors of −6 are ±1, ±2 ±3, ±6.

By trial f(1) = 0

So $x = 1$ is one of the roots.

> Note that the numbers must be factors of the constant in the equation, in this case −6.

> Using the factor theorem is an efficient way to find a root of an equation. However, it does not find the quotient.

> So try here $x = 1$.

Divide $f(x)$ by $(x - 1)$ ◄ — By long division.

$f(x) = (x - 1)(x^2 + 5x + 6) = 0$

Gives $\Rightarrow f(x) = (x - 1)(x + 3)(x + 2) = 0$ ◄ — Solve the resulting quadratic.

$\Rightarrow$ The roots are $x = 1, -2, -3$ ◄

— Alternatively, try the other factors of −6 until you have found three that work.

In the next example, the equation only has one real root.
3 Solve the equation $f(x) = x^3 + 2x^2 - 3x - 10 = 0$.

Solution

— The possible numbers must be factors of 10.

The factors of −10 are ±1, ±2, ±5 and ±10. ◄

$f(2) = 8 + 8 - 6 - 10 = 0$ ◄ — By trial $f(2) = 0$.

So $x = 2$ is a root.

$f(x) = (x - 2)(x^2 + 4x + 5) = 0$

In the quadratic, $a = 1$, $b = 4$ and $c = 5$, so $b^2 - 4ac = -4$.

Since $b^2 - 4ac < 0$ the quadratic has no roots.

So $x = 2$ is the only root.

In the next example, the equation has three roots, only one of which is an integer.
4 Solve the equation $f(x) = x^3 - 7x^2 + 13x - 3 = 0$.

Solution

By trial $f(3) = 0$ so $(x - 3)$ is a factor. ◄ — Factors of −3 are ±1, ±3.

$f(x) = (x - 3)(x^2 - 4x + 1) = 0$ ◄ — By long division.

So either $x = 3$ or $x^2 - 4x + 1 = 0$.

$\Rightarrow x = \dfrac{4 \pm \sqrt{16 - 4}}{2} = \dfrac{4 \pm \sqrt{12}}{2}$ ◄ — By formula.

$\quad = 2 \pm \sqrt{3}$

the roots are $3, 2 + \sqrt{3}, 2 - \sqrt{3}$

or $3, 3.73, 0.268$.

Exam-style question

TESTED ☐

The function $f(x)$ is defined by $f(x) = x^3 - 3x^2 + 4$.

(i) Show that $(x + 1)$ is a factor of $f(x)$.
(ii) Solve the equation $f(x) = 0$.
(iii) Sketch the graph of $y = f(x)$.

Short answers on page 120

Full worked solution online

CHECKED ANSWERS ☐

4.4 Inequalities

Key facts

1 Inequality signs
- $x < a$ means x is less than a.
- $x > a$ means x is greater than a.
- $x \leqslant a$ means x is less than or equal to a, or x is at most a.
- $x \geqslant a$ means x is greater than or equal to a, or x is at least a.
- The symbol $\leqslant$ is often written $\leq$.
- The symbol $\geqslant$ is often written $\geq$.

2 Representing an inequality on a number line

If the inequality is $f(x) \geqslant 0$ then the end points are included but if the inequality is $f(x) > 0$ then they are not.

If the end points are included, the standard notation on a diagram is for a filled circle, as shown in the diagram, at the end of the line.

If the end points are not included then the circle is not shaded (it is 'open'), as shown here.

e.g. The inequality $1 < x \leqslant 9$ is represented as follows.

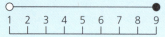

3 Linear inequalities

Inequalities can be manipulated in the same way as equations except for one rule: if both sides are multiplied or divided by a negative number then the direction of the inequality changes.

Subject to the above rule, linear inequalities can be solved in the same way as equations.

4 Solving quadratic inequalities graphically

The graph represents the function $y = f(x) = ax^2 + bx + c$ where $a > 0$.

If $f(x) < 0$ then the range of values of x is the range where the graph is below the axis.

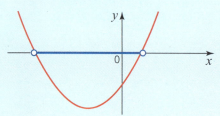

If $f(x) > 0$ then the range of values of x is the range where the graph is above the axis.

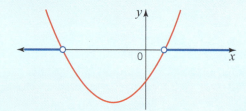

5 Solving quadratic inequalities algebraically

A quadratic inequality takes the form $f(x) < 0$ (with any of the four inequality signs), or can be manipulated into that form, where $f(x)$ is a quadratic function.

When $f(x)$ can be factorised you have two expressions whose product satisfies the inequality. You can then use the facts

$+ \times + = +$, $- \times - = +$, $+ \times - = -$ and $- \times + = -$

6 Test numbers

When you have found the solution to an inequality question, check it by trying one (or more) test numbers in the original inequality.

Worked examples

Linear inequalities

1 Solve the inequality $2 - x < 3(2x - 1)$.

Solution

$2 - x < 3(2x - 1)$ ← Expand the bracket.

$2 - x < 6x - 3$ ← Collect terms in x on one side and the numbers on the other. It does not matter which side.

$5 < 7x$ ← Divide by 7 to get x.

$x > \dfrac{5}{7}$ ← Logically $x > \dfrac{5}{7}$ is the same as $\dfrac{5}{7} < x$.

It is usual to write x on the left side of the inequality.

Sometimes you will get a double inequality. Apply the same rules throughout.

2 Find the range of values of x for which $3 < 2x - 7 < 9$.

Solution

$3 < 2x - 7 < 9$ ← Add 7 to all three parts.

$\Rightarrow 10 < 2x < 16$ ← Divide throughout by 2.

$\Rightarrow 5 < x < 8$

Worked example

Solving quadratic inequalities graphically

3 Solve the inequality $x^2 + 2x - 8 > 0$ graphically.

Solution

The graph of $y = x^2 + 2x - 8$ is shown below.

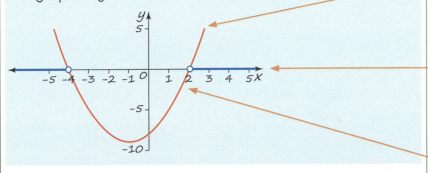

The graph is above the x-axis (i.e. $y > 0$) for values of x greater than 2 and less than -4.

Try a test number, e.g. $x = 0$. Does this satisfy the inequality? i.e. $0 + 0 - 8 < 0$ so no, it does not satisfy the inequality. It is not in the answer either, indicating that the solution is correct.

In this example the end points are not included so the end points are open circles.

Worked example

Solving quadratic inequalities algebraically

4 Solve the inequality $x^2 - 4x - 32 < 0$.

Solution

$$x^2 - 4x - 32 < 0$$
$$\Rightarrow (x - 8)(x + 4) < 0$$

Either $x - 8 < 0$ and $(x + 4) > 0$

i.e. $x < 8$ and $x > -4$

$\Rightarrow -4 < x < 8$

or $x - 8 > 0$ and $(x + 4) < 0$

i.e. $x > 8$ and $x < -4$, which is not possible

$\Rightarrow -4 < x < 8$

> Factorise the quadratic function

> You now have the product of two numbers which is negative. So one must be negative and the other positive.

> Try a test number, e.g. $x = 0$. Does this satisfy the inequality? i.e. $0 - 0 - 32 < 0$ so yes, it does satisfy the inequality. It is also in the answer region, also suggesting that the solution is correct.

> **Common mistake:** Inequality signs should be the same. It would be incorrect to write this answer as $8 > x < -4$.

Exam-style question

TESTED

(i) Solve algebraically the inequality $x^2 + 2x - 15 \geqslant 0$.
(ii) Illustrate your answer graphically.

Short answers on page 120

Full worked solution online

CHECKED ANSWERS

4.5 Index laws

REVISED

Key facts

1 The index laws

The three index laws, for any value of m and n, are as follows.

$$a^m \times a^n = a^{m+n}$$
$$a^m \div a^n = a^{m-n}$$
$$(a^m)^n = a^{mn}$$

You should also note the following that are derived from the above laws.

$$a^0 = 1$$
$$a^{-n} = \frac{1}{a^n}$$
$$\sqrt{a} = a^{\frac{1}{2}}$$
$$\sqrt[n]{a} = a^{\frac{1}{n}}$$

2 Disguised quadratic equations

Some more complicated looking equations can be written as a quadratic equation, e.g. $x^4 + x^2 - 6 = 0$ can be written as $(x^2)^2 + (x^2) - 6 = 0$, i.e. $y^2 + y - 6 = 0$ where $y = x^2$.

Worked examples

Using index laws

1 Write the term below as a single power of x.

$$\sqrt[3]{\frac{x^4 \times x^{\frac{1}{2}}}{x^3}}$$

Solution

$$\sqrt[3]{\frac{x^4 \times x^{\frac{1}{2}}}{x^3}} = \left(\frac{x^{4\frac{1}{2}}}{x^3}\right)^{\frac{1}{3}}$$

$$= \left(x^{\frac{3}{2}}\right)^{\frac{1}{3}} = x^{\frac{3}{2} \times \frac{1}{3}} = x^{\frac{1}{2}} = \sqrt{x}$$

> Remember that $\sqrt[3]{x}$ is the same as $x^{\frac{1}{3}}$.

Disguised quadratic equations

2 Solve the equation $x^6 + x^3 - 20 = 0$.

Solution

$x^6 + x^3 - 20 = 0$

Write $y = x^3$

$\Rightarrow \quad y^2 + y - 20 = 0$

$\Rightarrow (y + 5)(y - 4) = 0$

$\Rightarrow \qquad\qquad y = -5 \text{ or } 4$

i.e. $x^3 = -5$ or $x^3 = 4$

$\Rightarrow x = (-5)^{\frac{1}{3}}$ or $4^{\frac{1}{3}}$

$\Rightarrow x = -1.71$ or $x = 1.59$

> This equation only has two roots, but note that although this can be written as a quadratic equation it has power 6 and so there could be as many as six roots.

> Solve the quadratic in y in the usual way – this one factorises.

> Turn back into x^3 and take the cube roots.

Exam-style question

TESTED ☐

Solve the equation $x^4 - 11x^2 + 28 = 0$.

Short answer on page 120

Full worked solution online

CHECKED ANSWER ☐

4.6 Algebraic proof

Key fact

In an algebraic proof all the steps must be explained carefully.

Common mistake: Do not write the end result until you have got there from your starting point showing all the steps carefully. To write the answer will encourage you to miss out essential steps.

Hint: When you are asked to prove something you will get the demand 'prove that...' or 'show that...'.

You will be asked either of these when the end result is given to you.

'Show that ...' is often used for results that are mathematically more minor.

Worked examples

Algebraic proof

This example asks you to 'show that'.

1 Show that $x^2 + 4x + 1 = (x + 2)^2 - 3$.

In this course 'show that' is used rather more often than 'prove' but they mean essentially the same thing. You have to move from the original statement to the given end result. Since you know that this is true you need to demonstrate convincingly that you can derive the final result using correct mathematics.

Solution

$x^2 + 4x + 1 = x^2 + 4x + 4 - 4 + 1$

$\qquad = (x^2 + 4x + 4) - 4 + 1$

$\qquad = (x + 2)^2 - 3$

Don't forget to show all the steps.

2 Prove that $(n + 1)^2 - (n - 1)^2$ is a multiple of 4.

Solution

$(n + 1)^2 - (n - 1)^2$

$= (n^2 + 2n + 1) - (n^2 - 2n + 1)$

$= n^2 + 2n + 1 - n^2 + 2n - 1$

$= 4n$

So it is a multiple of 4.

Multiply out the brackets. Be careful of the signs.

This shows that the expression is a multiple of 4.

Exam-style question

Prove that the sum of three consecutive integers is a multiple of 3.

Short answer on page 120

Short answer on page 120

Full worked solution online

CHECKED ANSWER

4.7 Sequences

Key facts

1 Position to term sequences

Each term is defined by a rule that gives its value as a function of n, its position in the sequence,

e.g. $u_n = n^2 + 3$

So each term can be worked out directly,

e.g. $u_6 = 6^2 + 3 = 39$

> The nth term of a sequence is given the suffix n, e.g. a_n, u_n, x_n, etc.
>
> So, the first term is a_1, u_1, x_1, etc.

2 Term to term sequences

Each term is defined by a rule that gives its value as a function of the previous term (or terms).

In this case the first term is required as well as the rule,

e.g. $u_n = u_{n-1} + 3$ with $u_1 = 1$.

3 Linear sequences

A linear sequence is one where the nth term has the form $an + b$ where a and b are constants.

Sometimes linear sequences are called arithmetic sequences.

> The equivalent term to term form for a linear sequence is $u_{n+1} = u_n + a$ with $u_1 = b$.

4 Quadratic sequences

A quadratic sequence is a position to term sequence where the nth term has the form $an^2 + bn + c$ where a, b and c are constants.

> Linear and quadratic sequences have terms that diverge.

5 Limiting value of a sequence

Some sequences converge to a constant value. This is called the limiting value of the sequence,

e.g. if $a_n = 2 - \dfrac{1}{n}$

then $a_1 = 1$, $a_2 = 1.5$, $a_3 = 1.66...$, $a_4 = 1.75$,...

The terms are approaching the limiting value of 2.

> **Common mistake:** The limiting value of this sequence is 2. It would not be true to say, however, that $a_n = 2$.

Worked example

Position to term sequence

1 List the first 5 terms of the sequence given by $x_n = 3n^2 - 1$.

Solution

$x_1 = 3 \times 1^2 - 1 = 2$

$x_2 = 3 \times 2^2 - 1 = 11$

$x_3 = 3 \times 3^2 - 1 = 26$

$x_4 = 3 \times 4^2 - 1 = 47$

$x_5 = 3 \times 5^2 - 1 = 74$

> Each term is calculated by substituting for n, the position of the term in the sequence.
>
> So x_1 is found by substituting $n = 1$, x_2 by substituting $n = 2$, etc.

Worked example

Term to term sequence

2 List the first five terms of the sequence given by $x_1 = 1$ and $x_n = 3x_{n-1} - 1$.

Solution

$x_2 = 3 \times x_1 - 1 = 3 \times 1 - 1 = 2$

$x_3 = 3 \times x_2 - 1 = 3 \times 2 - 1 = 5$

$x_4 = 3 \times x_3 - 1 = 3 \times 5 - 1 = 14$

$x_5 = 3 \times x_4 - 1 = 3 \times 14 - 1 = 41$

> You are given x_1. Now use the rule to work out x_2, etc.

Worked examples

Finding a sequence

This example is the reverse process, asking for the formula for the sequence from the terms.

3 The first three terms of a linear sequence are 0, 1, 3.

(i) Find the term to term formula for the sequence.

(ii) Work out the next two terms.

Solution

(i) For a term to term sequence

$x_n = ax_{n-1} + b$

Take $x_1 = 0$

Then $x_2 = ax_1 + b$ giving $1 = b$

$x_3 = ax_2 + 1$ giving $3 = a + 1$

$\Rightarrow a = 2$ and $b = 1$

i.e. $x_n = 2x_{n-1} + 1$

(ii) For the next two terms

$x_n = 2x_{n-1} + 1$

$\Rightarrow x_4 = 2 \times 3 + 1 = 7$

$x_5 = 2 \times 7 + 1 = 15$

4 The nth term of a sequence is given by $a_n = \dfrac{2n+1}{3}$.

Work out the difference between the 13th and the 7th term.

Solution

$a_n = \dfrac{2n+1}{3} \Rightarrow a_{13} = \dfrac{2 \times 13 + 1}{3} = 9$

and $a_7 = \dfrac{2 \times 7 + 1}{3} = 5$

$\Rightarrow a_{13} - a_7 = 9 - 5 = 4$

> For each term substitute the value of n.

Worked example

Quadratic sequences

5 You are given that $a_n = 2n^2 + n - 5$.
Work out the first four terms of the sequence.

Solution

$a_n = 2n^2 + n - 5$

$\Rightarrow a_1 = 2 + 1 - 5 = -2$ ← Substitute for n in a_n.

$a_2 = 8 + 2 - 5 = 5$

$a_3 = 18 + 3 - 5 = 16$

$a_4 = 32 + 4 - 5 = 31$

Worked example

The limiting value of a sequence

6 In a sequence the nth term is given by $u_n = u_{n-1} + \dfrac{1}{2^{n-1}}$, $u_1 = 3$.

Find the limiting value of this sequence as $n \to \infty$.

Solution

$u_1 = 3$, $u_2 = u_1 + \dfrac{1}{2^1} = 3\dfrac{1}{2}$

$u_3 = u_2 + \dfrac{1}{2^2} = 3\dfrac{1}{2} + \dfrac{1}{4} = 3\dfrac{3}{4}$

$u_4 = u_3 + \dfrac{1}{2^3} = 3\dfrac{3}{4} + \dfrac{1}{8} = 3\dfrac{7}{8}$

Continuing this sequence gives

$u_5 = 3\dfrac{15}{16}$, $u_6 = 3\dfrac{31}{32}$, ... ← In the fractional part the denominator doubles with each new term and the numerator is always 1 less.

The limiting value of the fractional part is 1.

So the limiting value of the sequence is $3 + 1 = 4$.

Exam-style question

TESTED

The first three terms of a quadratic sequence are 2, 7, 16.

Find an expression for the nth term in the form $an^2 + bn + c$.

Short answer on page 120

Full worked solution online

CHECKED ANSWER

Review questions (Chapters 1–4)

1 The ratio of female : male employees in a company is 2 : 3. Given that the company employs 20 females, work out the number of male employees. [2]

2 John's annual salary of £27 000 is increased by 5%. Work out his new annual salary. [2]

3 Simplify $5(x + 3y) - 3(5x + y)$. [2]

4 Simplify $\dfrac{2cd^2}{3e} \div \dfrac{4c^2 d}{9e^2}$. [2]

5 Solve the equation $2x - 3 = 6 - x$. [2]

6 Solve the equation $\dfrac{2(x - 1)}{3} - 2 = \dfrac{3x + 1}{4}$. [3]

7 You are given that $a : b = 2 : 3$. Work out $a + b : b$. [2]

8 When £x is increased by 10% the amount is the same as when £y is decreased by 10%. Given that $x + y = £20 000$, find x and y. [3]

9 In how many ways can 6 people be chosen from a group of 11? [2]

10 A 'combination lock' consists of 4 wheels, each with the digits 1 to 6 on it. The wheels are rotated in such a way that when one particular number shows, the lock is released.
Paula tries each number in turn, starting with 0000.
It takes 3 seconds to try each number and move to the next if it is not the right combination.
What is the maximum time that it could take to try all the numbers? [3]

11 Simplify $(x - 3)(x + 2) + (x + 3)(x + 2)$. [2]

12 Expand and simplify $(x - 3)(x + 2)(x + 3)$. [2]

13 Expand $(2 - x)^5$. [2]

14 Work out the coefficient of x^3 in the expansion of $(3 + 2x)^6$. [3]

15 Express $(\sqrt{3} + 3\sqrt{5})(2\sqrt{3} - \sqrt{5})$ in the form $a + b\sqrt{c}$ where a, b and c are integers to be determined. [2]

16 Simplify $\dfrac{1}{2 + \sqrt{2}} + \dfrac{1}{2 - \sqrt{2}}$. [2]

17 Express $\dfrac{3\sqrt{3} - \sqrt{2}}{2\sqrt{3} + \sqrt{2}}$ in the form $a + b\sqrt{n}$ where a and n are integers and b is a rational number. [4]

18 Factorise $x^2 + 3x - 40$. [2]

19 Factorise $3x^2 - 11xy + 10y^2$. [3]

20 Rearrange the formula $S = 2\pi rh + 2\pi r^2$ so that h is the subject. [2]

21 Make x the subject of the formula $y = \dfrac{x + 2}{2x + 1}$. [4]

22 Write $\dfrac{3}{x(x + 1)} + \dfrac{2}{x(x - 1)}$ as a single fraction in its simplest form. [3]

23 Solve the equation $\dfrac{2}{x + 1} + 1 = x$. [3]

24 Write $x^2 + 6x + 7$ in the form $(x + a)^2 + b$ where a and b are to be determined. [3]

25 Work out the values of a, b and c such that $6 - 4x - x^2 = a + b(x + c)^2$. [4]

26 You are given that $f(x) = 2x + 1$ and $g(x) = x^2$. Find

(i) $fg(x)$

(ii) $gf(x)$. [4]

27 You are given that $f(x) = x^2$ and $g(x) = x + 2$.

Find the value of k such that $fg(k) = gf(k)$. **[4]**

28 You are given $f(x) = 2x - 3$. Find $f^{-1}(4)$. **[3]**

29 You are given that $f(x) = 2x + 1$.

(i) Work out $f^{-1}(x)$. **[2]**

(ii) On the same graph, sketch the curves $y = f(x)$ and $y = f^{-1}(x)$. **[2]**

30 You are given the function $f(x) = x^2 + 1$ in the domain $x \geqslant 0$.

On one graph, draw the curves $y = f(x)$, $y = f^{-1}(x)$ and $y = x$. **[4]**

31 (i) On one graph, sketch the curves $y = 2e^x$ and $y = e^{-x} + 2$. **[4]**

(ii) Hence find an approximate solution to the equation $2e^x = e^{-x} + 2$. **[1]**

32 Sketch the graph of $y = f(x)$ where

$y = x$ for $0 \leqslant x < 1$

$y = 1$ for $1 \leqslant x < 2$

$y = 1 - (x - 2)^2$ for $2 \leqslant x < 3$

$y = 0$ for $x \geqslant 3$ **[4]**

33 The graph of $y = f(x)$ is shown.

The domain of x is $0 \leqslant x < 5$.

Define $y = f(x)$. **[3]**

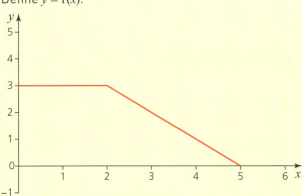

34 Solve the equation $x^2 - 4x - 7 = 0$.

Give your answers correct to three significant figures. **[3]**

35 Solve the equation $\dfrac{3}{x-1} + \dfrac{2}{x-2} = 1$ giving the roots exactly. **[4]**

36 Solve simultaneously the equations $2x + y = 1$, $3x - 2y = 12$. **[4]**

37 Solve simultaneously the equations $y = x + 6$ and $y = x^2 - x + 3$. **[4]**

38 You are given $f(x) = x^3 - 8x^2 + 5x + 14$.

(i) Show that $(x - 2)$ is a factor of $f(x)$. **[2]**

(ii) Solve the equation $f(x) = 0$. **[3]**

39 You are given that the cubic equation $x^3 + ax^2 + bx - 22 = 0$. has three distinct, positive roots.

By forming two equations in a and b, find the values of a and b. **[5]**

40 Solve the following inequalities.

(i) $3 - x > 5(x + 1)$. **[3]**

(ii) $x^2 - 5x < 6$. **[3]**

41 Find the set of integers that satisfy the inequality $-8 < 3x - 1 < 13$. [3]

42 Write $\sqrt{\dfrac{\left(x^{\frac{3}{2}}\right)^4 \times x^3}{x^5 \times x^2}}$ as a single power of x. [2]

43 Solve the equation $x^{\frac{3}{2}} = 8$. [2]

44 Prove that, for all integers, $n > 0$, $2(n + 1) - (3 - 2n)$ is positive. [3]

45 Prove that, for all integers, $n > 1$, $n^2 - n$ is not prime. [3]

46 On 1 January one year £2000 is deposited into a savings account which pays compound interest. On 1 January the next year the amount had become £2100.

   (i) What was the rate of interest in that year? [2]

   (ii) On the assumption that the rate of interest remains constant, find the amount in the account after 3 years. [2]

47 A sequence of positive integers, $u_1, u_2, u_3, \ldots$ is such that $u_{n+1} = 5u_n + 6$ with $u_1 = 1$. Work out u_4. [2]

48 A sequence of positive numbers, $u_1, u_2, u_3, \ldots$ is such that $u_n = \dfrac{2n}{3n + 1}$.

   Work out the limiting value of $u_n = \dfrac{2n}{3n + 1}$ as $n \to \infty$. [2]

Short answers on pages 120–121

Full worked solutions online

CHECKED ANSWERS

SECTIONS 2, 3 AND 4

Target your revision (Chapters 5–9)

Try answering each question below. If you get stuck, follow the page reference underneath to revise that topic.

1 The gradient of a line
Work out the gradient of the line joining the points (−1, 3) and (3, −2).

(see page 58)

2 The distance between two points
Find the distance between the points (2, 4) and (7, −2).

(see page 58)

3 The midpoint of a line joining two points
Find the midpoint of the line joining (12, −1) to (4, 7).

(see page 58)

4 Parallel and perpendicular lines
Write down the gradient of a line that is perpendicular to the line with gradient 2.

(see page 58)

5 The equation of a line
Find the equation of the line through (1, 3) with gradient −1.

(see page 60)

6 The intersection of two lines
Find the point of intersection of the lines with equations $y = 2x − 9$ and $2x + 3y = 5$.

(see page 60)

7 Dividing a line in a given ratio
The points A and B have coordinates (1, 1) and (4, 7). Find the coordinates of the point C which divides AB in the ratio 2 : 1.

(see page 60)

8 Equation of a circle with centre the origin
Write down the equation of a circle with centre the origin and radius 3.

(see page 62)

9 Equation of a circle with centre not at the origin
Work out the equation of the circle with centre (2, 3) and radius 5.

(see page 62)

10 Finding the centre and radius
Find the centre and radius of the circle with equation $x^2 + y^2 + 10x − 2y − 23 = 0$.

(see page 62)

11 Determining whether a point lies inside or outside a circle
Determine whether the point (3, 4) lies inside or outside the circle with equation $x^2 + y^2 − 14x + 2y + 25 = 0$.

(see page 62)

12 Perpendicular bisector of a chord
Points A(−5, 5) and B(7, 1) lie on the circle with equation $x^2 + y^2 = 50$.
Use coordinate geometry to show that the perpendicular bisector of the line AB passes through the centre of the circle.

(see page 64)

13 Pythagoras' theorem
Two sides of a right-angled triangle are 5 cm and 7 cm.
Work out the two possible values for the length of the third side.

(see page 70)

14 Angle properties of a polygon
Work out the size of each of the interior angles of a regular octagon.

(see page 70)

15 Special polygons
An exterior angle of a regular polygon is 20°.
Work out the number of sides of the polygon.

(see page 70)

16 The angle at the centre of a circle
C is the centre of a circle and A, B, D are points on the circumference.
Angle ADB = 52°. Work out angle ACB.

(see page 73)

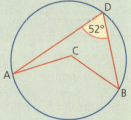

17 The angle in a semi-circle
A, B, C and D are points on a circle.
The angle at B is 90°.
Write down the angle at D.

(see page 73)

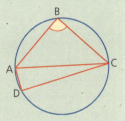

18 Angles in the same segment

The points A, B, C and D lie on a circle.
Angle ADC = 90°.
Angle ABD = 60°.
Work out angle CAD.

(see page 73)

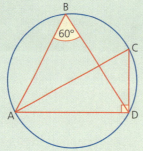

19 Angles in a cyclic quadrilateral

Points A(2, 6) and B(−6, 2) lie on the circle with equation $x^2 + y^2 = 40$ and centre O.
The tangents to the circle at A and B meet at a point T.
(i)   Show that OA and OB are perpendicular.
(ii) Hence deduce that AT and BT are perpendicular.

(see page 73)

20 Proof

Prove that the diagonals of a rhombus are perpendicular.

(see page 75)

21 Circle proof

The tangents from points R and S on a circle, centre O, meet at T.
Prove that OT is perpendicular to the chord RS.

(see page 75)

22 Special angles

In the right-angled triangle ABC the angles at A and B are 45° and AB = 7 cm.
Work out the lengths of AC and BC.

(see page 76)

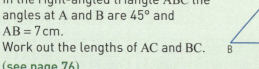

23 Angles of elevation and depression

A church tower stands vertically on horizontal ground. From the top of the tower the angle of depression to the gate is 41°. The height of the tower is 35 m. Work out the length of the path from the foot of the tower to the gate.

(see page 76)

24 Bearings

A ship is initially at point B due South of a lighthouse at L. The ship sails on a course of 040°. At its nearest point to the lighthouse, it is at S.
What is the bearing of the lighthouse from the ship when it is at S?

(see page 77)

25 Trigonometric ratios for angles of any size

Write down the following ratios:
sin 170°, cos 220°, tan 340°.

(see page 79)

26 Graphs of trigonometric functions

Sketch the graph of $y = \tan x$ for $0° \leqslant x \leqslant 360°$.

(see page 80)

27 Solution of trigonometric equations

Solve the equation $3 \sin \theta = 2$ for $0° \leqslant \theta \leqslant 360°$.

(see page 81)

28 Trigonometric identities

Show that $\cos\theta + \sin\theta \tan\theta = \dfrac{1}{\cos\theta}$.

(see page 81)

29 Trigonometric equations using identities

Solve the equation $3\cos^2\theta - \sin\theta - 2 = 0$ for $0° \leqslant \theta \leqslant 360°$.

(see page 81)

30 The area of a triangle

In the triangle PQR, QR = 6 cm, PR = 4 cm and the angle at R is 70°.
Work out the area of the triangle.

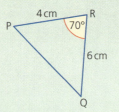

(see page 84)

31 The sine rule

In the triangle LMN, the angles at M and N are 35° and 65°. LN = 6 cm.
Find the length LM.

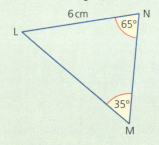

(see page 86)

32 The cosine rule

In the triangle ABC, AB = 7 cm, AC = 8 cm and angle A = 55°.
Work out the length of the side BC.

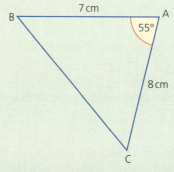

(see page 86)

33 The angle between a line and a plane

A box is a cuboid ABCDEFGH with horizontal base ABCD and top EFGH with E vertically above A.
$AE = 10\,cm$, $AB = 12\,cm$ and $AD = 15\,cm$.
Work out the angle that the diagonal AG makes with the base ABCD.

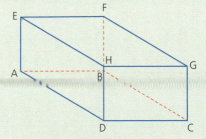

(see page 89)

34 Angle of greatest slope

A hillside can be modelled by a rectangle AEFD at an angle to a horizontal rectangle ABCD.
E and F are vertically above B and C.
$AD = 100\,m$, $CD = 60\,m$ and $CF = 20\,m$.
Find the angle of greatest slope.

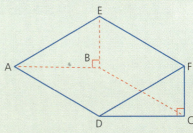

(see page 89)

35 Finding the gradient of a curve at a point

Find the gradient of the curve whose equation is $y = x^3 - 2x + 4$ at the point (2, 8).

(see page 94)

36 Tangent to a curve

Find the equation of the tangent to the curve $y = x^3 + x^2 + 2x - 5$ at the point (1, −1).

(see page 97)

37 Normal to a curve

Find the equation of the normal to the curve $y = x^3 - 3x^2 + 3x - 4$ at the point (3, 5).

(see page 97)

38 Increasing and decreasing functions

Determine the range of values for which $y = 2x^3 - 15x^2 + 24x + 10$ is a decreasing function.

(see page 98)

39 Turning points

You are given the equation of a curve $y = x^3 - 3x^2 - 9x + 4$.

(i) Determine the turning points on the curve.

(ii) For these values of x work out the value of $\dfrac{d^2 y}{dx^2}$.

(iii) Explain what these values indicate with respect to the nature of the turning points.

(see page 99)

40 Determining the nature of a turning point

Determine the nature of the turning points of the curve $y = 2x^3 - 3x^2 - 12x + 6$.

(see page 99)

41 Multiplying matrices

Given that $\mathbf{A} = \begin{pmatrix} 1 & 2 \\ -1 & 3 \end{pmatrix}$ and $\mathbf{B} = \begin{pmatrix} 2 & 1 \\ 1 & -4 \end{pmatrix}$, show that $\mathbf{AB} \neq \mathbf{BA}$.

(see page 103)

42 Equal matrices

Given that $\begin{pmatrix} 1 & 3 \\ 4 & -2 \end{pmatrix} = \begin{pmatrix} 1 & 3 \\ 4 & a \end{pmatrix}$, write down the value of a.

(see page 103)

43 Matrix transformations

Work out the image of the point (2, 3) for the transformation defined by the matrix $\begin{pmatrix} 2 & 1 \\ 1 & 4 \end{pmatrix}$.

(see page 105)

44 Transformations of the unit square

Describe the transformation defined by $\mathbf{M} = \begin{pmatrix} 0 & 1 \\ 1 & 0 \end{pmatrix}$.

(see page 105)

45 Successive transformations

Work out the matrix that defines a rotation of 90° clockwise followed by a reflection in the line $y = x$.

(see page 108)

Short answers on pages 121–122

Full worked solutions online

CHECKED ANSWERS

About this topic

Coordinate geometry provides the link between geometry and algebra. Being able to use algebra makes it much easier to solve many geometric problems. In this chapter, you meet the geometry of circles. You also have a reminder of the geometry of straight lines, met in Chapter 3.

Before you start, remember ...

- how to use coordinates
- Pythagoras' theorem
- how to solve linear equations.

5.1 Points and lines

REVISED

Key facts

1 The gradient of a line

The gradient of a line joining the points (x_1, y_1) and (x_2, y_2) is $\frac{y_2 - y_1}{x_2 - x_1}$.

$\left(\text{This is } \frac{\text{change in } y \text{ values}}{\text{change in } x \text{ values}}.\right)$

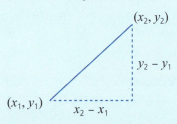

2 The distance between two points

Pythagoras' theorem gives the distance between two points (x_1, y_1) and (x_2, y_2) to be $\sqrt{(x_2 - x_1)^2 + (y_2 - y_1)^2}$.

3 The midpoint of a line joining two points

The midpoint of the line between (x_1, y_1) and (x_2, y_2) is $\left(\frac{x_1 + x_2}{2}, \frac{y_1 + y_2}{2}\right)$.

> The coordinates of the midpoint are the arithmetic means of the end points.

4 Parallel lines

Parallel lines have the same gradient. In this diagram $m_1 = m_2$.

m_1

m_2

5 Perpendicular lines

If two lines have gradients m_1 and m_2 respectively, then they are perpendicular if $m_1 \times m_2 = -1$.

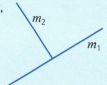

> **Common mistake:** Unless you scale the axes equally perpendicular lines may not look as if they are at right angles to each other.

Worked example

Gradients of lines

1 Find the gradient of the line ST joining
the two points S(1, 2) and T(7, 9).

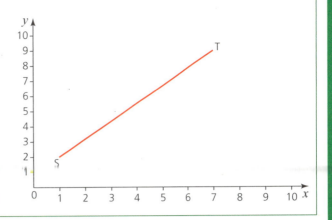

Solution

Taking S to be (x_1, y_1) and T to be (x_2, y_2),
and using $m = \dfrac{y_2 - y_1}{x_2 - x_1}$ gives

$m = \dfrac{9 - 2}{7 - 1} = \dfrac{7}{6}$.

Worked example

The distance between two points

2 Find the distance between the points
L(5, −3) and M(−2, 6).

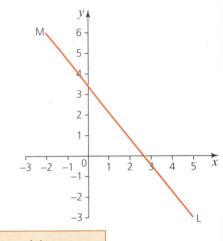

Solution

Taking L to be (x_1, y_1) and M to be (x_2, y_2),
and using

distance $= \sqrt{(x_2 - x_1)^2 + (y_2 - y_1)^2}$ gives

$LM = \sqrt{(-2 - 5)^2 + (6 - -3)^2}$

$\quad = \sqrt{(-7)^2 + (9)^2}$

$\quad = \sqrt{49 + 81} = \sqrt{130}$

> This is Pythagoras' theorem.

> Take care with the signs.

Worked example

The midpoint of a line

3 Find the midpoint of the line UV where the
coordinates of U and V are (−3, 7) and (4, −1).

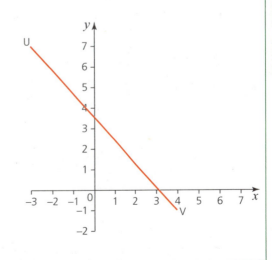

Solution

Taking U to be (x_1, y_1) and V to be (x_2, y_2),
and using

midpoint $= \left(\dfrac{x_1 + x_2}{2}, \dfrac{y_1 + y_2}{2} \right)$

gives the midpoint to be

$\left(\dfrac{-3 + 4}{2}, \dfrac{7 - 1}{2} \right) = \left(\dfrac{1}{2}, 3 \right)$.

Worked example

Parallel and perpendicular lines

4 The coordinates of A, B and C are (2, 1), (6, 3) and (1, 3) respectively. Show that the lines AB and AC are perpendicular.

Solution

Gradient of the line AB = m_1 = $\dfrac{3-1}{6-2}$

$\qquad = \dfrac{2}{4} = \dfrac{1}{2}$

Gradient of the line AC = m_2 = $\dfrac{3-1}{1-2}$

$\qquad = \dfrac{2}{-1} = -2$

$\Rightarrow m_1 \times m_2 = \dfrac{1}{2} \times -2 = -1$

So the lines are perpendicular.

> Find the gradient of each line.

> Use the rule $m_1 \times m_2 = -1$ to test whether the lines are perpendicular.

Exam-style question

TESTED ☐

Points A, B, C and D have coordinates (1, 0), (2, 2), (4, 1) and (2, −3) respectively.
(i) Show that the quadrilateral ABCD is a trapezium.

E is on the line CD with coordinates (3, −1).

(ii) Show that the quadrilateral ABCE is a parallelogram.
(iii) Show additionally that the quadrilateral ABCE is a rhombus.
(iv) Prove that ABCE is a square.

Short answers on page 122

Full worked solution online

CHECKED ANSWERS ☐

5.2 The equation of a line and intersections

REVISED ☐

Key facts

1 **The equation of a line**
The following three formulae give the equation of a straight line.
(a) Gradient m and intercept on y-axis of (0, c): $y = mx + c$
(b) Gradient m and through a given point (x_1, y_1): $y - y_1 = m(x - x_1)$
(c) through two points (x_1, y_1), (x_2, y_2):
$$\frac{y - y_1}{x - x_1} = \frac{y_2 - y_1}{x_2 - x_1}$$

2 **The intersection of two lines**
- Unless two lines are parallel, they meet in a point.
- To find that point, solve the equations simultaneously.
- Two lines are parallel if they have the same gradient.

> See Chapter 4.

3 **Dividing a line in a given ratio**
If the coordinates of A and B are A (x_1, y_1) and B (x_2, y_2), then the coordinates of a point C which divides the line AB in the ratio $p : q$ are
$$\left(\frac{qx_1 + px_2}{p+q}, \frac{qy_1 + py_2}{p+q} \right).$$

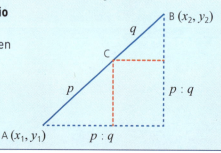

Worked example

The equation of a line

1 The equation of line *l* is $y + 2x - 4 = 0$.
   Find the equation of the line through (2, 3) which is perpendicular to *l*.

Solution

The gradient of the line *l* is –2.

So the gradient of the required line is $\frac{1}{2}$.

It passes through the point (2, 3) so the equation is

$$y - 3 = \frac{1}{2}(x - 2)$$

$$\Rightarrow 2y - 6 = x - 2$$

$$\Rightarrow 2y = x + 4$$

> Write the equation $y + 2x - 4 = 0$ in the form $y = -2x + 4$ to determine the gradient.
> Use $m_1 m_2 = -1$.

> Now you have a line with given gradient through a given points, so use the form
> $y - y_1 = m(x - x_1)$.

> Always leave the equation of a line with only three terms.

Worked example

The intersection of two lines

2 Find the coordinates of the point where the lines $x + 3y = 7$ and $y = 4x - 2$ intersect.

Solution

$$x + 3y = 7$$

$$y = 4x - 2$$

$$\Rightarrow x + 3(4x - 2) = 7$$

$$\Rightarrow x + 12x - 6 = 7$$

$$\Rightarrow 13x = 13$$

$$\Rightarrow x = 1$$

$$\Rightarrow y = 4 \times 1 - 2 = 2$$

The coordinates are (1, 2).

> Because *y* is the subject of one of the equations the easiest method to solve these equations simultaneously is by elimination.

> Don't forget the *y* coordinate.

> **Common mistake:** Don't forget to answer the question. The last line is important.

Worked example

Dividing a line in a given ratio

3 A and B have coordinates (1, 4) and (9, 8).
   The point C divides AB in the ratio 3 : 1.

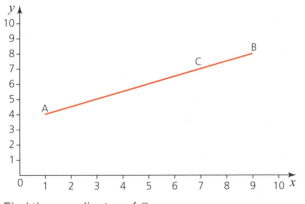

Find the coordinates of C.

Solution

Using $\left(\dfrac{qx_1 + px_2}{p+q}, \dfrac{qy_1 + py_2}{p+q}\right)$ with $p = 3$, $q = 1$, (x_1, y_1) is $(1, 4)$ and (x_2, y_2) is $(9, 8)$, gives C to be $\left(\dfrac{1+27}{4}, \dfrac{4+24}{4}\right)$, i.e. $(7, 7)$.

> Use the formula given where $p : q = 3 : 1$ and so $p = 3$, $q = 1$ and $p + q = 4$.

Exam-style question

TESTED

Points A and B have coordinates $(-2, -3)$ and $(4, 9)$. Point C lies on AB and divides AB in the ratio $2 : 1$.

Find the equation of the line through C which is perpendicular to AB.

Short answer on page 122

Full worked solution online

CHECKED ANSWER

5.3 The circle

REVISED

Key facts

1 **Definition**
A circle is the locus of all points that are a fixed distance (the radius) from a fixed point (the centre).

> A circle is a two-dimensional shape. In three dimensions this would be the definition of a sphere.

2 **The equation of a circle**
The equation of a circle with centre (a, b) and radius r is $(x - a)^2 + (y - b)^2 = r^2$.

> This is Pythagoras' theorem.

3 **Equation of a circle with centre at the origin**
The equation of a circle with centre the origin and radius r is $x^2 + y^2 = r^2$.

> This is a special case of the general result when (a, b) is $(0, 0)$.

4 **Finding the centre and radius of a circle given its equation**
The equation $(x - a)^2 + (y - b)^2 = r^2$ may be written as
$x^2 + y^2 - 2ax - 2by + a^2 + b^2 - r^2 = 0$
When you are given the equation of a circle, you can find the values of a and b and so its centre (a, b) by comparing it with this general form. You can then go on to find its radius r.

> **Common mistakes:** if you try plotting this curve on your graphic calculator, it will not look like a circle if the axes have different scales. Also, your calculator may only show half the circle.

Worked example

The equation of a circle

1 Find the equation of the circle with centre $(1, 2)$ and radius 4.

Solution

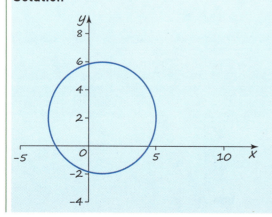

Applying the formula gives

$(x - 1)^2 + (y - 2)^2 = 4^2 = 16$

If you are asked for the equation of a circle in an examination, then either form will be acceptable unless one specific form is demanded.

If this is expanded

$(x - 1)^2 + (y - 2)^2 = 4^2 = 16$

$\Rightarrow x^2 - 2x + 1 + y^2 - 4y + 4 = 16$

$\Rightarrow x^2 + y^2 - 2x - 4y + 4 + 1 = 16$

$\Rightarrow \qquad x^2 + y^2 - 2x - 4y = 16 - 1 - 4$

$= x^2 + y^2 - 2x - 4y - 11 = 0$

Worked example

Finding the centre and radius

In the previous example you were given the centre and radius of a circle and asked to find its equation. The next example is the other way round; you are given the equation of a circle and are asked to find its centre and radius.

2   Find the centre and radius of the circle $x^2 + y^2 - 4x + 6y + 4 = 0$.

Solution

$\qquad\qquad x^2 + y^2 - 4x + 6y + 4 = 0$

$\Rightarrow \qquad (x^2 - 4x) + (y^2 + 6y) + 4 = 0$

$\Rightarrow (x^2 - 4x + 4) + (y^2 + 6y + 9) + 4 = 4 + 9$

$\Rightarrow \qquad\qquad (x - 2)^2 + (y + 3)^2 = 9$

So the centre of the circle is (2, –3) and the radius is 3.

Complete the square for the x and y terms.

Worked example

Whether a point is inside or outside a circle

3   Determine whether the point (4, −6) lies inside or outside the circle with equation $x^2 + y^2 - 6x + 8y + 19 = 0$.

Solution

Comparison with the general equation for a circle
$x^2 + y^2 - 2ax - 2bx + a^2 + b^2 - r^2 = 0$ gives $a = 3$ and $b = -4$.

So, the centre is (3, −4).

The radius, r, is given by $a^2 + b^2 - r^2 = 19$.

So $r^2 = 6$ and $r = \sqrt{6}$.

The distance, d, from the centre of the circle (3, −4) to the point (4, −6) is given by

$d^2 = (4 - 3)^2 + (-6 - -4)^2 = 1 + 4 = 5$

$\Rightarrow d = \sqrt{5}$

Since $\sqrt{5} < \sqrt{6}$ the point is inside the circle.

Find the centre of the circle and the radius.
Note that r^2 is satisfactory here.

Use Pythagoras to find distance of the point from the centre.

The distance is less than the radius, so it lies inside the circle.

Exam-style question

A circle has centre (0, 1) and radius 3.

(i)   Find the equation of the circle.
(ii) The circle cuts the x-axis at A and B. Find the length AB.
(iii) Determine whether the point (2, 3.5) lies inside, outside or on the circle.

Short answer on page 122

Full worked solution online

5.4 Circle geometry

Key facts

The following geometry facts for a circle need to be known and may be used in an examination.

1   Angle in a semi-circle
The angle in a semi-circle is 90°.

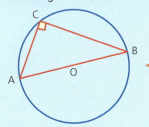

Angle ACB = 90°.

2   Bisection of a chord
The perpendicular from the centre to a chord bisects the chord.

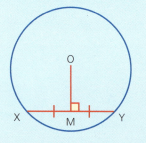

XM = YM.

3   Angle between a tangent and radius at a point
The angle between a tangent and a radius at a point on the circumference is 90°.

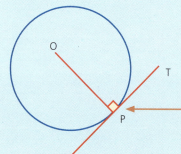

Angle OPT = 90°.

4 Two tangents from a point to a circle

Two tangents from a point to a circle are equal in length.

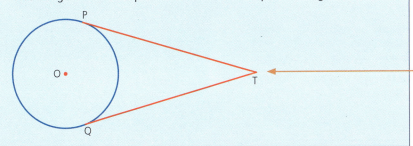

TP = TQ.

Worked example

Angle in a semi-circle

1 Points P, Q and R have coordinates (1, 2), (4, 5) and (5, 4) and lie on a circle.

(i) Show that the lines PQ and QR are perpendicular.

(ii) Hence find the centre of the circle.

Solution

(i)

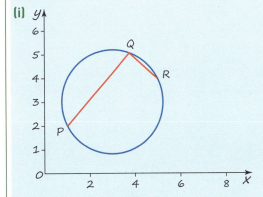

Gradient PQ = $\dfrac{5-2}{4-1} = \dfrac{3}{3} = 1$

Find the gradients.

Gradient QR = $\dfrac{4-5}{5-4} = -1$

Since $1 \times -1 = -1$, the lines are perpendicular.

Show that they satisfy $m_1 m_2 = -1$.

(ii) Since PQ and QR are perpendicular and the points lie on a circle it means that the angle PQR is an angle in a semi-circle.

Hence PR is a diameter of the circle.

So the centre is the midpoint of PR,

i.e. $\left(\dfrac{1+5}{2}, \dfrac{2+4}{2}\right)$ which is (3, 3).

Use the formula for the midpoint.

Worked example

Bisection of a chord

2 A circle has equation $x^2 + y^2 = 50$ and points A(−7, 1) and B(5, 5) lie on the circumference.
  (i) Find the equation of the line through the centre which is perpendicular to the chord AB.
  (ii) Determine the coordinates of the point where this line cuts AB.

Solution

(i)

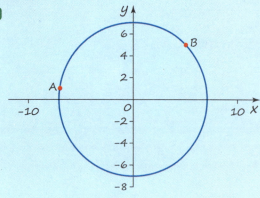

Gradient of chord AB $= \dfrac{5 - 1}{5 - (-7)}$

$\qquad\qquad\qquad\quad = \dfrac{4}{12} = \dfrac{1}{3}$

$\Rightarrow$ Gradient of perpendicular line $= -3$

So equation is $y - 0 = -3(x - 0) \Rightarrow y + 3x = 0$.

(ii) The line goes through the midpoint of AB

which is $\left(\dfrac{-7 + 5}{2}, \dfrac{1 + 5}{2}\right)$, i.e. (−1, 3).

> Find the gradient of AB.

> Use $m_1 m_2 = -1$.

> The circle has centre the origin.

Worked example

Angle between a tangent and radius at a point

3 (i) Show that the point T(2, 3) lies on the circle with equation $x^2 + y^2 = 13$.
  (ii) Find the equation of the tangent to the circle at T.

Solution

(i) Substituting (2, 3) into $x^2 + y^2$ gives $2^2 + 3^2 = 4 + 9 = 13$.
  So, the point (2, 3) lies on the circle $x^2 + y^2 = 13$.

(ii)

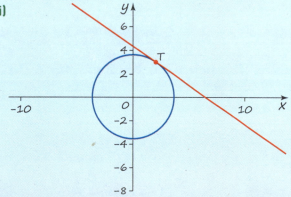

The gradient of OT $= \dfrac{3-0}{2-0} = \dfrac{3}{2}$.

$\Rightarrow$ Gradient of tangent $= -\dfrac{2}{3}$

$\Rightarrow$ Equation of tangent is $y - 3 = -\dfrac{2}{3}(x - 2)$

$\qquad\qquad \Rightarrow \quad 3y - 9 = 4 - 2x$

$\qquad\qquad \Rightarrow 3y + 2x = 13$

> Find the gradient of the radius from the centre to T.

> Use $m_1 \times m_2 = -1$.

> Use $y - y_1 = m(x - x_1)$.

> Your final equation should have only three terms.

Worked example

Two tangents from a point to a circle

4 A circle has equation $x^2 + y^2 = 34$.
   (i) Find the equations of the tangents from the points P(3, 5) and Q(−5, 3).
   The two tangents meet at T.
   (ii) Find the coordinates of T.
   (iii) Find the lengths of TP and TQ and show that they are equal.
   (iv) What theorem does this result demonstrate?

Solution

(i) The centre is at O(0, 0).

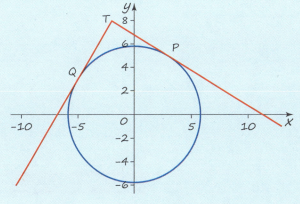

Finding the equation of the tangent PT:

Gradient of OP $= \dfrac{5-0}{3-0} = \dfrac{5}{3}$

> Start by finding the equation of each tangent in the same way as in Example 3.

$\Rightarrow$ Gradient of tangent $= -\dfrac{3}{5}$

$\Rightarrow$ Equation of tangent at P is $y - 5 = -\dfrac{3}{5}(x - 3)$

$\qquad\qquad \Rightarrow 5y - 25 = 9 - 3x$

$\qquad\qquad \Rightarrow 5y + 3x = 34$

Finding the equation of the tangent QT:

Gradient of OQ $= \dfrac{3-0}{-5-0} = \dfrac{3}{-5}$

$\Rightarrow$ Gradient of tangent $= \dfrac{5}{3}$

$\Rightarrow$ Equation of tangent at Q is $y - 3 = \dfrac{5}{3}(x + 5)$

$\qquad\qquad \Rightarrow \quad 3y - 9 = 5x + 25$

$\qquad\qquad \Rightarrow 3y - 5x = 34$

(ii) T is the intersection of the two tangents:

$5y + 3x = 34$ ①

$3y - 5x = 34$ ②

$5 \times ① + 3 \times ②: 34y = 272$

$\Rightarrow y = 8$ giving $3x = 34 - 40 = -6$

$\Rightarrow x = -2$

T has coordinates $(-2, 8)$.

> Find the intersection of the two lines by solving the equations simultaneously.

> Start by eliminating x.

> Then find x for the value of y.

(iii) Finding the distances TP and TQ:

$$TQ = \sqrt{(-2 - -5)^2 + (8 - 3)^2}$$
$$= \sqrt{3^2 + 5^2} = \sqrt{34}$$

$$TP = \sqrt{(-2 - 3)^2 + (8 - 5)^2}$$
$$= \sqrt{5^2 + 3^2} = \sqrt{34}$$

So TP = TQ.

> Note: this could have been proved by geometry. The triangles TPO and TQO are congruent (the same shape and the same size) and so the two sides TP and TQ are equal.

(iv) This result demonstrates the theorem that the two tangents from a point to a circle are equal in length.

Exam-style question

A(−1, 3) and B(5, 11) are the end points of a diameter of a circle.

(i) Work out the centre and radius of the circle.

C is the point (−2, 10).

(ii) Show that C lies on the circle.

(iii) Find the gradients of AC and BC and hence show that they are perpendicular.

(iv) What theorem does this result demonstrate?

Short answer on page 122

Full worked solution online

Chapter 6: Geometry I

About this topic

This chapter covers revision and extension of GCSE geometry, including calculations of length, area and volume of a variety of shapes. Angles and circle theorems are also covered.

You have already met the trigonometric ratios sin, cos and tan, for angles between 0° and 90°, and used them to solve right-angled triangles. In this chapter the definition is extended to include angles of any size, and you learn to solve any triangle. Working with angles of any size allows you to think of $\sin\theta$, $\cos\theta$ and $\tan\theta$ as mathematical functions in their own right. This important step is illustrated by the final section on identities.

Before you start, remember ...

- Pythagoras' theorem
- the circle theorems from GCSE
- the three trigonometric ratios, sin, cos and tan (see page 76)
- bearings.

6.1 GCSE revision on mensuration and angles

REVISED

Key facts

1 Formulae for length, area and volume

These formulae need to be known. You have met them in GCSE.

Area of triangle $= \frac{1}{2} \times$ base $\times$ height $= \frac{1}{2}bh$

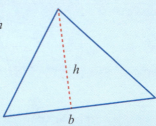

Area of parallelogram $=$ base $\times$ height $= bh$

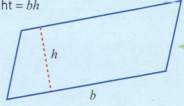

Common mistake: The 'height' of a parallelogram is not the length of the sides that are not the base.

Area of trapezium $= \frac{1}{2} \times$ sum of parallel sides $\times$ distance between them

$= \frac{1}{2} \times (a + b)h$

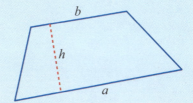

Circumference of circle = $\pi d = 2\pi r$

Area of circle = πr^2

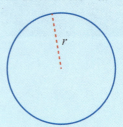

Volume of prism = area of cross section × length

= Al

A 3D shape is a prism if there is a constant cross-section. This might be a triangle or a circle or any shape of quadrilateral.

2 Pythagoras' theorem
For a right-angled triangle in which the two adjacent sides have lengths a and b with hypotenuse length c

$c^2 = a^2 + b^2$

3 Angles at a point
- Vertically opposite angles are equal.

  $a = c$ and $b = d$.
- Adjacent angles on a straight line add up to 180°.

  $a + b = c + d = 180°$.

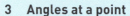

4 Angles relating to parallel lines
- Alternate angles are equal.
- Corresponding angles are equal.
- Interior angles add up to 180°.

Angles a and b are alternate angles and are equal.
Angles a and c are corresponding angles and are equal.
Angles a and d are interior angles and add up to 180°.

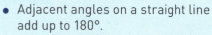
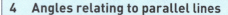

5 Angle properties of a polygon
- The angle sum of a triangle is 180°.
- The angle sum of a quadrilateral is 360°.
- The angle sum of a polygon with n sides is $(n-2) \times 180°$.

6 Special polygons
- A parallelogram is a quadrilateral with opposite sides parallel.
- A rhombus is a parallelogram with all four sides equal.
- A trapezium is a quadrilateral with a pair of sides parallel.
- A kite is a quadrilateral with one line of symmetry and the diagonals intersecting at 90°.
- A regular polygon is a polygon with all sides equal and all angles equal.

Worked examples

Area of plane figures

1 Two sides of a parallelogram have length 6 cm. The area is 30 cm². Work out the height of the parallelogram.

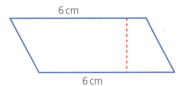

Solution

$$\text{Area} = \text{base} \times \text{height}$$

$$\Rightarrow \quad 30 = 6 \times \text{height}$$

$$\Rightarrow \text{height} = \frac{30}{6} = 5 \text{ cm}$$

2 The diagram shows a trapezium with area 24 cm². BC is perpendicular to DC. Work out the length of CD.

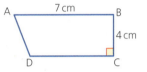

Solution

Because BC is perpendicular to the parallel sides its length is the height.

$$\text{So area} = \frac{1}{2}(AB + CD) \times BC$$

$$\Rightarrow \quad 24 = \frac{1}{2}(7 + CD) \times 4$$

$$\Rightarrow 7 + CD = 12$$

$$\Rightarrow \quad CD = 5 \text{ cm}$$

Worked example

Pythagoras' theorem

3 A ladder of length 10 metres is placed so that it leans against a vertical wall with its foot 2.5 metres from the base of the wall on horizontal ground.
   How far up the wall does the ladder reach?

Solution

$$h^2 + 2.5^2 = 10^2$$

$$\Rightarrow \quad h^2 = 100 - 6.25 = 93.75$$

$$\Rightarrow \quad h = \sqrt{93.75} = 9.68$$

The ladder reaches 9.68 m up the wall.

Worked examples

Angles in a polygon

4 Work out the size of one of the interior angles of a regular pentagon.

Solution

Angle sum of a pentagon = $(5 - 2) \times 180 = 540°$

$\Rightarrow$ One angle $= \dfrac{540}{5} = 108°$

5 **(i)** Work out the size of an exterior angle of a regular hexagon.
   (ii) Hence find the interior angle of a regular hexagon.

Solution

(i) A hexagon has 6 sides.
The sum of the exterior angles is 360°.
So each exterior angle is $360° \div 6 = 60°$.

(ii) Sum of an interior angle and an exterior angle is 180°.
So one interior angle is 120°.

> As you go round a polygon you do a complete rotation of 360° and this is the sum of the exterior angles.

6 An interior angle of a regular polygon is 140°.
Work out the number of sides of the polygon.

Solution

Interior angle = 140°

So exterior angle = $180° - 140° = 40°$

$\dfrac{360}{40} = 9$ so there are 9 sides.

> The sum of an interior angle and an exterior angle is 180° as they form a straight line.

Exam-style question

TESTED ☐

The diagram shows a regular hexagon and a regular pentagon with a common side.

Work out the angle x.

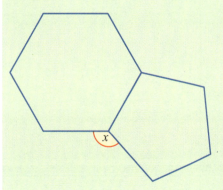

Short answer on page 122

Full worked solution online

CHECKED ANSWER ☐

6.2 Circle theorems

Key facts

Note: other circle theorems have been covered in Chapter 5.

1 The angle at the centre of a circle
The angle at the centre of a circle is twice the angle at the circumference.

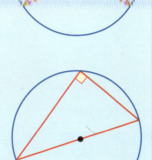

> You can write this more formally as 'The angle subtended by a chord at the centre is double the angle subtended at a point on the circumference.'

2 The angle in a semi-circle
The angle in a semi-circle is 90°.

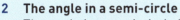

3 Angles in the same segment
Angles in the same segment are equal.

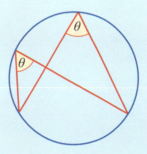

4 Angles in a cyclic quadrilateral
Opposite angles of a cyclic quadrilateral add up to 180°.

> A cyclic quadrilateral is one where all four vertices lie on a circle.

5 The alternate segment theorem
The angle between a chord and a tangent is equal to the angle in the alternate segment.

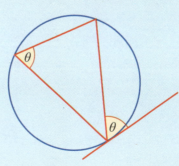

Worked example

Angle in a semi-circle

1 AB is a diameter of the circle and C is a point on the circumference. Angle BAC = 28°. Work out the angle x.

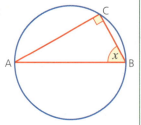

> This uses the fact that the angle in a semi-circle is 90°.

> **Warning:** In these questions you should not try to measure angles with a protractor, but use the theorems to calculate answers.

Solution

Angle C = 90° (angle in a semi-circle)

x = 180° – 90° – 28° (angle sum of triangle)

= 62°

Worked example

Angles in the same segment

2 AB is a chord of the circle and C and D are points on the circumference.
AD = BD. Angle C = 40°.
Work out the angle DBA.

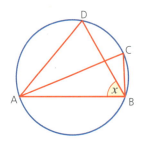

Solution

D = C = 40° (angles in the same segment)

Triangle ADB is isosceles (as AD = BD).

So angles DAB and DBA are equal.

Their sum is 180° – 40° = 140° so angle DBA = 70°.

> This uses the facts that the angles in the same segment are equal and also that the base angles of an isosceles triangle are equal.

Exam-style question

TESTED

In the diagram, the circle ABCD has centre O.

Angle DOB = 160°.

Work out the angle at A.

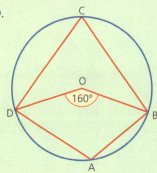

Short answer on page 122

Full worked solution online

CHECKED ANSWER

6.3 Geometric proof

Key facts

In a geometric proof reasons must be given for all statements.
The reasons can use the standard results given earlier in this chapter.

Worked examples

Geometric proofs

1 Prove that the tangent and the radius at a point on a circle are perpendicular.

Proof

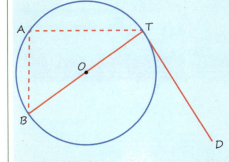

In the diagram, TD is the tangent to the circle at T. OT is the radius through T and BT is the diameter.

Construct the triangle BAT where A is a point on the circumference.

Angle DTB = angle BAT (alternate segment theorem).

Angle BAT = 90° (angle in a semi-circle).

Therefore angle DTB = 90°.

So lines TD and TB are perpendicular.

> Show your own constructions carefully – these are the dashed lines in the diagram.

> State the theorem you use to make a deduction.

> This is also covered in Chapter 5.

2 In the diagram, AC is a diameter of a circle. B and D are points on the circumference of the circle on either side of AC.
Angle CAB = x and angle BDA = y.
Prove that $x + y = 90°$.

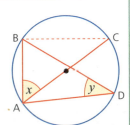

Proof

Join BC.

Angle BCA = angle BDA = y (angles in the same segment)

Angle ABC = 90° (angle in a semi-circle)

$\Rightarrow x + y + 90° = 180°$ (angle sum of triangle)

$\Rightarrow x + y = 90°$, as required.

> State your construction.

> State the theorems you have used to make your deductions.

Exam-style question

In the diagram, A, B and D are points on the circumference of a circle, centre C.

Angle CAB = x and angle ADB = y.

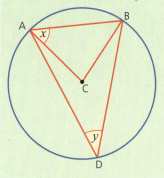

Prove that $x + y = 90°$.

Short answer on page 122

Full worked solution online

6.4 Trigonometry in two dimensions

Key facts

1 The definitions of the three trigonometrical ratios

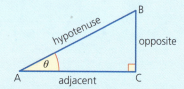

$$\sin\theta = \frac{\text{opposite}}{\text{hypotenuse}}, \quad \cos\theta = \frac{\text{adjacent}}{\text{hypotenuse}}, \quad \tan\theta = \frac{\text{opposite}}{\text{adjacent}}$$

2 Special angles

$\sin 45° = \cos 45° = \frac{1}{\sqrt{2}}, \tan 45° = 1$

$\sin 30° = \frac{1}{2}, \cos 30° = \frac{\sqrt{3}}{2}, \tan 30° = \frac{1}{\sqrt{3}}$

$\sin 60° = \frac{\sqrt{3}}{2}, \cos 60° = \frac{1}{2}, \tan 60° = \sqrt{3}$

3 Angles of elevation and depression

The angle between the horizontal and the line of sight to the top of the tower is the angle of elevation.

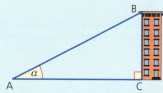

α is the angle of elevation from A to B.

The line AC is horizontal.

If the angle is below the horizontal then it is an angle of depression.

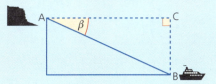

β is the angle of depression from A to B.
The line AC is horizontal.

4 Bearings

The bearing of a direction is the angle measured from north in a clockwise direction. It is usually given as a 3 digit number between 000° and 360°. So north-east is 045° and north-west is 315°.

Worked example

Using the sin ratio

1 In the triangle ABC, C is a right angle. BC = 6 cm, AB = 10 cm. Work out the angle A.

Solution

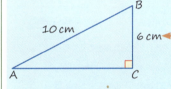

> BC is the side opposite the angle A. AB is opposite the right angle at C and so is the hypotenuse.

$$\sin A = \frac{opposite}{hypotenuse} = \frac{6}{10} = 0.6$$

$$\Rightarrow A = 36.9°$$

Worked example

Using the cos ratio

2 In the triangle PQR, R is a right angle. QR = 5 cm. Angle $Q = 42°$. Work out the length of PQ.

Solution

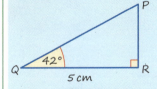

> This example involves the cos ratio as the two sides used are the adjacent and the hypotenuse.

$$\cos Q = \frac{adjacent}{hypotenuse} = \frac{5}{PQ}$$

$$\Rightarrow \frac{5}{PQ} = 0.7431$$

$$\Rightarrow PQ = \frac{5}{0.7431} = 6.728...$$

$$PQ = 6.73 \text{ cm (to 3 s.f.)}.$$

Worked example

Angle of elevation

3 John stands 30 metres away from a 10 metre pole that stands vertically on horizontal ground. Work out the angle of elevation of the top of the pole from John.

Solution

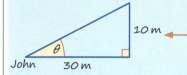

In the diagram the angle of elevation is θ.

$$\tan\theta = \frac{\text{opposite}}{\text{adjacent}}$$

$$\tan\theta = \frac{10}{30} = \frac{1}{3}$$

$$\Rightarrow \quad \theta = 18.4°$$

The angle of elevation is 18.4°.

Note: Always draw a diagram when answering questions like this.

In this case it is easy to see that the sides in the right-angled triangle are opposite and adjacent so it is the tan ratio that is required.

Exam-style question

TESTED ☐

Jean is walking on horizontal ground from a point A to a point C.

She can either walk directly over rough ground or she can take a straight path from A to a point B and then another straight path from B to C.

B is 300 metres due east of A and C is 500 metres due north from B.

Find

(i)   the distance Jean walks if she walks in a straight line from A to C,
(ii) the bearing on which she must walk,
(iii) the bearing of her return walk, from C to A.

Short answer on page 122

Full worked solution online

CHECKED ANSWERS ☐

6.5 Trigonometric functions for angles of any size

REVISED ☐

Key facts

1 Trigonometric ratios for angles of any size

The first diagram shows a point in the 1st quadrant with coordinates (x, y). The line joining it to the origin has length r and is at angle θ to the positive x-axis.

The trigonometric ratios are defined as follows:

$$\sin\theta = \frac{y}{r}, \cos\theta = \frac{x}{r}, \tan\theta = \frac{y}{x}$$

> These definitions give sin, cos and tan for any angles.

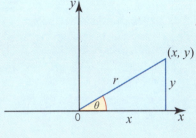

In the second diagram the point is in the 2nd quadrant so that the angle θ is obtuse and the value of x is negative. The same definitions for sin, cos and tan apply.

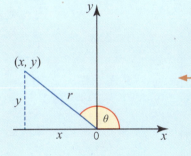

> You can draw equivalent diagrams for points in the 3rd and 4th quadrants.

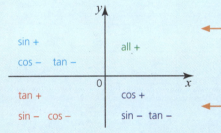

> In the 1st quadrant all three ratios are positive. In the second quadrant (angles 90°–180°) the sine of an angle is positive, the others are negative, and so on for the 3rd and 4th quadrants.

> Example: $\cos 100° = -0.174$, $\sin 100° = 0.985$, $\tan 100° = -5.671$

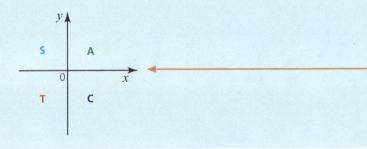

> One way to remember in which quadrant ratios are positive and negative is the mnemonic CAST.

2 Principal angles

When you use your calculator to find an angle, only one angle can be shown, although there are two angles in the range [0°, 360°]. The angle shown on your calculator is called the **principal angle** and the other angle has to be worked out.

In general, calculators will give the principal angle in the following ranges.

$\cos\theta$: $0° < \theta < 180°$
$\sin\theta$ and $\tan\theta$: $-90° < \theta < 90°$

Note that this means that for any value of the ratio there are two angles in the range $0° \leqslant \theta \leqslant 360°$.

Note: You will also meet angles labelled x rather than θ although x is usually reserved for a length.

3 Graphs of trigonometric functions

$y = \sin\theta$

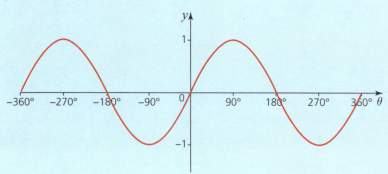

$y = \cos\theta$

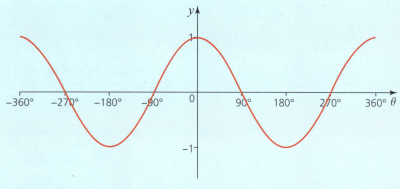

You should know the graphs and be able to use them whenever necessary.

$y = \tan\theta$

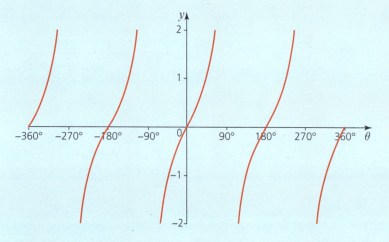

4 Drawing graphs

Drawing graphs is helpful in determining the behaviour of trigonometric functions.

A sketch of the functions gives you an indication of all possible answers.

5 Solution of trigonometric equations

Drawing graphs is useful in determining the roots of trigonometrical equations. Your calculator will give the principal angle from which you need to work out other roots within the required range.

5 Trigonometric identities

An identity is true for all permissible values of the variable. The symbol ≡ is sometimes used.

By contrast an equation is only true for certain values. The symbol = is used.

These identities should be known.

$$\tan\theta = \frac{\sin\theta}{\cos\theta}$$

$$\sin^2\theta + \cos^2\theta = 1$$

> Note: There are many more trigonometric identities that you will meet in more advanced work.

> For example
> $\tan^2\theta + 1 \equiv \dfrac{1}{\cos^2\theta}$ is true for all values of θ except 90°, 270°, … when $\tan\theta$ is not defined. It is an identity.
> $x^2 - 3x - 40 = 0$ is an equation. It is true for only two values of x, 8 and −5.

7 Trigonometric equations

Trigonometric identities are often used to simplify and solve more complicated equations.

> For example, the equation $2\sin^2\theta + \cos\theta = 2$ can be solved by using the identity $\sin^2\theta + \cos^2\theta = 1$ to make it into a quadratic equation in $\cos\theta$.

Worked examples

Angles greater than 90°

1 Given that $\cos\theta = \cos 230°$ with θ in the range $0° \leqslant \theta \leqslant 180°$, write down the other possible value of θ.

Solution

230° is in the 3rd quadrant and so cos 230° is negative.

So the other angle is in the 2nd quadrant.

230° = 180° + 50°

So the other angle is 180° − 50° = 130°.

> You need to be able to produce this answer without using your calculator to find the actual value of cos 230°.
> However, note that if you find cos 230° and then the inverse it will give you the principal angle which is 130°.

2 You are given that $\sin\theta = 0.4$.
   Find the two values of θ in the range $0° \leqslant \theta \leqslant 360°$.

Solution

The principal angle is $\theta = 23.6°$.

The other angle is in the 2nd quadrant.

So it is $\theta = 180° - 23.6° = 156.4°$.

Worked examples

Solving direct trigonometrical equations

3 Solve the equation $\sin\theta = 0.7$ for $0° \leqslant \theta \leqslant 360°$.

Solution

$\sin\theta = 0.7$

$\Rightarrow \theta = 44.4°$

> This is the principal angle.

There is also a root in the 2nd quadrant,

i.e. 180° − 44.4° = 135.6°.

4 Solve the equation $\cos\theta = 0.4$ for $0° \leqslant \theta \leqslant 360°$.

Solution

The principal angle is $x = 66.4°$.
The other angle is in the 4th quadrant and so is
$x = 360° - 66.4° = 293.6°$.

5 Solve the equation $\tan\theta = -0.8$ for $0° \leqslant \theta \leqslant 360°$.

Solution

$\tan\theta = -0.8$

$\Rightarrow \theta = -38.7°$ (this is the principal angle)

$\Rightarrow \theta = 180° - 38.7° = 141.3°$

or $\theta = 360° - 38.7° = 321.3°$

Worked examples

Solution of equations requiring the use of identities

6 Solve the equation $\sin\theta = 2\cos\theta$ for $0° \leqslant \theta \leqslant 360°$.

Solution

$\sin\theta = 2\cos\theta$

$\Rightarrow \dfrac{\sin\theta}{\cos\theta} = 2$ ← Divide by $\cos\theta$.

$\Rightarrow \tan\theta = 2$ ← Using the identity $\tan\theta = \dfrac{\sin\theta}{\cos\theta}$.

$\Rightarrow \theta = 63.4°$ or $243.4°$

7 Solve the equation $2\cos^2\theta + \sin\theta = 1$ for $0° \leqslant \theta \leqslant 360°$.

Solution

$2\cos^2\theta + \sin\theta = 1$

$\Rightarrow 2(1 - \sin^2\theta) + \sin\theta = 1$ ← Use the identity $\cos^2\theta = 1 - \sin^2\theta$ and arrange into a quadratic equation in $\sin\theta$.

$\Rightarrow 2 - 2\sin^2\theta + \sin\theta = 1$

$\Rightarrow 2\sin^2\theta - \sin\theta - 1 = 0$

$\Rightarrow (2\sin\theta + 1)(\sin\theta - 1) = 0$ ← Solve by factorising.

$\Rightarrow \sin\theta = 1$ or $-\dfrac{1}{2}$ ← Your calculator will probably tell you that $\theta = -30°$ which is not within the range required. If the value of $\sin\theta$ is negative then θ is in the 3rd and 4th quadrants.

$\Rightarrow \theta = 90°, 210°$ or $330°$

Worked example

Trigonometric identities

8 **(i)** Show that $\dfrac{\sin\theta}{\cos\theta} + \dfrac{\cos\theta}{\sin\theta} = \dfrac{1}{\sin\theta\cos\theta}$.

(ii) Show that the equation $\sin\theta\cos\theta = \dfrac{1}{2}$ is equivalent to
$\tan^2\theta - 2\tan\theta + 1 = 0$.

(iii) Hence solve the equation $\sin\theta\cos\theta = \dfrac{1}{2}$ for values of θ in the
range $0° \leqslant \theta \leqslant 360°$.

Solution

(i) $\text{LHS} = \dfrac{\sin\theta}{\cos\theta} + \dfrac{\cos\theta}{\sin\theta}$

> In this proof LHS stands for Left Hand Side and RHS stands for Right Hand Side.

$\quad = \dfrac{\sin^2\theta}{\sin\theta\cos\theta} + \dfrac{\cos^2\theta}{\sin\theta\cos\theta}$

> Find common denominator.

$\quad = \dfrac{\sin^2\theta + \cos^2\theta}{\sin\theta\cos\theta}$

> Now use the identity $\sin^2\theta + \cos^2\theta = 1$.

$\quad = \dfrac{1}{\sin\theta\cos\theta} = \text{RHS}$

(ii) Since $\sin\theta\cos\theta = \dfrac{1}{2}$ it follows that $\dfrac{1}{\sin\theta\cos\theta} = 2$

$\dfrac{\sin\theta}{\cos\theta} = \tan\theta$ and so $\dfrac{\cos\theta}{\sin\theta} = \dfrac{1}{\tan\theta}$,

> Remember the identity $\tan\theta = \dfrac{\sin\theta}{\cos\theta}$.

from part **(i)** $\dfrac{\sin\theta}{\cos\theta} + \dfrac{\cos\theta}{\sin\theta} = \dfrac{1}{\sin\theta\cos\theta}$

> Use the identity from **(i)**.

$\Rightarrow \qquad \tan\theta + \dfrac{1}{\tan\theta} = 2$

> Multiply by $\tan\theta$.

$\Rightarrow \qquad \tan^2\theta - 2\tan\theta + 1 = 0$

(iii) Using $t = \tan\theta$ gives

$t^2 - 2t + 1 = 0$

$\Rightarrow (t-1)^2 = 0$

> Solve the equation as a quadratic.

$\Rightarrow \qquad t = 1$

> Solve to find the values of t, i.e. $\tan\theta$.

$\Rightarrow \qquad \theta = 45°$

or $\theta = 180° + 45° = 225°$

> There are two values in the range.

Exam-style question

(i) Prove that $1 + \cos^2\theta + 4\sin^2\theta = 5 - 3\cos^2\theta$.

(ii) Solve the equation $1 + \cos^2\theta + 4\sin^2\theta = 3$ for $-180° \leqslant \theta \leqslant 180°$.

Short answer on page 122

Full worked solution online

Chapter 7: Geometry II

About this topic

You are familiar with calculations on right-angled triangles, finding the lengths of their sides and their areas. This chapter extends this work to triangles that do not contain a right angle.

Before you start, remember …

- the trigonometrical ratios for any angle
- the area of a triangle.

7.1 The area of a triangle

Key facts

1 **Standard convention**
   The three vertices of a triangle and the associated angles are labelled A, B, C and the sides opposite are labelled a, b, c.

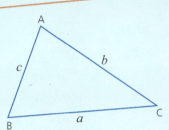

> Note: other letters may be used, e.g. L, M, N with associated lengths l, m, n.

2 **The area of a triangle**

   Area $= \frac{1}{2}bc\sin A = \frac{1}{2}ca\sin B = \frac{1}{2}ab\sin C$

Worked examples

Area of a triangle

1 In triangle ABC, AB = 6 cm, AC = 7 cm and angle A = 30°. Work out the area of the triangle.

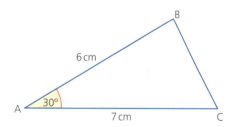

Solution

In the formula for the area, $c = 6$, $b = 7$, A = 30°.

Area $= \frac{1}{2}bc\sin A$

$= \frac{1}{2} \times 7 \times 6 \times \sin 30$

$= 21 \times \frac{1}{2}$

$= 10.5 \, cm^2$

> Apply the formula. You should remember that $\sin 30° = \frac{1}{2}$.

In the next example, the triangle has an obtuse angle and so if you draw a perpendicular from L to P on the base, it lies outside the triangle. However, the mathematics is still the same.

2   In the triangle LMN, LN = 5 cm, MN = 2 cm and angle N = 20°.
    Work out the area of the triangle.

Solution

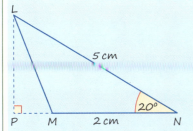

In the formula for the area, l = 2, m = 5, N = 20°.

$$\text{Area} = \frac{1}{2} lm \sin N$$

$$= \frac{1}{2} \times 2 \times 5 \times \sin 20$$

$$= 1.71 \, cm^2$$

> Apply the formula.

In this example the area is given and one of the angles is required.

3   In a triangle PQR, QR = 4.1 cm, PR = 5.2 cm. The area of the triangle is 9 cm².
    Work out the size of angle R.

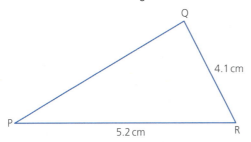

Solution

In the formula for the area, p = 4.1 and q = 5.2.

$$\text{Area} = \frac{1}{2} pq \sin R$$

> Apply the formula.

$$\Rightarrow \quad 9 = \frac{1}{2} \times 4.1 \times 5.2 \times \sin R$$

$$\Rightarrow \sin R = \frac{18}{4.1 \times 5.2} = 0.8442$$

> Make sin R the subject. Use your calculator to find R.

$$\Rightarrow \quad R = 57.6°$$

> Notice that there is another solution of the equation in the range 0° to 180° that your calculator does not give you.

However, there are two possible solutions.

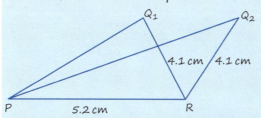

The two values for the angle R are 57.6° and 180° − 57.6° = 122.4°. Q_1Q_2 is parallel to PR and so the areas of the triangles PQ_1R and PQ_2R are the same.

Exam-style question

In the triangle ABC shown, BC = 3 cm, AC = 7 cm and the angle C = 40°.

(i) Work out the area of the triangle ABC.

(ii) The line AP is the perpendicular from A to the line BC. Work out the lengths of AP and CP and use them to check your answer to part **(i)**.

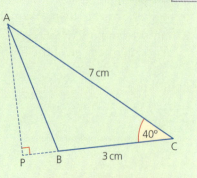

Short answers on page 122

Full worked solution online

7.2 The sine and cosine rules

Key facts

1 The sine rule

In the triangle ABC,

$$\frac{a}{\sin A} = \frac{b}{\sin B} = \frac{c}{\sin C}$$

which can be written

$$\frac{\sin A}{a} = \frac{\sin B}{b} = \frac{\sin C}{c}$$

2 The cosine rule

In the triangle ABC,

$$a^2 = b^2 + c^2 - 2bc\cos A$$

The angle can also be made the subject of the formula:

$$\cos A = \frac{b^2 + c^2 - a^2}{2bc}$$

> The cosine rule can use the other angles. In these cases, the formula becomes $b^2 = c^2 + a^2 - 2ca\cos B$ and $c^2 = a^2 + b^2 - 2ab\cos C$.

3 Choosing which rule to use

A triangle has six measurements – the three sides and the three angles.

You can use one of the rules to find out an unknown measurement if you know three independent measurements:

- 3 sides: use the cosine rule to work out one (or more) of the angles.

- 2 sides and the included angle: use the cosine rule to work out the length of the third side.

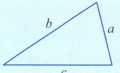

- 2 sides and a non-included angle: Use the sine rule to work out a second angle (and therefore third by the angle sum of a triangle).

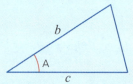

- 2 angles and one side: Use the sine rule to work out another side.

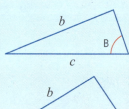

4 The ambiguous case

If you are given the lengths of two sides with a non-included angle then there could be two answers for the other two angles.

If the sides b and c are given with the (non-included) angle θ then the position of B can be in one of two places as shown (B and B').

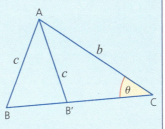

Note that the two angles at B' add to 180°. So if, for instance, you find that angle B = 40° then the angle B' = 140°. This is consistent with the solutions of, for example, sin B = 0.43 in the range 0° to 180°. One solution of this equation is an angle in the first quadrant and the other is in the second quadrant.

Worked example

The sine rule

1 In a triangle ABC, angle A = 56°, angle B = 47° and $a = 5$. Find b.

Solution

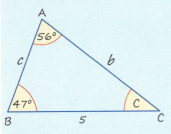

Using the sine rule

$$\frac{a}{\sin A} = \frac{b}{\sin B}$$

Choose the appropriate part of the sine rule and substitute.

$$\frac{5}{\sin 56} = \frac{b}{\sin 47}$$

$$\Rightarrow b = \frac{5\sin 47}{\sin 56} = 4.41$$

Make b the subject and solve.

Worked examples

The cosine rule

2 In a triangle ABC, angle A = 37°, $b = 4$ and $c = 6$. Find a.

Solution

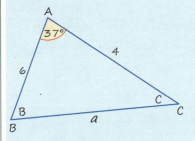

Using the cosine rule

$$a^2 = b^2 + c^2 - 2bc\cos A$$

$$a^2 = 4^2 + 6^2 - 2 \times 4 \times 6\cos 37$$

$$= 13.67$$

$$\Rightarrow a = 3.70$$

Common mistake: Take care to calculate the 3rd term before subtracting from the sum of the first two terms. Incorrect entry on your calculator will give an incorrect answer!

It will help if you are able to estimate an approximate answer for a.

---

3 Work out the size of the largest angle in the triangle ABC in the diagram.

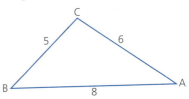

Solution

The largest angle is opposite the largest side.

So in this triangle the largest angle is C.

$$\cos C = \frac{a^2 + b^2 - c^2}{2ab}$$

$$\Rightarrow \cos C = \frac{5^2 + 6^2 - 8^2}{2 \times 5 \times 6} = -0.05$$

$$\Rightarrow \quad C = 92.9°$$

> **Common mistake:** Make sure that you substitute the correct values.

> Use the appropriate form of the cosine rule.

> Substitute and use your calculator to find angle C.

Worked example

Using both rules in the same question

4 In triangle ABC angles B and C are 70° and 50° respectively.
CB = 6 cm.
Work out the lengths of AC and AB.

Solution

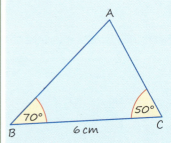

Angle A = (180° − 70° − 50°) = 60°.

Using the sine rule:

$$\frac{a}{\sin A} = \frac{b}{\sin B}$$

$$\Rightarrow \frac{b}{\sin 70°} = \frac{6}{\sin 60°}$$

$$\Rightarrow \quad b = \frac{6\sin 70°}{\sin 60°} = 6.51 \text{ cm}$$

Using the cosine rule:

$$c^2 = 6^2 + 6.51^2 - 2 \times 6 \times 6.51 \times \cos 50°$$
$$= 78.38 - 78.12\cos 50°$$
$$= 78.38 - 50.21 = 28.17$$

$$\Rightarrow c = 5.31 \text{ cm}$$

> Use the angle sum of the triangle to find angle A.

> Use the sine rule to find one side.

> Note: This could be done by using the sine rule twice.

> **Common mistake:** Work out the third term completely before subtracting. Make sure you enter the values correctly.

> Use the cosine rule to find the other side.

Exam-style question

A field is a quadrilateral with corners ABCD as shown in the diagram.

There is a straight footpath from D to B.

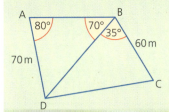

Work out the length of the side DC.

Short answer on page 122

Full worked solution online

7.3 Problems in three dimensions

Key facts

1   True shape diagrams

Polygons, for example triangles, are 2-dimensional shapes.

Drawings of 3-dimensional objects are always distorted; for example, right angles do not always appear to be 90°. Always start by drawing true shape diagrams of any polygons with which you are going to work.

2   The angle between a line and a plane

This diagram shows a line AB crossing a plane at A. The point C is also on the plane and angle ACB = 90°. The angle between the line and the plane is BAC.

This is an example of a sketch of a 3D shape when right angles are not drawn as 90°.

You will always find it helpful to draw 'true shape' diagrams, such as the triangle ABC.

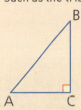

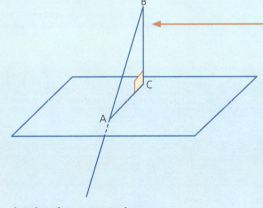

3   Angle of greatest slope

The diagram shows a sloping plane meeting the horizontal in a straight line. The line EF is perpendicular to that line and lies in the sloping plane. It is a **line of greatest slope** and the angle θ is the **angle of greatest slope**.

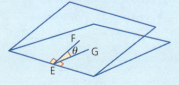

Worked example

Angle between a line and a plane

1 The diagram shows a cuboid with horizontal base.
   AB = 5 cm and BC = 4 cm.
   The top is 3 cm above the base; E is vertically above A.
   Find the angle that the diagonal AG makes with
   (i) the base ABCD   **(ii)** the face ABFE.

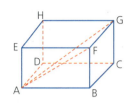

Solution

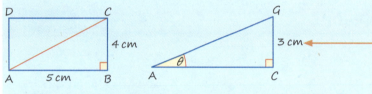

> These are true shape diagrams of triangles ABC and ACG. In part **(i)** the required angle is GAC.

(i) The required triangle is ACG. The right angle is at G and the required angle, θ, is GAC.

In triangle ABC, $AC^2 = 4^2 + 5^2 = 41$ so $AC = \sqrt{41}$

> AC is the diagonal of the rectangular base, ABCD.

In triangle AGC, $\tan\theta = \dfrac{CG}{AC}$

So $\tan\theta = \dfrac{3}{\sqrt{41}} \Rightarrow \theta = 25.1°$

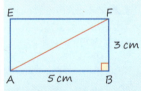
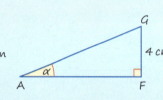

> AF is the diagonal of the vertical rectangular face AEFB.

(ii) The required angle, α, is GAF.

In triangle ABF, $AF^2 = 5^2 + 3^2 = 34$

So $\tan\alpha = \dfrac{4}{\sqrt{34}} \Rightarrow \alpha = 34.4°$

Worked example

Angle of greatest slope

2 The diagram shows a hillside of uniform slope.

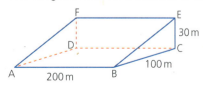

> A hillside with uniform slope may be modelled as a wedge as shown in the diagram. ABCD is a horizontal rectangular plane, the hillside is the rectangle ABEF such that FDCE is a vertical rectangular plane.

AB = 200 m, BC = 100 m and EC = 30 m.
Find
(i) the line of greatest slope,
(ii) the angle between the line AE on the slope and the horizontal.

Solution

(i) An angle of greatest slope is EBC. ◄

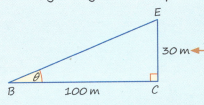

Angle EBC = $\tan^{-1}\left(\dfrac{30}{100}\right)$ = 16.7°

> BC and BE are perpendicular to the common line, AB, and so EBC is the angle of greatest slope.

> Another angle of greatest slope is FAD.

(ii) The triangle to be used is CAE.

> The angle that AE makes with the horizontal is angle CAE.

> In the triangle EAC, the line AC must be found.

However, first the length of AC must be found.
This is done from the triangle ABC in the rectangle ABCD.

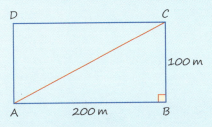

AC = $\sqrt{AB^2 + BC^2}$

   = $\sqrt{200^2 + 100^2}$ = 223.6... ◄

In triangle CAE,

Angle CAE = $\tan^{-1}\left(\dfrac{EC}{AC}\right)$

$\Rightarrow$ Angle CAE = $\tan^{-1}\left(\dfrac{30}{223.6...}\right)$ = 7.6° (to 1 d.p.)

> **Common mistake:** If you approximate too soon then the errors produced will accumulate. So, keep the unrounded number on your calculator until the very end.

Exam-style question

TESTED ☐

A pyramid PQRSV has a square horizontal base of side 5 cm. The vertex V is vertically above the centre of the base, O. The height of the pyramid is 8 cm.

Find the angle that the sloping edge VP makes with the horizontal.

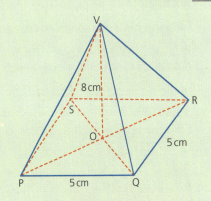

Short answer on page 122

Full worked solution online

CHECKED ANSWER ☐

Review questions (Chapters 5–7)

1 The points A, B, C and D have coordinates (−5, −6), (−2, 3), (1, 2) and (3, 8).

  (i) Show that AB and CD are parallel. [2]

  (ii) Show that AB and BC are perpendicular. [2]

2 The points A, B, C and D have coordinates (1, 3), (2, 5), (4, 4) and (3, 2).

  (i) Show that ABCD is a square. [4]

  (ii) Use coordinate geometry to show that the midpoint of the diagonal AC is the same point as the midpoint of the diagonal BD. [2]

3 A printing firm advertises the supply of flyers, to include creating the artwork, printing and paper costs.

  The price charged for differing numbers of flyers are as shown in the table.

Quantity	Cost
50	£14
100	£15
250	£18

  (i) Plot these points on a graph and draw a straight line through them. [2]

  (ii) Use the line to determine

    (a) the cost of producing the artwork and administration (a fixed charge for any number of flyers), [1]

    (b) the cost of paper and printing per 200 flyers. [2]

  (iii) Work out how much the firm would charge at these rates for printing 2000 flyers. [2]

4 (i) A circle has equation $x^2 + y^2 - 4x - 6y - 12 = 0$. Find the coordinates of its centre, C, and its radius. [3]

  (ii) Find the coordinates of the points, A and B, where the line $y = x - 6$ cuts the circle. [5]

5 The points A and B have coordinates (1, 2) and (5, 8) respectively.

  (i) Find the equation of the line AB. [3]

  (ii) C is the midpoint of AB. Find the coordinates of C. [1]

  (iii) The line through C perpendicular to AB meets the y-axis at D. Find the coordinates of D. [4]

  (iv) Find the equation of the circle which has D as its centre and for which AB is a tangent. [3]

6 (i) Find the centre and radius of the circle with equation $x^2 + y^2 - 2x - 2y - 23 = 0$. [3]

  (ii) Show that the point T(4, −3) lies on the circle. [1]

  (iii) Find the equation of the tangent to the circle at T. [4]

7 P(1, 3), Q(3, 5) and R(7, 1) are joined by straight lines. Find the angle PQR and hence work out the centre of the circle which passes through P, Q and R. [4]

8 In the circle shown, A, B and C are on the circumference and O is the centre of the circle. Angle OBC = 50°.

Find the angles OCB and CAB. Give your reasoning. [4]

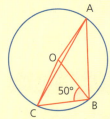

9 In the circle shown, AT and BT are tangents at A and B, meeting at T. O is the centre and C is on the circumference. AOC is a straight line.

Angle ATO = 25°.

Prove that CB is parallel to OT. **[4]**

10 From a point on the ground 23 m from the base of a vertical tower the angle of elevation to the top of the tower is 48°. Find the height of the tower. **[2]**

11 A hillwalker is initially 2 km due south of a tower. He then walks on a bearing of 035°. Work out the shortest distance he is from the tower. **[2]**

12 Find the four values of x in the range $0° \leqslant x \leqslant 360°$ that satisfy the equation $\sin 2x = 0.5$. **[4]**

13 Find the value of x in the range $0° \leqslant x \leqslant 360°$ that satisfies both $\tan x = 0.75$ and $\cos x = -0.8$. **[3]**

14 Find all the values of x in the range $0° \leqslant x \leqslant 360°$ that satisfy $\sin x = -2\cos x$. **[3]**

15 In the triangle EFG, EF = 4 cm, EG = 7 cm and the angle at E is 55°. Work out the area of the triangle EFG. **[2]**

16 In the triangle ABC, AB = 5 cm and BC = 8 cm. The area of the triangle is 19 cm². Work out the angle at B. **[2]**

17 In the triangle ABC shown in the diagram, AB = 6 cm and CB = 8 cm. The angle at C = 32°. Find the angle at A. **[3]**

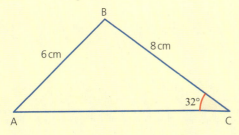

18 A triangular piece of land ABC is such that AB = 10 m, BC = 12 m and the angle ABC = 50°. Find the length of the side AC. **[3]**

19 A garden shed may be modelled by a cuboid with base 3 m by 4 m and height 2 m. Assuming that it is possible to get it into the shed, what is the maximum length of wood that can be stored? **[2]**

20 In the diagram, the rectangle ABEF lies on a plane hillside which slopes at an angle of 20° to the horizontal. ABCD is a horizontal rectangle. E and F are 100 m vertically above C and D respectively. AB = DC = FE = 500 m.

AE is a straight path.

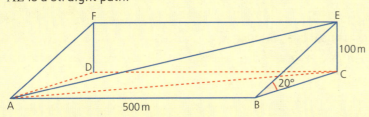

(i) Find the distance BE. **[3]**

(ii) Find the angle that the path AE makes with the horizontal. **[4]**

Short answers on page 123

Full worked solutions online

CHECKED ANSWERS

About this topic

Calculus is one of the really big ideas of mathematics and has wide ranging applications. This course includes differentiation as an introduction to calculus.

This allows you to use the equation of a curve to find its gradient at any point.

In this chapter differentiation is applied to finding the equations of tangents and normals, and to determining the turning points of a curve.

Before you start, remember ...

- how to find the equation of a straight line
- how to draw and sketch curves of the form $y = f(x)$ for a given function $f(x)$
- the relationship between the gradients of tangents and normals.

8.1 Differentiation and gradients of curves

REVISED

> ### Key facts
>
> **1   The gradient of a curve at a particular point**
> The gradient of a curve at a point is the gradient of the tangent at that point.
>
>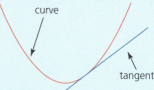
>
> **2   Finding the gradient function**
> The gradient function, denoted $\dfrac{dy}{dx}$, is used to find the gradient of a curve.
>
> **3   Finding the gradient of a curve at a point**
> The gradient of the curve at a point $x = x_1$ is found by substituting $x = x_1$ into the gradient function.
>
> **4   Differentiating**
> For a curve $y = ax^n$ where a is a constant and n is an integer (positive or negative), $\dfrac{dy}{dx} = nax^{n-1}$.
>
> The process is called differentiation and the result is sometimes called the derivative.
>
> **5   The sums and differences of terms**
> Each term of an expression is treated separately.
> The gradient function of a constant is 0.
>
> **6   More complicated functions**
> Any brackets of fractions must be cleared before differentiating,
> e.g. $y = x^2(x - 3) = x^3 - 3x^2 \Rightarrow \dfrac{dy}{dx} = 3x^2 - 6x.$

Worked example

Finding the gradient function

Each term is differentiated separately.

1 Find the gradient function of the curve $y = x^4 - 3x^2$.

Solution

$$y = x^4 - 3x^2$$

$$\Rightarrow \frac{dy}{dx} = 4x^3 - 6x$$

> There are two terms here. Each should be differentiated separately.

> Apply the rule
> $y = ax^n \Rightarrow \frac{dy}{dx} = nax^{n-1}$ to each term. In the first term $a = 1$ but in the second $a = -3$.

Worked example

Finding the gradient of a curve at a point

2 Find the gradient of the curve $y = x(x^2 + 2)$ at the point (1, 3).

Solution

$$y = x(x^2 + 2) = x^3 + 2x$$

$$\Rightarrow \frac{dy}{dx} = 3x^2 + 2$$

When $x = 1$, $\frac{dy}{dx} = 3 + 2 = 5$.

So the gradient is 5.

> First multiply out the brackets to give two terms added together.

> Differentiate each term separately

> Notice that the second term becomes $2x^0$ and $x^0 = 1$.

> Substitute $x = 1$ into the gradient function to find the gradient at a particular point.

> The question asks for the gradient. It is good practice to finish the question with the answer relating to the question.

Worked example

Negative powers

3 Find the gradient of the curve $y = 2x^2 + \dfrac{1}{x^2}$ at the point (1, 3).

Solution

$$y = 2x^2 + \frac{1}{x^2} = 2x^2 + x^{-2}$$

$$\Rightarrow \frac{dy}{dx} = 4x - 2x^{-3} = 4x - \frac{2}{x^3}$$

When $x = 1$, $\frac{dy}{dx} = 4 - 2 = 2$.

At (1, 3), the gradient is 2.

> Write the second term in the form x^n where $n = -2$.

> Differentiate. Note that when $n = -2$, $n - 1 = -3$.

> Substitute $x = 1$ to find the gradient of the curve at that point.

Worked examples

Finding the coordinates of a point with a given gradient

In the following example you work backwards to find x when $\frac{dy}{dx}$ is a quadratic expression.

4 Find the coordinates of the point on the curve $y = x^2 + 2x + 5$ at which the gradient is 4.

Solution

$$y = x^2 + 2x + 5$$

$$\Rightarrow \frac{dy}{dx} = 2x + 2$$

When $\frac{dy}{dx} = 4$, $2x + 2 = 4$

$$\Rightarrow \qquad\qquad 2x = 2 \Rightarrow x = 1$$

$$y = 1^2 + 2 \times 1 + 5 = 8$$

So the coordinates are (1, 8).

> Remember that 5 differentiated is 0 and that $2x$ differentiated is 2.

> Differentiate each term.

> Set $\frac{dy}{dx} = 4$ and solve the resulting equation in x.

> Don't forget to find the y-coordinate as well!

The next example involves a cubic function; there are two points with the same gradient.

5 Find the points on the curve $y = x^3 - 3x^2 - 7x + 4$ at which the gradient is 2.

Solution

$$y = x^3 - 3x^2 - 7x + 4$$

$$\Rightarrow \frac{dy}{dx} = 3x^2 - 6x - 7$$

When $\frac{dy}{dx} = 2$, $3x^2 - 6x - 7 = 2$

$$\Rightarrow \qquad\qquad 3x^2 - 6x - 9 = 0$$

$$\Rightarrow \qquad\qquad x^2 - 2x - 3 = 0$$

$$\Rightarrow \qquad\qquad (x + 1)(x - 3) = 0$$

$\Rightarrow$ Either $x = -1$ and so $y = -1 - 3 + 7 + 4 = 7$

or $x = 3$ and so $y = 27 - 27 - 21 + 4 = -17$

The coordinates are (–1, 7) and (3, –17).

> Differentiate term by term.

> Set $\frac{dy}{dx} = 2$. Because the curve is a cubic the gradient function will be a quadratic.

> Divide by the common factor of 3.

> Factorise the resulting quadratic.

> For each x-coordinate find the y-coordinate.

Exam-style question

TESTED ☐

The equation of a curve is $y = x^3 - 2x^2 - 3x + 5$.

(i) Find the gradient of the curve when $x = 1$.

(ii) Find the coordinates of the points where the gradient is 1.

Short answers on page 123

Full worked solution online

CHECKED ANSWERS ☐

8.2 Tangents and normals

Key facts

1. The gradient of the tangent at a point (x_1, y_1) on a curve is the value of $\dfrac{dy}{dx}$ at that point

   If $\dfrac{dy}{dx} = m$ then the equation of the tangent is $y - y_1 = m(x - x_1)$.

2. The normal to the curve at a point is the line through the point perpendicular to the tangent

   If the gradient of the tangent is m_1 and the gradient of the normal is m_2 then $m_1 \times m_2 = -1$.

Worked example

The tangent to a curve

1. Find the equation of the tangent to the curve $y = x^2 - 5x + 2$ at the point $(1, -2)$.

Solution

$$y = x^2 - 5x + 2$$

$$\Rightarrow \frac{dy}{dx} = 2x - 5$$

> Find the gradient function and then the gradient at that point.

When $x = 1$, $\dfrac{dy}{dx} = 2 \times 1 - 5 = -3$.

The tangent is the line through $(1, -2)$ with gradient $m = -3$.

> Use the standard form for the equation of a straight line through a given point with given gradient.

$$\Rightarrow 3x + y = 1$$

> Express your equation with three terms only, so combine the numbers.

Worked example

The normal to a curve

A 'normal' is a line through the point of intersection of the tangent with the curve and is perpendicular to it.

2. Find the equation of the normal to the curve $y = x^2 - 3x + 2$ at the point $(3, 2)$.

Solution

$$y = x^2 - 3x + 2$$

$$\Rightarrow \frac{dy}{dx} = 2x - 3$$

> Differentiate and substitute the value of x of the point to find the gradient of the tangent.

When $x = 3$, $\dfrac{dy}{dx} = 2 \times 3 - 3 = 3$

So gradient of tangent at $(3, 2)$ is 3.

So gradient of normal at that point is $-\dfrac{1}{3}$.

> Use $m_1 m_2 = -1$ to find the gradient of the normal.

So equation of normal is

$$y - 2 = -\frac{1}{3}(x - 3)$$

> Then find the equation of the normal in the usual way, simplifying it to three terms.

$$\Rightarrow 3y - 6 = 3 - x$$

$$\Rightarrow 3y + x = 9$$

A curve has equation $y = x^2 - 4x - 2$.
(i) Find the equation of the normal to the curve at the point A(−1, 3).

The normal at A meets the curve again at B.
(ii) Find the coordinates of B.

Short answers on page 123

Full worked solution online

8.3 Increasing and decreasing functions and the second derivative

REVISED ☐

Key facts

1 Increasing and decreasing functions

The value of the gradient function of a curve at a point determines whether the curve is increasing, decreasing or is stationary.

If $\dfrac{dy}{dx} > 0$ the function is increasing.

If $\dfrac{dy}{dx} < 0$ the function is decreasing.

If $\dfrac{dy}{dx} = 0$ the function is stationary.

2 The second derivative

If you differentiate a function twice then you obtain what is called the second derivative.

This is denoted by $\dfrac{d^2 y}{dx^2}$.

E.g. $y = x^3 + 2x \Rightarrow \dfrac{dy}{dx} = 3x^2 + 2 \Rightarrow \dfrac{d^2 y}{dx^2} = 6x$.

It represents the rate of change of the gradient function.

If $\dfrac{d^2 y}{dx^2} > 0$ the gradient function is increasing.

If $\dfrac{d^2 y}{dx^2} < 0$ the gradient function is decreasing.

If $\dfrac{d^2 y}{dx^2} = 0$ the gradient function is stationary.

Worked examples

Increasing and decreasing functions

1 You are given the curve with equation $y = x^3 + 2x^2 - 5$.
Determine whether the curve is increasing or decreasing when
(i) $x = -1$ (ii) $x = 5$.

Solution

(i) $y = x^3 + 2x^2 - 5$

$\Rightarrow \dfrac{dy}{dx} = 3x^2 + 4x$ ←

When $x = -1$, $\dfrac{dy}{dx} = 3 - 4 = -1 < 0$ ←

So the function is decreasing.

> Differentiate: remember that when differentiated, 5 becomes 0.

> Substitute the value of x and find the value of $\dfrac{dy}{dx}$.

(ii) When $x = 5$, $\frac{dy}{dx} = 3 \times 25 + 4 \times 5 = 95$ so the function is increasing.

2 For what values of x is the curve $y = x^3 - 3x - 1$ increasing?

Solution

$y = x^3 - 3x - 1$

$\Rightarrow \dfrac{dy}{dx} = 3x^2 - 3$ ← ⎤

$\dfrac{dy}{dx} > 0$ when $3x^2 - 3 > 0$ ← ⎦   Differentiate and form an inequality.

i.e. when $3x^2 > 3 \Rightarrow x^2 > 1$

So the curve is increasing when $x > 1$ and when $x < -1$. ←   Solve the quadratic inequality.

Exam-style question   TESTED ☐

You are given the curve with equation $y = 4x + \dfrac{1}{x}$.

(i) Determine the range of values for which the curve is increasing.

(ii) Determine the range of values for which the gradient function is increasing.

(iii) (a) Sketch the graph of $y = 4x + \dfrac{1}{x}$.

    (b) Comment on your answers to parts **(i)** and **(ii)** in relation to the graph.

Short answers on page 123

Full worked solution online   CHECKED ANSWERS ☐

8.4 Stationary points   REVISED ☐

Key facts

1 Stationary points

A stationary point on a curve is a point where the gradient is zero.

2 The nature of stationary points

There are three types of stationary points:

- maximum
- minimum
- stationary point of inflection.

Maxima and minima are often called turning points.

Maximum   Minimum   Stationary point of inflection

Note: 'maxima' is the plural of 'maximum' and 'minima' is the plural of 'minimum'.

3 Sketching curves

When sketching a curve the essential details should be shown. This includes the intercepts on the axes and if you know the coordinates of any turning points, then these should also be included.

4 Determining the nature of a turning point

There are three ways to decide the nature of a turning point.

(i) Use values of the function $y = f(x)$ either side of the turning point.

(ii) Use values of the gradient, $\dfrac{dy}{dx}$, either side of the turning point.

(iii) Use the value of the second derivative, $\dfrac{d^2y}{dx^2}$, at the turning point.

Worked example

Finding turning points

1 **(i)** Find the coordinates of the turning points of the curve
$y = 2x^3 - 3x^2 - 12x + 24$.

(ii) Sketch the curve, showing the turning points.

Solution

(i)
$$y = 2x^3 - 3x^2 - 12x + 24$$

$$\Rightarrow \frac{dy}{dx} = 6x^2 - 6x - 12$$ ← Differentiate.

$$= 0 \text{ when } 6x^2 - 6x - 12 = 0$$ ← Set $\frac{dy}{dx}$ equal to 0. Note that a common factor of 6 is extracted, making the algebra a little easier.

$$\Rightarrow x^2 - x - 2 = 0 \Rightarrow (x - 2)(x + 1) = 0$$

So either $x = 2$ giving $y = 16 - 12 - 24 + 24 = 4$

or $x = -1$ giving $y = 31$

$\Rightarrow$ Coordinates of the turning points are (2, 4) and (−1, 31).

(ii)

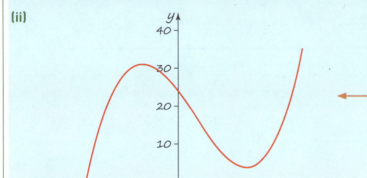

Turning points at (−1, 31) and (2, 4).

Worked example

Using values of the function

2 Use the values of the function $y = 2x^3 - 3x^2 - 12x + 24$ to verify the result from Example 1 that (2, 4) is a minimum point.

Solution

$f(1.9) = 4.088 > 4$ ← Try values either side of $x = 2$, so 1.9 and 2.1 will do.

$f(2) = 4$

$f(2.1) = 4.092 > 4$

So the turning point at (2, 4) is a minimum.

Worked example

Use values of the gradient

3 Use the values of the gradient function of the curve
$y = 2x^3 - 3x^2 - 12x + 24$ to verify the result from Example 1 that
$(-1, 31)$ is a maximum point.

Solution

$$y = 2x^3 - 3x^2 - 12x + 24$$

$$\Rightarrow \frac{dy}{dx} = 6x^2 - 6x - 12$$

The points with $x = -1.1$ and $x = -0.9$ are either side of the
turning point $(-1, 31)$.

x	-1.1	-1	-0.9
$\dfrac{dy}{dx}$	1.86	0	-1.74

Since the gradient goes from positive to 0 to negative the
turning point is a maximum.

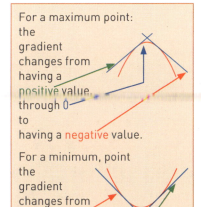

For a maximum point:
the gradient changes from having a positive value, through 0 to having a negative value.

For a minimum, point the gradient changes from having a negative value, through 0 to having a positive value.

Worked example

Use the second derivative

4 Use the values of $\dfrac{d^2 y}{dx^2}$ for the function $y = 2x^3 - 3x^2 - 12x + 24$ to
verify the results from Example 1 that $(2, 4)$ is a minimum point and
$(-1, 31)$ is a maximum point.

Solution

$$y = 2x^3 - 3x^2 - 12x + 24$$

$$\Rightarrow \frac{dy}{dx} = 6x^2 - 6x - 12$$

$$\Rightarrow \frac{d^2 y}{dx^2} = 12x - 6$$

Differentiate twice to find $\dfrac{d^2 y}{dx^2}$.

When $x = 2$, $\dfrac{d^2 y}{dx^2} = 24 - 6 = 18 > 0$

So this turning point is a minimum.

When $x = -1$, $\dfrac{d^2 y}{dx^2} = -12 - 6 = -18 < 0$

So this turning point is a maximum.

The condition for a minimum point is $\dfrac{d^2 y}{dx^2} > 0$ and for a maximum point $\dfrac{d^2 y}{dx^2} < 0$.

Exam-style question

TESTED

You are given a curve with equation $y = x^3 - 3x^2 + 6$.

(i) Find the coordinates of the stationary points on the curve and determine the nature of each point.
(ii) Sketch the curve with equation $y = x^3 - 3x^2 + 6$, showing the turning points.

Short answers on page 123

Full worked solution online

CHECKED ANSWERS

Review questions (Chapter 8)

1 Work out the gradient function for each the following functions.

   (i) $y = \dfrac{2}{x^2}$,

   (ii) $y = x^3 - 2x^2 + 6x - 2$. [4]

2 A curve has equation $y = x^2 - 2x + 7$.

   Find the coordinates of the point on the curve where the gradient is 4. [3]

3 The gradient of a curve at a point (x, y) is given by $\dfrac{dy}{dx} = 2x^3 - x^2 + 5$.

   Find the equation of the normal to the curve at the point P(1, 2). [5]

4 A curve has equation $y = x^3 + 2x^2 - 5x + 5$.

   (i) Find the equation of the tangent at P(1, 3). [4]

   (ii) This tangent cuts the curve again at point R. Find the coordinates of R. [4]

   (iii) Determine whether the tangent is a normal to the curve at R. [3]

5 Find the range of values of x for which the function $y = 2x + \dfrac{1}{x^2}$ is increasing. [3]

6 Show that $y = x^3 + x^2 + x - 5$ is an increasing function for all values of x. [4]

7 You are given the curve $y = x^3 + 3x^2 + 2$.

   For $x = 2$, work out the values of

   (i) y

   (ii) the gradient

   (iii) the rate of change of the gradient. [5]

8 (i) Show that there is a stationary point at (1, 9) on the curve $y = x^3 - 6x^2 + 9x + 5$ and determine the nature of this stationary point. [5]

   (ii) Find the coordinates of the other stationary point. [2]

   (iii) Hence sketch the curve. [2]

9 (i) Find the stationary points on the curve $y = x^3 + \dfrac{3}{2}x^2 - 6x + 4$. [4]

   (ii) Determine the nature of each stationary point. [4]

   (iii) Sketch the curve. [2]

Short answers on page 124

Full worked solutions online

CHECKED ANSWERS

About this topic

An $m \times n$ matrix is a rectangular array of numbers with m rows and n columns.

So it contains mn numbers.

The numbers are usually called elements. The plural of matrix is matrices.

In this course you only meet 2×1 and 2×2 matrices. Matrices are used extensively in many branches of more advanced mathematics, as well as in engineering and science.

Before you start, remember ...

- basic transformations from your GCSE course
- vectors
- solving equations.

9.1 Multiplying matrices

REVISED ☐

Key facts

1   Multiplication by a number
Each element is multiplied by that number.

e.g. $2\begin{pmatrix} 1 & 2 \\ 3 & 4 \end{pmatrix} = \begin{pmatrix} 2 & 4 \\ 6 & 8 \end{pmatrix}$.

2   Multiplying matrices
When 2 matrices of different sizes are multiplied then the rule is $(a \times b) \times (b \times c)$ gives $(a \times c)$.
If the two numbers on the inside are not the same then multiplication is not possible.

For example, a $(3 \times 2) \times (2 \times 5)$ multiplication gives a 3×5 matrix.

It is not possible to multiply the other way round. i.e. $(2 \times 5) \times (3 \times 2)$ is not possible because $5 \neq 3$.

3   Multiplying a 2 × 2 matrix by a 2 × 2 matrix
The elements of each row are multiplied by the corresponding elements of each column.
The answer is a 2×2 matrix.

4   The identity matrix
The matrix $\begin{pmatrix} 1 & 0 \\ 0 & 1 \end{pmatrix}$ is the identity 2×2 matrix and denoted by $\mathbf{I}$.

For any 2×2 matrix $\mathbf{A}$, $\mathbf{AI} = \mathbf{IA} = \mathbf{A}$.

For a 2×1 matrix $\begin{pmatrix} a \\ b \end{pmatrix}$, $\begin{pmatrix} 1 & 0 \\ 0 & 1 \end{pmatrix}\begin{pmatrix} a \\ b \end{pmatrix} = \begin{pmatrix} a \\ b \end{pmatrix}$.

e.g. $\begin{pmatrix} 4 & 6 \\ -1 & 5 \end{pmatrix}\begin{pmatrix} 1 & 2 \\ 3 & 4 \end{pmatrix}$

$= \begin{pmatrix} 4 \times 1 + 6 \times 3 & 4 \times 2 + 6 \times 4 \\ -1 \times 1 + 5 \times 3 & -1 \times 2 + 5 \times 4 \end{pmatrix}$

$= \begin{pmatrix} 22 & 32 \\ 14 & 18 \end{pmatrix}$

5   Equal matrices
If two matrices are equal then their corresponding elements are equal.

A 2×1 matrix can be used to define the position of a point in two dimensions and is then called a position vector.

Worked example

Multiplication by a number

1 You are given the matrix $\mathbf{A} = \begin{pmatrix} 1 & 2 \\ 3 & -1 \end{pmatrix}$.
Work out $4\mathbf{A}$.

Solution

$$4A = \begin{pmatrix} 4\times1 & 4\times2 \\ 4\times3 & 4\times-1 \end{pmatrix} = \begin{pmatrix} 4 & 8 \\ 12 & -4 \end{pmatrix}$$

> Each element is multiplied by 4.

Worked example

Multiplying a 2 × 2 matrix by a 2 × 1 matrix

2 You are given the matrix $\mathbf{A} = \begin{pmatrix} 1 & 2 \\ 3 & -1 \end{pmatrix}$ and the matrix $\mathbf{B} = \begin{pmatrix} 4 \\ 5 \end{pmatrix}$.
Work out $\mathbf{AB}$.

Solution

$$AB = \begin{pmatrix} 1 & 2 \\ 3 & -1 \end{pmatrix}\begin{pmatrix} 4 \\ 5 \end{pmatrix} = \begin{pmatrix} 1\times4+2\times5 \\ 3\times4-1\times5 \end{pmatrix} = \begin{pmatrix} 14 \\ 7 \end{pmatrix}.$$

> Multiply 'across' the row and 'down' the column.

Worked example

Multiplying a 2 × 2 matrix by a 2 × 2 matrix

3 You are given the matrix $\mathbf{A} = \begin{pmatrix} 2 & -2 \\ 4 & 1 \end{pmatrix}$ and the matrix $\mathbf{B} = \begin{pmatrix} 3 & 5 \\ -1 & 6 \end{pmatrix}$.
Work out $\mathbf{AB}$.

Solution

$$AB = \begin{pmatrix} 2 & -2 \\ 4 & 1 \end{pmatrix}\begin{pmatrix} 3 & 5 \\ -1 & 6 \end{pmatrix} = \begin{pmatrix} 2\times3-2\times-1 & 2\times5-2\times6 \\ 4\times3+1\times-1 & 4\times5+1\times6 \end{pmatrix}$$

$$= \begin{pmatrix} 8 & -2 \\ 11 & 26 \end{pmatrix}$$

> Note that $\mathbf{BA}$ can be calculated and is also a 2 × 2 matrix but it is not be the same as $\mathbf{AB}$. In general $\mathbf{AB} \neq \mathbf{BA}$.

Worked example

The identity matrix

4 You are given the matrices $\mathbf{A} = \begin{pmatrix} 3 & 4 \\ 1 & 5 \end{pmatrix}$ and $\mathbf{B} = \begin{pmatrix} 5 & -4 \\ -1 & 3 \end{pmatrix}$.
Show that $\mathbf{AB} = k\mathbf{I}$.
State the value of k.

Solution

$$\begin{pmatrix} 3 & 4 \\ 1 & 5 \end{pmatrix}\begin{pmatrix} 5 & -4 \\ -1 & 3 \end{pmatrix} = \begin{pmatrix} 11 & 0 \\ 0 & 11 \end{pmatrix}$$

$$\begin{pmatrix} 11 & 0 \\ 0 & 11 \end{pmatrix} = 11\begin{pmatrix} 1 & 0 \\ 0 & 1 \end{pmatrix} = 11I$$

> The leading diagonal elements have the same value and the other elements are 0. Extract this as a factor leaving the identity matrix.

So it has the required form and $k = 11$.

Worked example

Equal matrices

5   Given that $\begin{pmatrix} 1 & 3 \\ -1 & p \end{pmatrix} = \begin{pmatrix} q & 3 \\ -1 & 2 \end{pmatrix}$, write down the values of p and q.

Solution

Because corresponding elements are equal, $p = 2$ and $q = 1$. ◄—— Equal matrices have equal elements.

Exam-style question   TESTED ☐

Given that $\mathbf{A} = \begin{pmatrix} 1 & 2 \\ -1 & n \end{pmatrix}$ and $\mathbf{B} = \begin{pmatrix} 2 & m \\ 1 & -2 \end{pmatrix}$, find values of m and n for which $\mathbf{AB} = \begin{pmatrix} 4 & 1 \\ 1 & -11 \end{pmatrix}$.

Short answer on page 124

Full worked solution online   CHECKED ANSWER ☐

9.2 Transformations   REVISED ☐

Key facts

1   **A point may be represented by a position vector**
    The point with coordinates (a, b) may be represented by the
    position vector $\begin{pmatrix} a \\ b \end{pmatrix}$.

2   **Matrices can define transformations**
    Example:

$$\begin{pmatrix} a & c \\ b & d \end{pmatrix}\begin{pmatrix} 1 \\ 0 \end{pmatrix} = \begin{pmatrix} a \\ b \end{pmatrix} \text{ and } \begin{pmatrix} a & c \\ b & d \end{pmatrix}\begin{pmatrix} 0 \\ 1 \end{pmatrix} = \begin{pmatrix} c \\ d \end{pmatrix}$$

    The point $(1, 0)$ is transformed to the point (a, b) and the point $(0, 1)$ is transformed to the point (c, d). ◄—— The transformed point is called the **image point**.

3   **The unit square**
    The unit square has vertices O(0, 0), A(1, 0), B(1, 1) and C(0, 1).

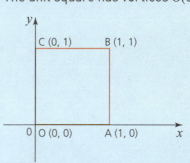

4   **Transformations of the unit square**
    To find the transformation represented by a 2×2 matrix it is often helpful to multiply it by the position vectors of the vertices of the unit square.

Worked example

Finding the image of a point under a transformation

1 (i) Write the point (3, 2) as a position vector.

   (ii) The point (3, 2) is transformed by the matrix $\begin{pmatrix} 2 & 0 \\ 1 & 2 \end{pmatrix}$.

   Find the image of the point.

Solution

(i) The position vector for the point (3, 2) is $\begin{pmatrix} 3 \\ 2 \end{pmatrix}$.

(ii) $\begin{pmatrix} 2 & 0 \\ 1 & 2 \end{pmatrix}\begin{pmatrix} 3 \\ 2 \end{pmatrix} = \begin{pmatrix} 2 \times 3 + 0 \times 2 \\ 1 \times 3 + 2 \times 2 \end{pmatrix} = \begin{pmatrix} 6 \\ 7 \end{pmatrix}$

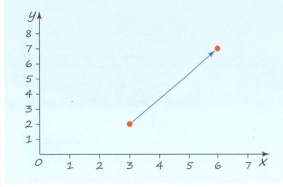

Worked examples

Transformations defined by matrices

2 (i) Multiply the matrix $\begin{pmatrix} 0 & 1 \\ 1 & 0 \end{pmatrix}$ by the position vectors of the points of the unit square.

   (ii) Draw a graph illustrating the unit square and its image under the transformation $\begin{pmatrix} 0 & 1 \\ 1 & 0 \end{pmatrix}$.

   (iii) Describe the transformation defined by the matrix $\begin{pmatrix} 0 & 1 \\ 1 & 0 \end{pmatrix}$.

Solution

(i) $\begin{pmatrix} 0 & 1 \\ 1 & 0 \end{pmatrix} \times \begin{pmatrix} 0 \\ 0 \end{pmatrix} = \begin{pmatrix} 0 \\ 0 \end{pmatrix}$

$\begin{pmatrix} 0 & 1 \\ 1 & 0 \end{pmatrix} \times \begin{pmatrix} 1 \\ 0 \end{pmatrix} = \begin{pmatrix} 0 \\ 1 \end{pmatrix}$

$\begin{pmatrix} 0 & 1 \\ 1 & 0 \end{pmatrix} \times \begin{pmatrix} 1 \\ 1 \end{pmatrix} = \begin{pmatrix} 1 \\ 1 \end{pmatrix}$

$\begin{pmatrix} 0 & 1 \\ 1 & 0 \end{pmatrix} \times \begin{pmatrix} 0 \\ 1 \end{pmatrix} = \begin{pmatrix} 1 \\ 0 \end{pmatrix}$

(ii)

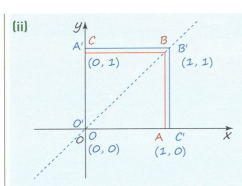

(iii) The transformation is a reflection in the line $y = x$.

3 Determine the transformation defined by the matrix $\begin{pmatrix} 2 & 0 \\ 0 & 2 \end{pmatrix}$.

Solution

$$\begin{pmatrix} 2 & 0 \\ 0 & 2 \end{pmatrix} \times \begin{pmatrix} 0 \\ 0 \end{pmatrix} = \begin{pmatrix} 0 \\ 0 \end{pmatrix}$$

$$\begin{pmatrix} 2 & 0 \\ 0 & 2 \end{pmatrix} \times \begin{pmatrix} 1 \\ 0 \end{pmatrix} = \begin{pmatrix} 2 \\ 0 \end{pmatrix}$$

$$\begin{pmatrix} 2 & 0 \\ 0 & 2 \end{pmatrix} \times \begin{pmatrix} 1 \\ 1 \end{pmatrix} = \begin{pmatrix} 2 \\ 2 \end{pmatrix}$$

$$\begin{pmatrix} 2 & 0 \\ 0 & 2 \end{pmatrix} \times \begin{pmatrix} 0 \\ 1 \end{pmatrix} = \begin{pmatrix} 0 \\ 2 \end{pmatrix}$$

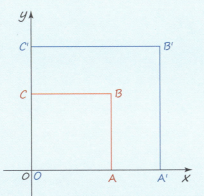

This transformation represents an enlargement, scale factor 2. ◀━━━━━━

> Enlargement can be by any scale factor, including a negative value.

Transforming the points (1, 0) and (0, 1) by a 2×2 matrix gives you sufficient information to describe the transformation. This is shown in the next example.

4 Determine the transformation defined by the matrix $\begin{pmatrix} 0 & -1 \\ 1 & 0 \end{pmatrix}$.

Solution

$$\begin{pmatrix} 0 & -1 \\ 1 & 0 \end{pmatrix} \times \begin{pmatrix} 1 \\ 0 \end{pmatrix} = \begin{pmatrix} 0 \\ 1 \end{pmatrix}$$

$$\begin{pmatrix} 0 & -1 \\ 1 & 0 \end{pmatrix} \times \begin{pmatrix} 0 \\ 1 \end{pmatrix} = \begin{pmatrix} -1 \\ 0 \end{pmatrix}$$

> Note that these examples have covered reflection, enlargement and rotation.
>
> Translations cannot be represented by matrices.

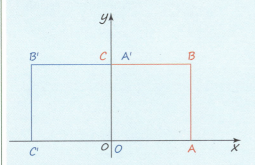

The transformation represents a rotation of 90° anticlockwise about the origin. ◀━━━━━━

> Note that in this course only rotations of 90°, 180° and 270° will be used.

Exam-style question

The unit square, OABC, has coordinates (0, 0), (1, 0), (1, 1) and (0, 1).

The vertices of the square are transformed by the matrix **P**, where $\mathbf{P} = \begin{pmatrix} 0 & 1 \\ -1 & 0 \end{pmatrix}$.

(i) Write down the coordinates of the images A′, B′ and C′ of the points A, B, and C.
(ii) Draw on a grid the unit square OABC and the transformed image OA′B′C′.
(iii) Hence describe the transformation defined by **P**.

Short answers on page 124

Full worked solution online

9.3 Combining transformations

Key facts

1 **Notation**
   When a transformation represented by the matrix **P** is followed by a transformation represented by the matrix **Q**, the matrix for the combined transformation is **QP**.

Worked examples

Successive transformations

1 The point X has coordinates (2, 3).

   When X is transformed by $\mathbf{P} = \begin{pmatrix} -1 & 0 \\ 0 & 1 \end{pmatrix}$ its image is X′.

   When X′ is transformed by $\mathbf{Q} = \begin{pmatrix} 1 & 0 \\ 0 & -1 \end{pmatrix}$ its image is X″.

   (i) Find the coordinates of X′ and X″.
   (ii) Find the matrix which represents the combined transformation.
   (iii) Use this matrix to check your answer for X″.
   (iv) Describe geometrically the two transformations and the combined transformation.

Solution

(i) $\begin{pmatrix} -1 & 0 \\ 0 & 1 \end{pmatrix}\begin{pmatrix} 2 \\ 3 \end{pmatrix} = \begin{pmatrix} -2 \\ 3 \end{pmatrix}$ so X′ is (−2, 3).

$\begin{pmatrix} 1 & 0 \\ 0 & -1 \end{pmatrix}\begin{pmatrix} -2 \\ 3 \end{pmatrix} = \begin{pmatrix} -2 \\ -3 \end{pmatrix}$ so X″ is (−2, −3).

(ii) $QP = \begin{pmatrix} 1 & 0 \\ 0 & -1 \end{pmatrix}\begin{pmatrix} -1 & 0 \\ 0 & 1 \end{pmatrix} = \begin{pmatrix} -1 & 0 \\ 0 & -1 \end{pmatrix}$

(iii) $\begin{pmatrix} -1 & 0 \\ 0 & -1 \end{pmatrix}\begin{pmatrix} 2 \\ 3 \end{pmatrix} = \begin{pmatrix} -2 \\ -3 \end{pmatrix}$

This confirms that X″ is (−2, −3).

(iv)

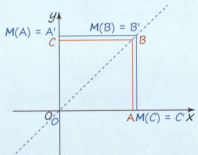

P represents a reflection in the y-axis.

Q represents a reflection in the x-axis.

QP represents a rotation of 180° centre (0, 0).

2 The matrix **L** represents a rotation of 90° anticlockwise and **M** represents a reflection in the line $y = x$.

(i) Write down the matrices **L** and **M**.

(ii) Draw diagrams to show the transformations **LM** and **ML** applied to the unit square, OABC.

(iii) Hence show that the two transformations **LM** and **ML** are not the same.

Solution

(i) $L = \begin{pmatrix} 0 & -1 \\ 1 & 0 \end{pmatrix}$

$M = \begin{pmatrix} 0 & 1 \\ 1 & 0 \end{pmatrix}$

(ii) The transformation **LM** means transform using **M** first followed by **L**.

M is a reflection in the line $y = x$.

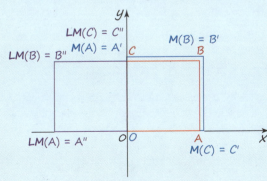

L is a rotation through 90° about the origin anticlockwise.

The diagram shows that the combined transformation **LM** is a reflection in the y-axis.

The combined transformation **ML** means transform using **L** first and then **M**.

L is a rotation through 90° about the origin anticlockwise.

M is a reflection in the line $y = x$.

The combined transformation **ML** is shown in the diagram below.

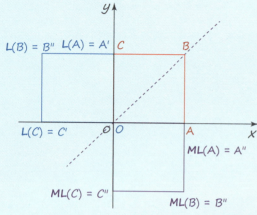

The diagram shows that the combined transformation **ML** is reflection in the x-axis.

(iii) The images under **LM** and **ML** are not the same, so the transformations are not the same.

Exam-style question

(i) Write down the matrix **P** that represents a transformation of a rotation clockwise of 90° centre the origin.

(ii) Work out $\mathbf{P}^2$ and state what transformation it represents.

(iii) Deduce the matrix $\mathbf{P}^4$ and state what transformation it represents.

Short answers on page 124

Full worked solution online

Review questions (Chapter 9)

1 Multiply out $\begin{pmatrix} 1 & 2 \\ -1 & 4 \end{pmatrix}\begin{pmatrix} 3 \\ -1 \end{pmatrix}$. [2]

2 You are given that $\mathbf{A} = \begin{pmatrix} 1 & -1 \\ 2 & 3 \end{pmatrix}$, $\mathbf{B} = \begin{pmatrix} 2 & 0 \\ 1 & 1 \end{pmatrix}$ and $\mathbf{C} = \mathbf{AB} = \begin{pmatrix} 1 & -1 \\ a & b \end{pmatrix}$.

   Work out the values of a and b. [3]

3 You are given that $\mathbf{A} = \begin{pmatrix} 2 & -3 \\ 4 & a \end{pmatrix}$, $\mathbf{B} = \begin{pmatrix} 1 & 3 \\ -4 & b \end{pmatrix}$ and $\mathbf{AB} = k\mathbf{I}$.

   Work out the values of a, b and k. [4]

4 Work out the image of the point (1, 4) for the transformation represented by the matrix $\begin{pmatrix} 1 & 2 \\ 3 & -2 \end{pmatrix}$. [2]

5 The matrix $\mathbf{M}$ represents an anticlockwise rotation through 90° about the origin.

   (i) Write down $\mathbf{M}$.

   (ii) Work out $\mathbf{M}^2$ and say what it represents. [2]

6 The unit square OABC is transformed by the matrix $\begin{pmatrix} 3 & 0 \\ 0 & 3 \end{pmatrix}$ to OA'B'C'.

   Show the image on a diagram and describe the transformation. [3]

7 The point P with coordinates (1, 2) is transformed to the point P' by $\begin{pmatrix} 2 & 3 \\ 1 & -1 \end{pmatrix}$ followed by a

   further transformation $\begin{pmatrix} 1 & 0 \\ 1 & -1 \end{pmatrix}$ to P''.

   Find the coordinates of P''. [4]

8 The unit square is transformed by a reflection in the y-axis followed by an anticlockwise rotation of 90° about the origin.

   (i) Work out the matrix for the combined transformation. [2]

   (ii) Use your answer to part (i) to find the image of (3, 5) under the combined transformation. [2]

9 $\mathbf{A}$ is the matrix that represents a reflection in the line $y = x$.

   (i) Write down the matrix $\mathbf{A}$. [1]

   (ii) Work out $\mathbf{A}^2$. [2]

   (iii) Comment on your answer. [1]

Short answers on pages 124–125

Full worked solutions online

CHECKED ANSWERS

Exam preparation

During the course
- Do not leave your revision until the last few weeks – little and often is the best way. As you progress through your course continue to think about and revise topics already covered.

Before your exam
- Be organised and disciplined – don't get distracted by other things.
- Turn your phone off while revising – you won't do it if you keep talking to friends.
- Use this book to ensure you have everything covered and can understand your notes.
- Do as many questions as you can, particularly on the topics with which you do not feel comfortable.
- Aim to put aside some past and practice papers for a final check on your ability to deal with questions in the given time.
- The paper is 1 hour 45 minutes long and carries 80 marks. A very rough guide is a mark a minute with approximately 20 minutes for consolidation and checking.

Formulae you need to learn

All GCSE is assumed knowledge.

Polynomials and binomial expansions	**The factor theorem**
	If $f(a) = 0$ then $(x - a)$ is a factor of $f(x)$ and $x = a$ is a root of the equation $f(x) = 0$.
	The binomial expansion
	$(a + x)^n = (a)^n + {}^nC_1(a)^{n-1}(x) + {}^nC_2(a)^{n-2}(x)^2 + \ldots + {}^nC_r(a)^{n-r}(x)^r + \ldots + (x)^n$
	Factorial notation
	$n! = n(n-1)(n-2)\ldots3.2.1$
	Coefficients of the binomial expansion
	$${}^nC_r = {}_nC_r = \binom{n}{r} = \frac{n!}{r!(n-r)!} = \frac{n(n-1)(n-2)\ldots(n-r+1)}{1.2.3.4\ldots r}$$
	Pascal's triangle
	Can be used for small values for n to determine the coefficients.

	<pre> 1
	          1 1
	        1 2 1
	      1 3 3 1
	    1 4 6 4 1
	  1 5 10 10 5 1</pre>

Coordinate geometry	The equation of a straight line with gradient m and intercept, c, on the y-axis is $y = mx + c$.
	The equation of a straight line with gradient m and passing through the point (x_1, y_1) is $y - y_1 = m(x - x_1)$.
	For a line joining the points (x_1, y_1), (x_2, y_2):
	• Distance between the points, $d = \sqrt{(x_1 - x_2)^2 + (y_1 - y_2)^2}$
	• Midpoint of the line segment is $\left(\dfrac{x_1 + x_2}{2}, \dfrac{y_1 + y_2}{2}\right)$
	• Gradient of the line $= \dfrac{y_2 - y_1}{x_2 - x_1}$
	• The equation of the line is $\dfrac{y - y_1}{y_2 - y_1} = \dfrac{x - x_1}{x_2 - x_1}$
Circles	The equation of a circle, centre (a, b) and radius r is $(x - a)^2 + (y - b)^2 = r^2$.

Trigonometry	**Pythagoras' theorem** In a right-angled triangle where $A = 90°$ $a^2 = b^2 + c^2$ $\sin^2\theta + \cos^2\theta = 1$ $\dfrac{\sin\theta}{\cos\theta} = \tan\theta$ **Sine rule** $\dfrac{a}{\sin A} = \dfrac{b}{\sin B} = \dfrac{c}{\sin C}$ or $\dfrac{\sin A}{a} = \dfrac{\sin B}{b} = \dfrac{\sin C}{c}$ **Cosine rule** $a^2 = b^2 + c^2 - 2bc\cos A$ or $\cos A = \dfrac{b^2 + c^2 - a^2}{2bc}$	
Calculus	$y = ax^n \Rightarrow \dfrac{\mathrm{d}y}{\mathrm{d}x} = nax^{n-1}$ for $n \neq 0$ $y = \mathrm{f}(x) \pm \mathrm{g}(x) \Rightarrow \dfrac{\mathrm{d}y}{\mathrm{d}x} = \mathrm{f}'(x) \pm \mathrm{g}'(x)$	
Laws of indices	$a^n . a^m = a^{n+m}, (a^n)^m = a^{nm}, \dfrac{a^n}{a^m} = a^{n-m},$ $a^{-n} = \dfrac{1}{a^n}, a^0 = 1$	

During your exam

Watch out for these words

Exact... Do not use your calculator to calculate an answer. Leave it with included surds or powers.

For example, $2\sqrt{3}$ or 5π.

Show that... The answer has been given to you so you must show every step of the working.

Prove that... This is the same as 'show that...' except that it is likely to be algebraic rather than arithmetic.

Hence... You should follow a statement or a previous result.

Calculate... Unless there is a specific instruction (such as 'show all your working') you may use your calculator to do the arithmetic.

Determine... Expect to give some working to justify your result.

Give, state, write down... You should be able to write down the answer without working anything out.

Sketch, plot, draw...

- A **sketch** is a diagram that is not necessarily to scale showing the main features of the curve being drawn, such as turning points, intersections with the axes, etc. It often will involve the general shape for large positive or negative values of x.
- **Draw** means a diagram that is slightly more accurate for the context.
- **Plot** means that you have to work out values of y for a number of values of x (which will be given to you in the question) and then to mark them accurately on graph paper, joining them with a smooth curve.

Remember

If you give an answer that is correct with no working, it is possible that you will receive full marks (unless you have used a method that has been specifically prohibited). However, if you give an incorrect answer with no working then you will receive no marks at all. If you show some working, you may well receive some intermediate marks even though your final answer is wrong.

If you cannot do part (i) of a question it does not mean that you cannot do part (ii), so don't ignore the rest of the question just because you cannot do the first part. If the first part is a 'show that...', then you may use this in part(ii) even if you were not able to show it.

Pace yourself: 105 minutes for 80 marks means roughly a mark a minute plus 20 minutes reading and checking. If you spend 10 minutes on the first question which only carries 2 marks, then you will be putting pressure on yourself later in the exam.

If you get stuck on a question

- Don't panic!
- Move on to the next question and give yourself time to return to it.
- Re-read the question – have you missed an essential piece of information?
- Would a sketch help?

Finishing off

Try to pace yourself so that you have time to check your answers.

Check that your answer to each question is
- to the correct accuracy
- in the right form
- complete.

Answers

SECTION 1 ALGEBRA

Target your revision (Chapters 1–4) (pages 1–3)

1 (i) 650 : 250

   (ii) £18

2 $\dfrac{9}{8}$

3 $2(7n - m)$

4 $\dfrac{b}{12}$

5 $x = \dfrac{2}{3}$

6 7 : 2

7 $p = 120\%$ of q

8 20

9 720

10 330

11 $8x - 7y$

12 $2x^2 + 3xy - 2y^2$

13 $2x^3 - 3x^2 - 11x + 6$

14 1 8 28 56 70 56 28 8 1

15 $-40x^3$

16 90 720

17 $8\sqrt{2}$

18 $8\sqrt{2}$

19 $\dfrac{1}{7}(3 + \sqrt{2})$

20 $2^2 \times 5 \times 7$

21 $3x(2x + 3y)$

22 (i) $(x - 3)(x - 5)$

   (ii) $(x + 3)(2x - 5)$

23 $(2y - 5)(2y + 5)$

24 $x = \dfrac{2y + 1}{y - 1}$

25 $r = \sqrt[3]{\dfrac{3V}{4\pi}}$

26 $\dfrac{-x^2 + x - 2}{x^2 - 1}$

27 $x = 21$

28 $a = -6$, $b = -5$

29 Minimum value is 3 when $x = 2$

30 $-2 \leqslant x \leqslant 2$

31 $\mathrm{fg}(x) = 2 - x^2$

   $\mathrm{gf}(x) = (2 - x)^2$

32 -10

33 (i) $4y = 5x + 13$

   (ii)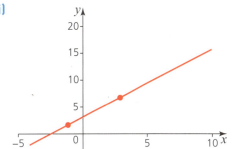

34 (i) $y = 4 - (x - 3)^2$

   (ii)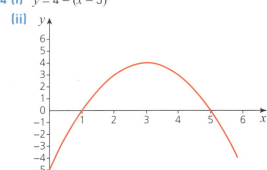

35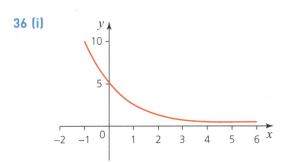

36 (i)

   (ii) From the graph the curve drops below $y = 1$ at approximately $x = 2.3$

37

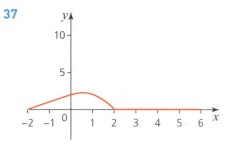

38

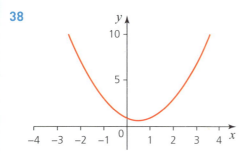

39 $y = 4, -2$

40 $x = -3 \pm 3\sqrt{2}$

41 $x = \frac{1}{2}(1 \pm \sqrt{33})$

42 $b^2 - 4ac < 0$, so no

43 11

44 $x = \pm\sqrt{2}$ or $\pm\sqrt{3}$

45 $x = 2, y = -5$

46 $(0, 4), (1, 2)$

47 $x = 2, y = -1, z = -3$

48 $x^2 + 4x + 7, 19$

49 (ii) $1, -2, 3$

50 (i) $-1 \leqslant x < 3$

(ii)

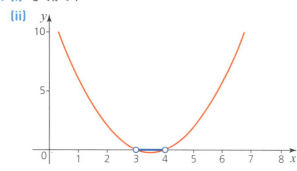

51 (i) $3 < x < 4$

(ii)

52 x^{-2}

53 $(n + 2)^2 - (n - 2)^2 = n^2 + 4n + 4 - (n^2 - 4n + 4)$
$$= 8n$$
So for all n this is a multiple of 8.

54 $a_{34} = 101$

55 45

56 $a_n = n^2 + 3n - 2$

57 1

Chapter 1A GCSE-level revision for number and algebra

1.1 Numbers and the number system Practice question (GCSE level) (page 5)

Adam receives $14 \times 700 = £9800$

Beryl receives $15 \times 700 = £10\,500$

Chloe receives $16 \times 700 = £11\,200$

1.2 Simplifying algebraic expressions Practice question (GCSE level) (page 7)

$\dfrac{-2x}{y^2}$

1.3 Solving linear equations Practice question (GCSE level) (page 8)

$x = -\dfrac{11}{4}$

1.4 Algebra and number Practice question (GCSE level) (page 9)

Initially there were 20 red and 15 blue marbles.

Chapter 1B Further revision for number and algebra

1.5 The product rule for counting Exam-style question (page 11)

(i) 720   **(ii)** 120   **(iii)** 48

1.6 Expanding brackets Exam-style question (page 13)

(i) (a) $6x^3 + 23x^2 + 25x + 6$

    (b) $22x^2 + 58x + 34$

(ii) When $x = 1$, volume = 60 cubic units, total surface area = 114 square units.

1.7 Binomial expansions – using Pascal's triangle Exam-style question (page 15)

$243x^5 - 1620x^4y + 4320x^3y^2 - 5760x^2y^3 + 3840xy^4 - 1024y^5$

1.8 Manipulating surds Exam-style question (page 16)

$\dfrac{25 + \sqrt{3}}{22}$

Chapter 2 Algebra II

2.1 Factorising Exam-style question (page 19)

$(3x + 5y)(2x - 3y)$

2.2 Rearranging formulae Exam-style question (page 20)

$C = \dfrac{5}{9}(F - 32)$

2.3 Simplifying algebraic fractions and solving equations Exam-style question (page 21)

$x = \dfrac{22}{19}$

2.4 Completing the square Exam-style question (page 24)

$\dfrac{1}{2}$

Chapter 3 Algebra III

3.1 Function notation and composite functions Exam-style question (page 27)

$x = 1$

3.2 Inverse functions Exam-style question (page 29)

$\text{f}(x) = x + 3$ and $\text{g}(x) = x^2 + 1$

$\Rightarrow \text{fg}(x) = \text{f}(x^2 + 1) = x^2 + 4$

Let $\text{fg}(x) = y$

$\Rightarrow y = x^2 + 4 \Rightarrow x = \sqrt{y - 4}$

$\Rightarrow (\text{fg})^{-1}(x) = \sqrt{x - 4}$

It is a one-one function for all values of $x \geqslant 4$

3.3 Drawing graphs Exam-style question (page 32)

(i)

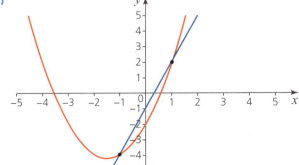

(ii) (1, 2) and (−1, −4)

3.4 Graphs of exponential functions Exam-style question (page 33)

(i)

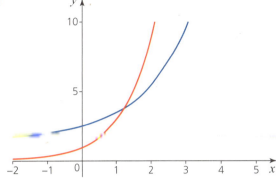

(ii) 1.2

(iii) For $x = 1.2$, $y = 3.74$ so $x = 1.2$ is a reasonable estimate for the root, given that it is read from the graph.

3.5 Graphs of functions with more than one part to their domains (page 35)

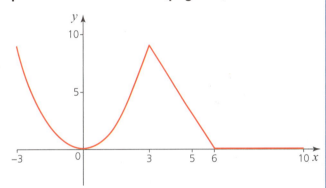

Chapter 4 Algebra IV

4.1 Quadratic equations Exam-style question (page 39)

(i) $\left(x - \dfrac{3}{2}\right)^2 - 3.25$

(ii) (1.5, −3.25)

(iii)

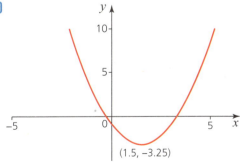

(iv) $x = 3.303$ or -0.303

4.2 Simultaneous equations Exam-style question (page 41)

(i)

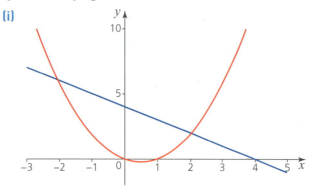

(ii) $x = 2$, $y = 2$ and $x = -2$, $y = 6$

(iii) The solution is $x = 2$, $y = 2$ and $x = -2$, $y = 6$

4.3 Solving cubic equations Exam-style question (page 43)

(ii) $x = -1, 2, 2$

(iii)

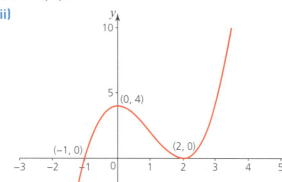

4.4 Inequalities Exam-style question (page 46)

(i) x satisfies $x \geqslant 3$ or $x \leqslant -5$

(ii)

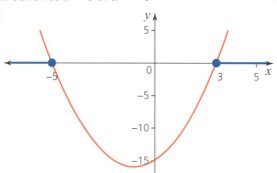

4.5 Index laws Exam-style question (page 47)

$x = \pm 2$ or $x = \pm\sqrt{7} = \pm 2.65$

4.6 Algebraic proof Exam-style question (page 48)

Go online for the full worked solution.

4.7 Sequences Exam-style question (page 51)

$u_n = 2n^2 - n + 1$

Review questions (Chapters 1–4) (pages 52–54)

1. 30
2. £28 350
3. $2(6y - 5x)$
4. $\dfrac{3de}{2c}$
5. $x = 3$
6. $x = -35$
7. $5 : 3$
8. $x = 9000$, $y = 11\,000$
9. 462
10. 1 h 4 min 48 s
11. $2x(x + 2)$
12. $x^3 + 2x^2 - 9x - 18$
13. $32 - 80x + 80x^2 - 40x^3 + 10x^4 - x^5$
14. 4320
15. $-9 + 5\sqrt{15}$
16. 2
17. $2 - \dfrac{1}{2}\sqrt{6}$
18. $(x + 8)(x - 5)$
19. $(x - 2y)(3x - 5y)$
20. $h = \dfrac{S - 2\pi r^2}{2\pi r}$
21. $x = \dfrac{2 - y}{2y - 1}$
22. $\dfrac{5x - 1}{x(x^2 - 1)}$
23. $x = \pm\sqrt{3}$
24. $(x + 3)^2 - 2$, $a = 3$, $b = -2$
25. $10 - (x + 2)^2$, $a = 10$, $b = -1$, $c = 2$
26. **(i)** $2x^2 + 1$ **(ii)** $(2x + 1)^2$
27. $k = -\dfrac{1}{2}$
28. 3.5
29. **(i)** $\dfrac{x - 1}{2}$

    (ii)

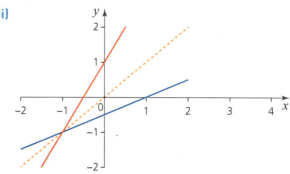

30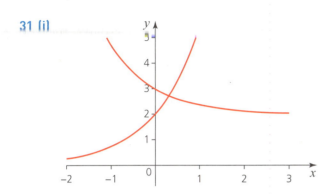

31 (i)

(ii) $x = 0.3$

32

33 $y = 3, 0 \leqslant x \leqslant 2$
$x + y = 5, 2 \leqslant x \leqslant 5$

34 $x = 5.32$ or $x = -1.32$

35 $x = 4 \pm \sqrt{6}$

36 $x = 2$
$y = -3$

37 $x = 3, y = 9$
and $x = -1, y = 5$

38 (ii) $x = -1, 2$ or 7

39 $a = -14, b = 35$

40 (i) $x < -\frac{1}{3}$ **(ii)** $-1 < x < 6$

41 Set of integers $\{-2, -1, 0, 1, 2, 3, 4\}$

42 x^1

43 $x = 4$

44 Go online for the full worked solution.

45 Go online for the full worked solution.

46 (i) 5%   **(ii)** £2315.25

47 $u_4 = 311$

48 $\frac{2}{3}$

SECTION 2 GEOMETRY

Target your revision (Chapters 5–9) (pages 55–57)

1 $-\frac{5}{4}$

2 $\sqrt{61} \approx 7.81$

3 $(8, 3)$

4 $-\frac{1}{2}$

5 $x + y = 4$

6 $(4, -1)$

7 $(3, 5)$

8 $x^2 + y^2 = 9$

9 $x^2 + y^2 - 4x - 6y - 12 = 0$

10 Centre $(-5, 1)$, radius 7

11 Outside

12 For AB midpoint is $(1, 3)$, gradient $-\frac{1}{3}$
For perpendicular gradient $= 3$ passing through $(1, 3)$
$\Rightarrow y - 3 = 3(x - 1) \Rightarrow y = 3x$
which passes through the centre $(0, 0)$.

13 $d = \sqrt{74} \approx 8.6$ or
$d = \sqrt{24} \approx 4.9$

14 $135°$

15 18

16 $104°$

17 $90°$

18 $30°$

19 Go online for the full worked solution.

20 Go online for the full worked solution.

21 Go online for the full worked solution.

22 AC $=$ BC $= 7\sin 45 = \dfrac{7}{\sqrt{2}} = \dfrac{7}{2}\sqrt{2}$

23 40.3 m

24 $310°$

25 $0.1736, -0.7660, -0.3640$

26

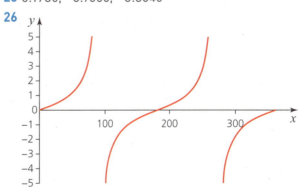

27 41.8°, 138.2°

28 Go online for the full worked solution.

29 25.7°, 154.3°, 230.1°, 309.9°

30 11.3 cm^2

31 9.48 cm

32 6.98 cm

33 27.5°

34 18.4°

35 10

36 $y = 7x - 8$

37 $12y + x = 63$

38 $1 \leqslant x \leqslant 4$

39 (i) $(-1, 9)$ and $(3, -23)$

    (ii) $-12, 12$

    (iii) At $x = -1$ the turning point is a maximum.
        At $x = 3$ the turning point is a minimum.

40 Minimum at $(2, -14)$, maximum at $(-1, 13)$

41 $\mathbf{AB} = \begin{pmatrix} 4 & -7 \\ 1 & -13 \end{pmatrix}, \mathbf{BA} = \begin{pmatrix} 1 & 7 \\ 5 & -10 \end{pmatrix}$

    So $\mathbf{AB} \neq \mathbf{BA}$

42 $a = -2$

43 $\begin{pmatrix} 7 \\ 14 \end{pmatrix}$ so coordinates of image are $(7, 14)$

44 Reflection in the line with equation $y = x$

45 $\begin{pmatrix} 1 & 0 \\ 0 & -1 \end{pmatrix}$

Chapter 5 Coordinate geometry

5.1 Points and lines Exam-style question (page 60)

(i) Go online for the full worked solution.

(ii) Go online for the full worked solution.

(iii) Go online for the full worked solution.

(iv) Go online for the full worked solution.

5.2 The equation of a line and intersections Exam-style question (page 62)

$x + 2y = 12$

5.3 The circle Exam-style question (page 64)

(i) $x^2 + (y - 1)^2 = 9$

(ii) AB is $4\sqrt{2} = 5.66$ (to 3 s.f.)

(iii) Outside

5.4 Circle geometry Exam-style question (page 68)

(i) Centre is at $(2, 7)$; radius is 5

(ii) Go online for the full worked solution.

(iii) Go online for the full worked solution.

(iv) The angle in a semi-circle is a right angle.

Chapter 6 Geometry I

6.1 GCSE Revision on mensuration and angles Exam-style question (page 72)

$x = 132°$

6.2 Circle theorems Exam-style question (page 74)

Angle DAB = 100°

6.3 Geometric proof Exam-style question (page 76)

Go online for the full worked solution.

6.4 Trigonometry in two dimensions Exam-style question (page 78)

(i) 583 m

(ii) 031°

(iii) 211°

6.5 Trigonometric functions for angles of any size Exam-style question (page 83)

(i) Go online for the full worked solution.

(ii) $\theta = 35.3°, 144.7°, -35.3°$ or $-144.7°$

Chapter 7 Geometry II

7.1 The area of a triangle Exam-style question (page 86)

(i) 6.75 cm^2

(ii) AP = 4.50 cm, CP = 5.36 cm, area = 6.75 cm^2. Go online for the full worked solution.

7.2 The sine and cosine rules Exam-style question (page 89)

DC = 42.1 m

7.3 Problems in three dimensions Exam-style question (page 91)

66.2°

Review questions (Chapters 5–7) (pages 92–93)

1 (i) Go online for the full worked solution.
   (ii) Go online for the full worked solution.

2 (i) Go online for the full worked solution.
   (ii) Go online for the full worked solution.

3 (i)

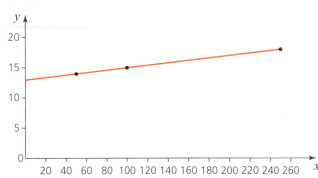

  (ii) (a) £13   (b) £4
  (iii) £53.00

4 (i) Centre is at (2, 3), radius = 5
  (ii) $x = 5$, $y = -1$
     and $x = 6$, $y = 0$

5 (i) $2y = 3x + 1$
  (ii) (3, 5)
  (iii) (0, 7)
  (iv) $x^2 + (y - 7)^2 = 13$

6 (i) Centre (1, 1), radius 5
  (ii) Go online for full worked solution.
  (iii) $3x - 4y = 24$

7 Centre (4, 2)

8 Angle OCB = 50° (base angles of isosceles triangle are equal)
  Angle COB = 80° (angle sum of triangle = 180°)
  Angle CAB = 40° (angle at circumference is half the angle at the centre)

9 Go online for full worked solution.

10 25.5 m

11 1.15 km

12 15°, 75°, 195°, 255°

13 216.9°

14 $x = 296.6°$ and 116.6°

15 11.47 cm²

16 71.8°

17 44.96°

18 AC = 9.47 m

19 $\sqrt{29} \approx 5.39$ m

20 (i) 292 m
  (ii) 9.9°

SECTION 3 CALCULUS

Chapter 8 Calculus

8.1 Differentiation and gradient of curves Exam-style question (page 96)

(i) $\dfrac{dy}{dx} = -4$

(ii) $\left(-\dfrac{2}{3}, 5\dfrac{22}{27}\right)$ and (2, −1)

8.2 Tangents and normals Exam-style question (page 98)

(i) $6y = x + 19$

(ii) $\left(\dfrac{31}{6}, \dfrac{145}{36}\right)$

8.3 Increasing and decreasing functions and the second derivative Exam-style question (page 99)

(i) $x > \dfrac{1}{2}$ and $x < -\dfrac{1}{2}$

(ii) $x > 0$

(iii) (a)

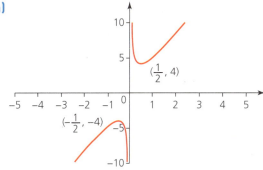

(b) Go online for full worked solution.

8.4 Stationary points Exam-style question (page 101)

(i) (0, 6) is a maximum.
   (2, 2) is a minimum.

(ii)

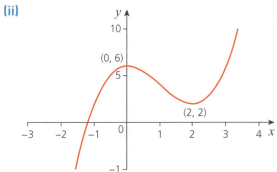

Review questions (Chapter 8) (page 102)

1 (i) $\dfrac{dy}{dx} = -4x^{-3}$

 (ii) $\dfrac{dy}{dx} = 3x^2 - 4x + 6$

2 $x = 3, y = 10$, or (3, 10)

3 $x + 6y = 13$

4 (i) $y = 2x + 1$

 (ii) $x = -4, y = -7$

 (iii) PR is not a normal

5 $x < 0$ or $x > 1$

6 Go online for full worked solution.

7 (i) 22

 (ii) 24

 (iii) 18

8 (i) Maximum at (1, 9)

 (ii) (3, 5)

 (iii)

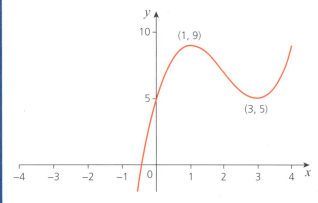

9 (i) $\left(1, \dfrac{1}{2}\right)$ and (−2, 14)

 (ii) $\left(1, \dfrac{1}{2}\right)$ is a minimum point

 (−2, 14) is a maximum point

 (iii)

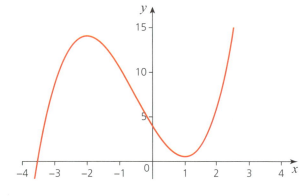

SECTION 4 MATRICES

Chapter 9 Matrices

9.1 Multiplying matrices Exam-style question (page 105)

$m = 5, n = 3$

9.2 Transformations Exam-style question (page 108)

(i) A′ is (0, −1), B′ is (1, −1), C′ is (1, 0)

(ii)

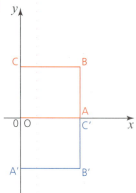

(iii) The transformation represents a rotation of 90° clockwise about the origin.

9.3 Combining transformations Exam-style question (page 110)

(i) $\mathbf{P} = \begin{pmatrix} 0 & 1 \\ -1 & 0 \end{pmatrix}$

(ii) $\mathbf{P}^2 = \begin{pmatrix} -1 & 0 \\ 0 & -1 \end{pmatrix}$

(iii) $\mathbf{P}^4 = \begin{pmatrix} 1 & 0 \\ 0 & 1 \end{pmatrix} = \mathbf{I}$, no transformation

Review questions (Chapter 9) (page 111)

1 $\begin{pmatrix} 1 \\ -7 \end{pmatrix}$

2 $a = 7, b = 3$

3 $k = 14, b = 2, a = 1$

4 (9, −5)

5 (i) $\mathbf{M} = \begin{pmatrix} 0 & -1 \\ 1 & 0 \end{pmatrix}$

(ii) $\mathbf{M}^2 = \begin{pmatrix} -1 & 0 \\ 0 & -1 \end{pmatrix}$, this represents a rotation of 180°, centre the origin.

6 $(1, 0)$ is transformed to $(3, 0)$, $(0, 1)$ is transformed to $(0, 3)$ and $(1, 1)$ is transformed to $(3, 3)$

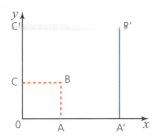

The transformation is an enlargement, centre the origin with scale factor 3.

7 $P'' = (8, 9)$

8 (i) $\begin{pmatrix} 0 & -1 \\ -1 & 0 \end{pmatrix}$

(ii) $\begin{pmatrix} -5 \\ -3 \end{pmatrix}$

9 (i) $\mathbf{A} = \begin{pmatrix} 0 & 1 \\ 1 & 0 \end{pmatrix}$

(ii) $\mathbf{A}^2 = \begin{pmatrix} 1 & 0 \\ 0 & 1 \end{pmatrix}$

(iii) $\mathbf{A}^2 = \mathbf{I}$. A reflection in $y = x$ followed by a reflection in $y = x$ transforms the unit square back to its original position.

Notes

Full worked solutions at www.hoddereducation.co.uk/myrevisionnotesdownloads

Hybrid SPECT/CT Imaging in Clinical Practice

Hybrid SPECT/CT Imaging in Clinical Practice

edited by

Ora Israel
Rambam Health Care Campus
B. Rapaport School of Medicine,
Technion-Israel Institute of Technology
Haifa, Israel

Stanley J. Goldsmith
New York-Presbyterian Hospital/New York Weill Cornell Medical Center
New York, New York, U.S.A.

Taylor & Francis
Taylor & Francis Group
New York   London

Taylor & Francis is an imprint of the
Taylor & Francis Group, an informa business

Published in 2006 by
Taylor & Francis Group
270 Madison Avenue
New York, NY 10016

International Standard Book Number-10: 0-8247-2854-8 (Hardcover)
International Standard Book Number-13: 978-0-8247-2854-0 (Hardcover)

Library of Congress Cataloging-in-Publication Data

Catalog record is available from the Library of Congress

Taylor & Francis Group
is the Academic Division of Informa plc.

Visit the Taylor & Francis Web site at
http://www.taylorandfrancis.com

We thank
    our contributors for their effort and collegiality,
    our staffs for their support and tolerance,
    our families for being there when we needed them,
    and each other for all of the above.

                             Ora Israel and Stanley J. Goldsmith

Preface

Nuclear medicine imaging in general, and single-photon emission computed tomography (SPECT) in particular, enable the diagnosis of a physiologic or pathophysiologic process by providing high target to background images that have high specificity and hence only sparse anatomic detail.

To improve the anatomic localization of the functionally active process identified by the nuclear medicine procedures, investigators have used a variety of methods to coregister scintigraphy with other imaging techniques. The history of fusion evolved from visual comparison of function and anatomy. Physicians with a primary interest in image interpretation, as well as those involved in patient care, have relied for decades on their mental capabilities to integrate information obtained from different diagnostic modalities, performed separately, exhibited side-by-side, or recorded in their memory. Further improvements in this crude type of fusion have been sought by drawing regions of interest on operator-visually-aligned slices.

First real image fusion was obtained by coregistering results of separately performed studies, usually nuclear medicine and computed tomography (CT), using dedicated software and workstations. Coregistration of functional and anatomic images is technically feasible and its clinical utility has been demonstrated in "fixed" organs such as the brain, but is more difficult in other regions of the body. The separate acquisition of SPECT and CT, at times days apart, makes superimposition of the two images difficult. Even with the use of fiducial markers, it is impossible to reproduce exact positioning during the gamma camera and CT acquisition. The need to manipulate the data using sophisticated software is also problematic. Discrepancies between functional and structural data that were detected following coregistration could therefore be attributed to either technical problems of coregistration or to clinically significant findings.

Because coregistration techniques were considered cumbersome, they did not gain wide application and were excluded from routine clinical use. The technique has been limited to a few centers with dedicated medical physics groups, and a few highly skilled clinical teams performing dedicated research involving image fusion. In order to be widely accepted, coregistration has to use simple acquisition and processing technology, it has to stand up to proven and accepted quality control criteria, and it needs to be readily accessible to review. Based on, and inspired by the pioneering work of Bruce Hasegawa and his team, hybrid imaging has become a welcomed reality.

In 1999, the first hybrid SPECT/CT imaging system (Millenium VG & Hawkeye™, General Electric Healthcare Technologies) was introduced. It consisted of a dual-detector gamma camera combined with a single-slice, low-power X-ray CT system. Each modality is imaged separately and sequentially during a single patient examination. The transmission data obtained with the CT component are useful for attenuation correction of the emission SPECT data and also as an anatomic template on which the scintigraphic images are superimposed in order to provide a fused image of the two modalities.

The concept of using X-ray–based attenuation correction of SPECT images was initiated by the need to improve imaging of the heart. CT-attenuation corrected myocardial perfusion scintigraphy is considered at present the state-of-the-art modality in this field. Within a short time, however, hybrid imaging has opened a whole new world of opportunities. It permits a better understanding of the role of imaging in the assessment of disease, as well as the impact it generates on patient care.

SPECT/CT, as well as positron emission tomography/computed tomography (PET/CT), have been rapidly proven their value in tumor imaging. Simultaneous assessment of SPECT and CT images provides previously unavailable information on the nature, size, and localization of tumors, and improves the diagnostic ability and accuracy of in vivo imaging of neoplasms. Hybrid imaging provides an explanation for several phenomena previously observed in cancer patients but not clearly understood, for example, the fact that a tumor mass is not necessarily the only finding seen in tumor imaging. Also, with the help of hybrid imaging, early detection of cancer relapse can be obtained before an abnormal mass is detected by conventional, anatomic imaging modalities. Hybrid imaging enhances the awareness to a possible dissociation between mass and disease in cancer, which may further alter management of the individual cancer patient. It is obvious today that hybrid imaging, in only a short period of six years, has led to a change in the routine concept of non-invasive assessment of cancer.

With hybrid imaging, unique physiologic information benefits from a precise topographic localization. Over many decades, nuclear medicine and anatomic imaging (such as plain X-rays, computed tomography, and ultrasound) have been viewed as competitive modalities and a tremendous effort has been invested to prove the advantages of one technique over the other. The medical community has now come to the understanding that none of these modalities provide the only answer to complex clinical questions. The need for simultaneous evaluation of anatomic and metabolic information about normal and disease processes, using registration of complementary data, has evolved from these important clinical requirements.

While this book was in the final editing stages, new, second-generation SPECT/CT devices have become available for clinical use and research, characterized mainly by improvements at the CT end of the system (Symbia, Siemens Medical Systems; Precedence, Philips Medical Systems, Hawkeye-4 & Infinia LS, GE Healthcare).

This volume includes chapters that illustrate the current state-of-the-art clinical SPECT/CT imaging. In many instances, the technique is essential for accurate anatomic localization of radiotracer accumulation. This is particularly relevant to tumor imaging using radiopharmaceuticals such as ^{111}In-octreotide, ^{111}In-capromab pendetide (ProstaScint®), ^{131}I-MIBG,^{131}I as NaI,^{67}Ga, and ^{99m}Tc-colloid agents for lymphoscintigraphy, as well as for non-oncologic examinations for diagnosis of parathyroid adenoma and infectious processes, or bone scintigraphy, all representing procedures with great pathophysiologic specificity. Paradoxically, this apparent advantage is also the limiting factor as it consequently results in very little nonspecific tracer distribution that may enable anatomically localizing a suspicious focus of uptake.

It is remarkable that even in cases where there is reasonable identification of anatomic structures on nuclear medicine procedures, the availability of hybrid fusion images improves diagnostic accuracy and raises the level of confidence of the inter-preting physician and referring clinician. An additional advantage of hybrid ima-ging, although difficult to quantitate, is the fact that combining anatomic mapping and radiotracer distribution reduces the gap between experienced nuclear medicine readers and trainees, as well as radiologists who do not specialize in nuclear imaging, and furthermore makes the doubts of the clinicians confronted with "scintigraphic dots" disappear.

This volume is also privileged to include contributions regarding future appli-cations of our specialty. The use of SPECT/CT for dosimetry calculations is of utmost importance in an era when radioimmunotherapy is becoming an important therapeutic tool in cancer patients. Also, small animal metabolic and hybrid imaging is a new and valuable tool in reaching an understanding of basic biological processes and subsequently in new drug developments.

Why did it take longer for SPECT/CT than for PET/CT to conquer the place it deserves in the world of modern imaging? The answer is partially related to technical limitations that are or will be, hopefully, solved in the near future. Also, nuclear medicine physicians have acquired a high level of expertise in reading planar and SPECT studies, and some of us harbored the conservative insight that there was not much more to add.

The editors of this book, as well as the authors of the various chapters, have had the fortunate opportunity to be among the first users of hybrid imaging with SPECT/CT. This initial work has helped make the difference. As always, success has come from teamwork, from joint efforts between scientists involved in basic research and engineering teams, between academic groups and industry, and between numerous clinical research teams in different corners of the world. None of the above could have been accomplished single-handedly.

We are on the right path, but there is still a great deal of work to do. We need to establish indications for performing SPECT/CT, and the optimal workflow. Risk-to-benefit ratios related to additional radiation also need to be considered.

The way of SPECT/CT from "toy to tool" has been a challenging adventure, met with doubts, high expectations, but also with a great amount of appreciation by our colleagues in the imaging and clinical communities. The availability of new SPECT/CT devices, further developments in technology and radiopharmacy, as well as the growing interest in the clinical potential of hybrid SPECT/CT, confirm our belief that there are exciting times ahead in nuclear medicine.

This project could not have been envisioned without the help of good friends in our clinical departments in particular and in the nuclear medicine community, in general. Working together and stirring the enthusiasm of each other has made this effort possible, and to all we extend our deepest gratitude.

Ora Israel
Stanley J. Goldsmith

Contents

Contributors

Naomi P. Alazraki Emory University School of Medicine and the VA Medical Center, Atlanta, Georgia, U.S.A.

Rachel Bar-Shalom Rambam Health Care Campus, B. Rapaport School of Medicine, Technion—Israel Institute of Technology, Haifa, Israel

Roland Bruno Bares Department of Nuclear Medicine, University Hospital Tuebingen, Tuebingen, Germany

Scott C. Bartley Emory University School of Medicine and the VA Medical Center, Atlanta, Georgia, U.S.A.

Dominique Delbeke Vanderbilt University Medical Center, Nashville, Tennessee, U.S.A.

Susanne Martina Eschmann Department of Nuclear Medicine, University Hospital Tuebingen, Tuebingen, Germany

Einat Even-Sapir Department of Nuclear Medicine, Tel Aviv Sourasky Medical Center, Sackler School of Medicine, Tel Aviv University, Tel Aviv, Israel

Alex Frenkel Rambam Health Care Campus, Haifa, Israel

Eric C. Frey Division of Medical Imaging Physics, Russell H. Morgan Department of Radiology and Radiological Science, Johns Hopkins University, Baltimore, Maryland, U.S.A.

Stanley J. Goldsmith New York-Presbyterian Hospital/New York Weill Cornell Medical Center, New York, New York, U.S.A.

Bruce H. Hasegawa UCSF Physics Research Laboratory, Department of Radiology, University of California, San Francisco, California, U.S.A.

Marius Stefan Horger Department of Diagnostic Radiology, University Hospital Tuebingen, Tuebingen, Germany

Ora Israel   Rambam Health Care Campus, B. Rapaport School of Medicine, Technion—Israel Institute of Technology, Haifa, Israel

Zohar Keidar   Rambam Health Care Campus, B. Rapaport School of Medicine, Technion—Israel Institute of Technology, Haifa, Israel

Carl-Martin Kirsch   Department of Nuclear Medicine, Saarland University Medical Center, Homburg/Saar, Germany

Lale Kostakoglu   New York-Presbyterian Hospital/New York Weill Cornell Medical Center, New York, New York, U.S.A.

Yodphat Krausz   Department of Medical Biophysics and Nuclear Medicine, Hadassah—Hebrew University Medical Center, Jerusalem, Israel

William H. Martin   Vanderbilt University Medical Center, Nashville, Tennessee, U.S.A.

Rosna M. Mirtcheva   New York-Presbyterian Hospital/New York Weill Cornell Medical Center, New York, New York, U.S.A.

Michael K. O'Connor   Division of Medical Physics, Department of Radiology, Mayo Clinic, Rochester, Minnesota, U.S.A.

1

Physics and History of SPECT/CT

Bruce H. Hasegawa
UCSF Physics Research Laboratory, Department of Radiology, University of California, San Francisco, California, U.S.A.

Alex Frenkel
Rambam Health Care Campus, Haifa, Israel

INTRODUCTION

The field of diagnostic radiology encompasses a wealth of imaging techniques that now are essential for evaluating and managing patients who need medical care. Imaging methods such as plain film radiography, X-ray computed tomography (CT), and magnetic resonance imaging (MRI) can be used to evaluate a patient's anatomy with submillimeter spatial resolution to discern structural abnormalities and to evaluate the location and extent of disease, whereas X-ray fluoroscopy and angiography can be used to evaluate the status of the cardiovascular, gastrointestinal, or genitourinary systems. In comparison, functional imaging methods including planar scintigraphy, single-photon emission computed tomography (SPECT), positron emission tomography (PET), and magnetic resonance spectroscopy (MRS) assess regional differences in the biochemical status of tissues.

In nuclear medicine, including SPECT and PET, functional measurements are performed by administering the patient with a radiopharmaceutical that is accumulated in response to its biochemical attributes. By design, the radiopharmaceutical has a targeted action, allowing it to be imaged to evaluate specific physiological processes in the body. Now there are many radiopharmaceuticals available for medical diagnosis, with additional radiotracers available for in vivo as well as in vitro biological experimentation. Because the amount of radioactivity that can be administered internally to the patient is limited by considerations of radiation dose, radionuclide images inherently have poor photon statistics, are produced with only modest spatial resolution, and require relatively long scan times. In addition, the visual quality of radionuclide images can be degraded by physical factors such as photon attenuation and scatter radiation (1–7). In contrast, anatomical imaging, especially with radiography and CT, can be performed quickly, and with excellent signal-to-noise characteristics and submillimeter spatial resolution. These considerations illustrate that X-ray (transmission) and radionuclide (emission) imaging provide different but complementary information about the medical status of the patient (8–11).

Several investigators have developed methods that attempt to improve the correlation between anatomical and physiological information from X-ray transmission and radionuclide emission imaging (12–14). Commonly, image registration techniques produce a single "fused" or "combined" image in which, for example, the radionuclide distribution is displayed in color over a gray-scale CT image of the same anatomical region. The simplest form of image registration performs "rigid-body" translation and rotation to match the two image data sets. These techniques can be applied most successfully to neurological studies (15,16), where the skull provides a rigid structure that maintains the geometrical relationship of structures within the brain. The situation is more complicated when image registration techniques are applied to other areas of the body—for example, the thorax and abdomen, where the body can bend and flex, especially when the X-ray and radionuclide data are captured using different machines in separate procedures, often on different days. The shape of the patient table, the orientation of the body and limbs during the imaging procedure, and the respiratory state of the patient can affect the geometrical relationships between different anatomical regions. In these cases, image registration might match the patient anatomy in one region of the body, but not in all anatomical regions. Image warping can improve registration over a larger region of the patient's anatomy (13,14), but in most cases, software-based image registration can be challenging and, at most institutions, is not used routinely for clinical procedures.

With the goal of improving the correlation of different types of image data, other investigators have developed instrumentation that integrate both X-ray and radionuclide imaging in a single device (17–32). This technique, often called dual-modality imaging, physically combines a PET or SPECT system with a CT scanner, using a common patient table, computer, and gantry so that both the X-ray and radionuclide image data are acquired sequentially without removing the patient from the scanner. This technique thereby produces anatomical and functional images with the patient in the same position and during a single procedure, which simplifies the image registration and fusion processes (21,22,25,29,31). Dual-modality imaging offers several potential advantages over conventional imaging techniques. Radionuclide data can identify areas of disease that are not apparent on the X-ray images alone (11,33–36). Conversely, X-ray images provide detailed anatomical information that helps the physician to differentiate normal radionuclide uptake from that indicating disease, and to help localize disease sites within the body. The X-ray data can be used to generate a patient-specific map of attenuation coefficients, which in turn is used to correct the radionuclide data for errors due to photon attenuation, scatter radiation, and other physical effects (17,18,21,22,24,37–39), thus improving both the visual quality and the quantitative accuracy of the correlated scintigraphic images.

BRIEF HISTORY OF DUAL-MODALITY IMAGING

Whereas the advent of dedicated dual-modality imaging systems is relatively recent, the potential advantages of combining anatomical and functional imaging has been recognized for several decades by radiological scientists and physicians. Many of the pioneers of nuclear medicine, including Mayneord (40–42), Anger (43,44), Cameron and Sorenson (45), and Kuhl et al. (46) recognized that a radionuclide imaging system could be augmented by adding an external radioisotope source to acquire transmission data for anatomical correlation of the emission image. For example, Kuhl et al. (47) incorporated an external radionuclide source on his Mark IV brain scanner to produce anatomical images useful for both localizing regions of radionu-clide uptake and to correct soft-tissue absorption in the radionuclide emission data. In a 1974 review of photon attenuation, Budinger and Gullberg (48) noted that a patient-specific attenuation map could be produced from transmission data acquired using an external radionuclide source or extracted from a spatially correlated CT scan of the patient. A specific implementation of a combined emission–transmission scanner was disclosed by Mirshanov (49) who produced an engineering concept dia-gram showing a semiconductor detector to record radionuclide emission data and a strip scintillator to record coregistered X-ray transmission data from a patient. In addition, Kaplan (50) proposed a high-performance scintillation camera to record both emission data from an internal radionuclide source and transmission data from an external X-ray source. However, these concepts were never reduced to practice or implemented in either an experimental or a clinical setting.

In the late 1980s and early 1990s, Hasegawa et al. (19,23,51,52), at the Univer-sity of California, San Francisco (UCSF), pioneered the development of dedicated emission–transmission imaging systems which could record both radionuclide and X-ray data for correlated functional and structural imaging. The first prototype, developed by Lang et al. (23,52), used an array of high-purity germanium detectors (HPGe) (Fig. 1) with sufficient energy discrimination and count-rate performance to discriminate γ-rays emitted by an internally distributed radiopharmaceutical from X-rays transmitted through the body from an external source. Phantom experiments with this first prototype led to the development of a second prototype (Fig. 2) having a 20-cm reconstruction diameter that could record emission and transmission data from a stationary animal or object using a single HPGe detector array. Kalki et al. (53,54) used this second prototype in animal studies both to demonstrate the capabil-ity of the system to facilitate image correlation and to test the feasibility of correcting the radionuclide data for photon attenuation using coregistered X-ray transmission data. Because the HPGe detector implemented in these first two prototypes was

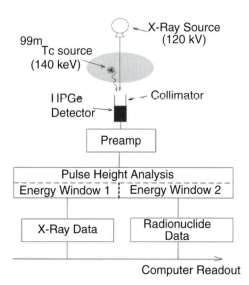

Figure 1 Schematic display of data acquisition of combined emission–transmission imaging system developed at UCSF using single high-purity germanium detector array with fast pulse-counting electronics for simultaneous emission–transmission imaging.

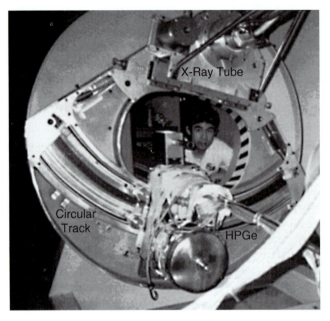

Figure 2 Early emission–transmission CT system at UCSF that had an HPGe detector was translated across a circular arc to simulate an entire detector array and was rotated around an isocenter for tomographic imaging. Detector was read-out using fast pulse-counting electronics and energy discrimination to separate the simultaneously acquired X-ray and radionuclide data. *Abbreviations*: CT, computed tomography; HPGe, high-purity germanium detector. *Source*: From Ref. 21.

expensive and impractical for clinical use, the UCSF group next implemented a SPECT/CT scanner (Fig. 3) for patient studies by siting a GE 9800 Quick CT scanner in tandem with a GE 400 XR/T SPECT system (17,18,37). This configuration allowed the patient to remain on a common patient table for radionuclide and X-ray imaging with separate subsystems that already had been optimized for clinical use both in terms of technical performance and cost-effectiveness. The investigators used this system to demonstrate that CT data could produce a patient-specific attenuation map that could be incorporated into an iterative reconstruction algorithm for attenuation correction of the correlated radionuclide data. The system was used for imaging studies with phantoms, animals, and patients and demonstrated that the use of combined emission and transmission data could improve both the visual quality and the quantitative accuracy of radionuclide data in comparison to SPECT alone (17,18,37).

Early independent work in SPECT/CT was also pioneered by investigators and engineers at Elscint Inc., Haifa, Israel, continued in Elgems Inc., a joint venture formed with GE Medical Systems in 1997, and since 1999 the global nuclear medicine business of GE Healthcare Technologies. This SPECT/CT system (56) was the first dual-modality imaging system that included a fully integrated gantry and patient table, with an integrated computer system and software for reconstruction, display, and analysis. It also included slip-ring capability, an innovation that allowed continuous scanning of both the X-ray and radionuclide imaging systems (Fig. 4). Early units of the SPECT/CT system were placed at the nuclear medicine departments at Vanderbilt University (Nashville, Tennessee) and at Rambam Medical Center in

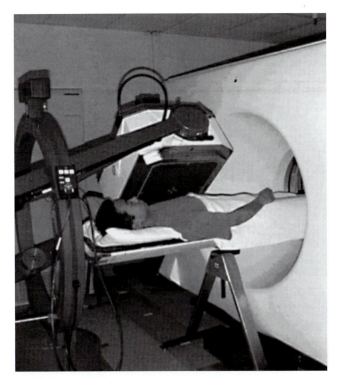

Figure 3 Combined SPECT/CT system configured at UCSF from a GE 9800™ CT Quick
CT scanner and a 600 XR/T SPECT system with a common patient table to translate the patient
between the reconstruction volumes of the CT and SPECT systems. *Abbreviation*: SPECT/CT,
single-photon emission computed tomography/computed tomography. *Source*: From Ref. 55.

Haifa, Israel, where the first clinical imaging studies were performed. This pioneering
work is notable because it represents the first dual-modality imaging system that was
developed for clinical dissemination and utilization by a corporate research
and development group, and presaged the rapid expansion of both PET/CT and
SPECT/CT dual-modality imaging from 2000 to the present.

   Although this chapter is devoted to advances in SPECT/CT, it is worth noting
that the first integrated PET/CT system was developed by Townsend et al. at the
University of Pittsburgh in 1998 (25,30–32) by combining the imaging chains from
a Somatom AR.SP (Siemens Medical Systems) CT system with an ECAT ART
(CTI/Siemens) PET scanner. Both the CT components and the PET detectors were
mounted on opposite sides of the rotating stage of the CT system, and a patient was
imaged using a common patient table translated between the centers of the two
tomographs, which are offset axially by 60 cm (30). The PET/CT prototype was
operational at the University of Pittsburgh from May 1998 to August 2001, during
which over 300 cancer patients were scanned. Reviews of dual-modality imaging,
including PET/CT, are presented in Chapter 2 by Hasegawa and Zaidi (57), and
in the journal articles by Townsend et al. (29,31,32).

   In 1999 and 2000, SPECT/CT and PET/CT dual-modality imaging systems
were introduced by the major medical equipment manufacturers for clinical use, with
approximately 500 systems of each type sold by mid- 2004. A significant success of
PET/CT has been the improved quality of 18F-fluorodeoxyglucose (FDG) images
for tumor localization (33,34,36,58). In comparison, the major use of SPECT/CT

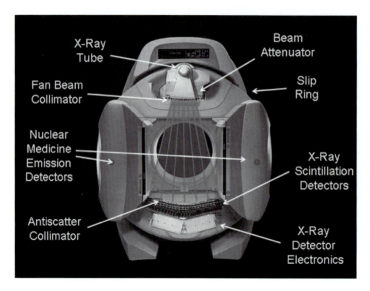

Figure 4 The first commercially available integrated SPECT/CT system was developed by a team of engineers and scientists at Elscint Ltd. (Haifa, Israel), was continued within Elgems Ltd. (joint venture between Elscint and GE Medical Systems), and was introduced by GE Medical Systems as the Millennium VG Hawkeye™. The SPECT data are acquired with a dual-headed SPECT system while the X-ray data are acquired with a dedicated X-ray tube and X-ray detector mounted on the same gantry. The system includes a slip ring for continuous rotation acquisition of both radionuclide and X-ray data. The system also features an integrated patient table, computer system, and software for acquisition, tomographic reconstruction, image processing, and image display. The X-ray data are used both for patient-specific attenuation correction and for anatomical localization of the radionuclide data. *Abbreviation*: SPECT/CT, single-photon emission computed tomography/computed tomography. *Source*: Elgems, GE Healthcare Technologies, Haifa, Israel.

has been in reducing attenuation artifacts and improving the quality of myocardial perfusion imaging with ^{99m}Tc-sestamibi (17,53,59–61). SPECT/CT also has demonstrated advantages for oncologic imaging with single-photon agents (9,22,35,36, 56,62–66). Both SPECT/CT and PET/CT have demonstrated their ability to facilitate attenuation correction using an X-ray-based patient-specific attenuation map that can be produced faster and more accurately than attenuation maps generated with external radionuclide sources (7,31). Clinical studies are underway to evaluate the applicability of X-ray-based correction of photon attenuation (67), and early results demonstrate improvement in sensitivity, specificity, and predictive accuracy in comparison to SPECT perfusion studies reconstructed without correction for photon attenuation errors. By providing high-resolution anatomical information from CT, dual-modality imaging also correlates functional and anatomical data to improve disease localization (11,33,34,36,58) and facilitates treatment planning for radiation oncology (9,68) or surgery (65,69,70).

GENERAL DESIGN FEATURES OF DUAL-MODALITY IMAGING SYSTEMS

Modern dual-modality imaging systems incorporate subsystems for radionuclide imaging and for X-ray-computed tomography that essentially use the same components as those in dedicated nuclear medicine and CT systems. Specifically, SPECT/CT systems use conventional dual-headed scintillation cameras suitable for planar scintigraphy or tomographic imaging of single-photon radionuclides, or coincidence imaging of PET radiopharmaceuticals. The first-generation clinical SPECT/CT scanners used a low-resolution CT detector (35,36,56,71) that offered relatively modest scan times. However, newer SPECT/CT scanners incorporate state-of-the-art multislice helical CT scanners identical to those used for diagnostic CT procedures.

The integration of the radionuclide and X-ray imaging chains in a dual-modality imaging system requires special considerations beyond those needed for scanners designed for single-modality imaging alone. One challenge is offered by the presence of X-ray scatter from the patient that has the potential to reach and possibly damage the radionuclide detectors, which are designed for the relatively low photon fluency rate encountered in radionuclide imaging (61). To avoid this possibility, the radionuclide detector in a dual-modality system is typically offset in the axial direction from the plane of the X-ray source and detector. This distance can be relatively small when a low-dose 140 kV, 1-mA X-ray tube is used (61), but can be 60 cm or more when a diagnostic CT scanner is coupled to a modern PET scanner operated without septa (25,27,28,31).

As noted above, all dual-modality systems rely on separate X-ray and radionuclide imaging chains that must be supported on a common mechanical gantry to maintain consistent spatial relationship between the two data sets, and allow the detectors to be rotated and positioned accurately for tomographic imaging. The requirements for translational and angular positioning accuracy are, of course, different for CT and SPECT. For example, CT requires approximately 1000 angular samples acquired with an angular position and center of rotation maintained with submillimeter accuracy. In comparison, SPECT and PET have spatial resolutions of a few millimeters, and therefore can be performed with an accuracy of slightly less than a millimeter for clinical imaging. To achieve these design goals, the mechanical requirements for a dual-modality imaging system can be implemented in several ways. In first-generation SPECT/CT systems (35,36,55,71), the SPECT detectors and CT imaging chain were mounted on the same rotating platform and were used sequentially while rotated around the patient. This limited the rotational speed of the X-ray and radionuclide imaging chains to approximately 20 seconds per rotation, but also had the advantage that it could be performed using a gantry similar to that used with a conventional scintillation camera. Second-generation SPECT/CT systems include high-performance diagnostic CT subsystems. This requires the heavy SPECT detectors to be mounted on a separate rotating platform from the CT imaging chain, which is rotated at speeds of 0.25 to 0.4 seconds per revolution. While this design obviously increases the performance of the CT subsystem, it also increases the cost and complexity of the gantry.

As discussed for PET/CT by Townsend et al. (31,32), the patient table is another seemingly simple, yet important element of a dual-modality scanner. Most imaging systems use cantilevered patient tables to support the patient in the bore of the imaging system. Patient tables are designed to support patients weighing up to 500 pounds, but obviously deflect to varying degrees when they are loaded and extended with normal adult patients. However, dual-modality systems use X-ray and radionuclide imaging chains in tandem and thereby require longer patient tables than conventional imaging systems. Moreover, the table extension and the degree of deflection can be different for the X-ray and radionuclide imaging chains, which can introduce inaccuracy in the image registration. This problem is overcome by several methods. A patient table supported in front of the scanner, with a secondary support between or at the far end of the X-ray and radionuclide imaging chains, can be used to minimize table deflection. Alternatively, the patient platform can be fixed to a base that is translated across the floor to move the patient into the scanner. As the patient platform is stationary relative to its support structure (which acts as a fulcrum), the deflection of the patient table is identical when the patient is positioned in the radionuclide imager or the CT scanner.

In currently available dual-modality scanners, the computer systems are well integrated in terms of system control, data acquisition, image reconstruction, image display, and data processing and analysis. The CT data have to be calibrated so that the data can be used as an attenuation map to correct the radionuclide data for photon attenuation (17,18,37). For physician review, the CT and radionuclide data are registered and presented with the radionuclide image as a color overlay on the gray-scale CT image. Available software functions allow the operator to utilize the dual-modality data in correcting, viewing, and interpreting, and are therefore important design elements in modern dual-modality imaging systems.

SPECT/CT IMAGING SYSTEMS

As noted earlier, the first SPECT/CT imaging systems were developed at the University of California, San Francisco, by Hasegawa et al. The first prototype instruments (Figs. 1 and 2) were configured as small-bore (i.e., approximately 20 cm reconstruction diameter) SPECT/CT systems with a small segmented HPGe operated in pulse-counting mode for simultaneous X-ray and radionuclide imaging (19,23,51,53). A clinical prototype SPECT/CT system (Fig. 3) also was configured at UCSF by installing a SPECT camera and CT scanner in tandem with a common patient table (17,18,37,66), and was used for studies on both large animals and patients.

The first dual-modality imaging system designed for routine clinical use was the Millennium VG™ introduced by GE Medical Systems (Milwaukee, Wisconsin) in 1999 (35,36,55,71). This system had both X-ray and radionuclide imaging chains that were integrated on a common gantry for "functional–anatomical mapping." X-ray projection data are acquired by rotating the X-ray detector and low-power tube 220° around the patient, with radiation generated during the image acquisition. As the patient is not removed from the patient table, the X-ray and radionuclide images are acquired with a consistent patient position in a way that facilitates accurate image registration. The image acquisition of the CT data requires approximately five minutes, and the radionuclide data requires a scan time of approximately 30 to 60 minutes. General Electric Healthcare has now introduced their current "Infinia" SPECT/CT system (Fig. 5) that integrates a dual-headed SPECT system with the same X-ray technology. This SPECT/CT system produces transmission data with significantly higher quality and better spatial resolution than those acquired with an external radionuclide source, and can be used for anatomical localization (11) and attenuation correction of SPECT for oncologic and cardiovascular applications.

In June 2004, both Siemens Medical Systems and Philips Medical Systems introduced SPECT/CT systems offering the performance available on state-of-the-art diagnostic CT systems (Fig. 5). The Symbia line of SPECT/CT systems (Siemens Medical Systems) has a single-slice, 2-slice, or 6-slice CT options with 0.6 second rotation speed coupled with a dual-headed variable-angle SPECT system. The 6-slice CT scanner will cover the abdomen in less than 6 seconds and allows both

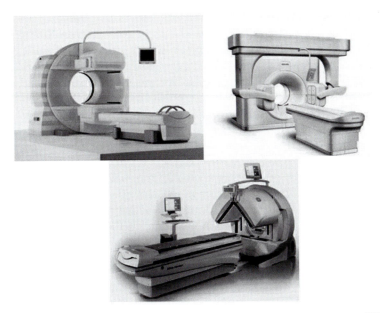

Figure 5 Current generation SPECT/CT systems include the Infinia™ SPECT/CT system (*bottom*) from General Electric Healthcare (Waukesha, Wisconsin), Symbia™ SPECT/CT system (*above left*) from Siemens Medical Solutions (Malvern, Pennsylvania and Erlangen, Germany), and Precedence™ SPECT/CT (*above right*) from Philips Medical Systems (Milpitas, California). *Abbreviation*: SPECT/CT, single-photon emission computed tomography/computed tomography. *Source*: General Electric Healthcare Technologies (Waukesha, Wisconsin), Siemens Medical Solutions (Malvern, Pennsylvania), and Philips Medical Systems (Milpitas, California).

X-ray-based attenuation correction and anatomical localization with the correlated diagnostic CT images. The Precedence SPECT/CT system (Philips Medical Systems) incorporates a dual-headed SPECT system with a 16-slice diagnostic CT scanner, and more recently with a 64-slice CT subsystem. Advanced CT capability has several potential benefits. The CT scan time is compatible with use of contrast media for improved lesion delineation with CT. The CT image quality and spatial resolution is similar to that offered by state-of-the-art diagnostic CT scanners. Advanced multi-slice CT scanners can be used for cardiac and coronary imaging. SPECT/CT scanners with multislice CT capabilities therefore offer improved performance for tumor imaging and have the potential to acquire and overlay radionuclide myocardial perfusion scans on a CT coronary angiogram showing the underlying arterial and cardiac anatomy in high spatial resolution.

CAPABILITIES OF DUAL-MODALITY IMAGING

Dual-modality techniques offer a critical advantage over separate CT and radionuclide imaging systems in acquiring and correlating functional and anatomical images without moving the patient (other than table translation). Dual-modality imaging also can account consistently for differences in reconstruction diameter, offsets in isocenter, image reconstruction coordinates, and image format (e.g., 128×128 vs. 512×512) between the CT and radionuclide image geometries to perform image coregistration and image fusion. Depending on the design of the system, image registration software also may be needed to account for table sag or for misalignment when the patient moves between the CT and scintigraphy. The coordinate systems in the radionuclide and CT image geometries are calibrated with respect to each other using fiducial markers that are scanned with both CT and scintigraphic imaging. The image registration must be confirmed to avoid misregistration errors in the dual-modality images or in the radionuclide image reconstructed using CT-derived attenuation maps.

Attenuation Correction

Dual-modality imaging provides a priori patient-specific information that is needed to correct the radionuclide data for photon attenuation and other physical effects. Traditionally, attenuation correction was performed by acquiring a transmission image from which a patient-specific map of attenuation coefficients was reconstructed and incorporated into the tomographic reconstruction of the radionuclide data (2,72–77). X-ray sources are now replacing most external radionuclide sources for acquisition of the transmission data. These data can be reconstructed to produce a patient-specific map of linear attenuation coefficients needed for the attenuation correction process for both SPECT and PET. Because the external radionuclide sources produce a limited fluence rate, the transmission scans often require several minutes and produce images that are noisy and photon-limited. In comparison, transmission data acquired using an X-ray source with a small focus have a significantly higher photon flux rate than a radionuclide source, and therefore produce tomographic reconstructions with 1 mm or better spatial resolution with excellent noise characteristics. The resulting CT images can then be calibrated to produce a patient-specific map of linear attenuation coefficients calculated for the energy of the radionuclide photons (17–19,26,31,37–39). The CT data from a dual-modality imaging system simplifies the process of correcting the radionuclide image for photon attenuation.

Several different CT-based attenuation correction techniques have been developed (17–19,26,31,37–39). The technique developed by Blankespoor et al. (18,78) obtains CT calibration measurements from a phantom with cylindrical inserts containing water, fat-equivalent (ethanol), and bone-equivalent material (K_2HPO_4). CT numbers extracted from each region are plotted against their known attenuation coefficients at the photon energy of the radionuclide to provide a piece-wise linear calibration curve (Fig. 6) that reflects the different combinations of tissue (i.e., soft-tissue and air vs. soft-tissue and bone) (18,79,80). During the dual-modality imaging study, both CT and radionuclide data of the patient are acquired. The calibration curve described above is used to convert the CT image of the patient into an object-specific attenuation map. The resulting attenuation map can then be incorporated into the reconstruction of the radionuclide data (81) using maximum likelihood expectation maximization (MLEM) or other iterative algorithms to correct the radionuclide data for perturbations due to photon attenuation. This or a similar process can be used to improve both the image quality and the quantitative accuracy of SPECT or PET images. Important clinical applications include attenuation correction of myocardial perfusion images (76,82) to resolve false-positive defects caused by soft-tissue attenuation. This has been demonstrated in a prospective study by Fricke et al. (83), who showed that X-ray-derived attenuation correction improved the accuracy of ^{99m}Tc-sestamibi imaging with SPECT in comparison to SPECT without attenuation correction and using ^{13}N-ammonia PET as a gold standard.

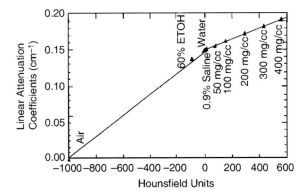

Figure 6 Calibration curve showing relationship between Hounsfield numbers measured from CT image to attenuation coefficients of various calibration materials at 140 keV, corresponding to the photon energy of ^{99m}Tc-labeled radiopharmaceuticals. *Abbreviation*: CT, computed tomography. *Source*: From Ref. 18.

Other studies have shown that attenuation correction improves contrast of ^{111}In-pentetreotide SPECT imaging of neuroendocrine lesions (84) and of ^{111}In-capromab pendetide imaging of prostate cancer (85). Although these studies are not definitive, they demonstrate that X-ray-based attenuation correction can improve assessment of myocardial perfusion, oncologic imaging, and other clinical applications of SPECT.

Image Registration

In addition to the energy calibration process described earlier, it is important to match the geometric characteristics of the CT-derived attenuation map and radionu-clide tomograms. This typically requires the CT data to be resampled so that it can be presented in slices that have the same pixel format (e.g., 128×128 or 256×256) and same slice width as the radionuclide image. Accurate spatial registration of the CT and radionuclide data is important because slight differences in the position of the attenuation map relative to the corresponding data can cause "edge artifacts," which produce bright and dark "rims" across the edges of regions where the CT and radio-nuclide data are misaligned. Takahashi et al. (60) performed a phantom study of myo-cardial perfusion imaging with ^{201}Tl-thallium to assess the effect of registration errors between CT and SPECT images from dedicated CT and SPECT scanners on the attenuation-corrected SPECT images. CT and SPECT images of the phantom with a breast attachment were acquired and then registered using software (86). Prior to attenuation correction and reconstruction of the SPECT data, the CT images were shifted one to three pixels in the three orthogonal directions, and were rotated 6° clockwise to simulate patient misregistration errors. In addition, the radionuclide data with soft-tissue attenuation from the breast and in the inferior wall were properly reconstructed using the attenuation maps derived from the unshifted X-ray CT images. However, as the pixel shifts increased, the attenuation-corrected SPECT images were reconstructed with increasing image artifacts and errors. The effect of misregistration between the CT and the SPECT data was also evaluated from patient data by Fricke et al. (59). In this retrospective study, the investigators evalu-ated 27 of 140 sets of myocardial perfusion images (acquired using a commercial

SPECT/CT system) that showed pronounced apical or anterior wall defects after CT-based attenuation correction of ^{99m}Tc-sestamibi SPECT. The body contours in the SPECT and CT data were evaluated in software to estimate the magnitude of shift needed to register the images, and the SPECT data were reconstructed with X-ray-based attenuation correction both before and after image registration was applied. In 15 of 27 patients, improved coregistration produced smaller and less-pronounced defects in the apex and anterior wall. In 6 of 27 patients, defects were defined as normal after image registration was applied. No improvement was seen in only four patients, apparently because the misregistration error was < 1 pixel (7 mm). These results show the impact and possibility of misregistration errors in attenuation-corrected data that can occur even when the patient is imaged with a dedicated SPECT/CT system. Many of these errors can be resolved if image registration software is applied to both the radionuclide and X-ray data before attenuation correction is performed on the SPECT or PET data. Misregistration errors also are minimized when the data are obtained with dual-modality imaging systems, rather than attempting to register images acquired during separate studies.

Presence of Iodinated Contrast Media

The process of generating attenuation maps as described earlier also assumes that the body is composed primarily of air (in the lung), soft tissue (essentially water-equivalent), and bone. However, diagnostic CT studies are obtained, as a rule, following the administration of intravenous and oral contrast media. Since contrast media in the body can attenuate photons emitted by a radiopharmaceutical during a PET or SPECT study, it is important to derive accurate attenuation maps that account for the presence of these materials in the body, regardless of whether they are administered orally and/or intravenously (37,66). It has been shown that PET attenuation can be overestimated in the presence of positive contrast agents, which can generate significant artifacts (87). One can account for attenuation differences between iodine versus bone using a technique that generates a calibration curve for contrast media, using a method similar to that described above for soft tissue and bone alone. In this method, calibration data are obtained experimentally using CT to image a phantom containing known concentrations of iodine contrast. The reconstructed CT values for each region in the phantom are extracted from the CT scan, and are related to the known linear attenuation coefficients as a function of CT number. Separate calibration curves are generated for each material (i.e., iodine vs. bone), for different X-ray potentials used to acquire CT scans, and for different photon energies (e.g., 140, 364, and 511 keV) encountered in nuclear medicine (37,66). It is important to use image processing or other techniques to segment bony and iodine-containing regions in the patient's CT image so that these can be scaled independently when forming the attenuation map. The attenuation maps with calibrated values for soft tissue, fat, bone, and iodine are then incorporated into an MLEM or other iterative reconstruction algorithm for attenuation compensation of the radionuclide image. The effect of contrast media and other high-density materials (e.g., metal implants, prosthetics) has been evaluated for PET images reconstructed with X-ray-derived attenuation maps (88–92). Robust correction methods for iodinated contrast media are being developed and it is likely that these issues will be further investigated as the clinical use of dual-modality imaging systems expands over the next few years.

Compensation for Scatter Radiation

The X-ray data from a dual-modality imaging system can also be used to compensate the radionuclide data for contamination by scatter radiation. One class of scatter compensation techniques characterizes the scatter distribution for different radionuclide and attenuation distributions with a spatially-dependent kernel, which can be used to convolve data from the photopeak window for estimating the scatter distribution from the acquired data. For example, the method proposed by Mukai et al. (93), assumes that the scatter radionuclide image ρ can be estimated from the "true" (i.e., "scatter-free") image η, as $\rho = P\eta$, where P is a matrix having elements $P(b,b')$ specifying the probability that a primary photon emitted from voxel b' is scattered in voxel b with an energy suitable for detection within the primary energy window, which can be calculated by integrating the Klein-Nishina equation over the primary acquisition energy window taking into account the energy resolution of the emission system. Obviously, the "true" or "scatter-free" image is unknown, making it necessary to approximate this using the currently available image estimated by the tomographic reconstruction algorithm. The matrix P can be approximated from the coregistered CT-derived attenuation map of the object provided by the dual-modality imaging system to obtain the total attenuation $[\Delta(b,b')]$ along a line from the center of voxel b to the center of voxel b'. All the parameters can be determined accurately from X-ray-based attenuation maps, illustrating how this process is facilitated by the availability of CT data from dual-modality imaging. Once the scatter distribution is calculated, it can be incorporated in the tomographic reconstruction algorithm to compensate the radionuclide data for this effect.

Correction for Partial Volume Errors

The visual quality and quantitative accuracy of the radionuclide data can be improved by using the methods described earlier for image fusion with CT and by use of X-ray data to correct the radionuclide data for photon attenuation and Compton scatter. In addition, the radionuclide image can be improved by correcting the data for the geometric response of the imaging system, which for SPECT is determined primarily by the geometry of the collimator (3,37,94,95). Traditionally, radionuclide quantification methods have focused on techniques that compensate the scintigraphic image reconstruction process for physical errors such as photon attenuation and scatter radiation as discussed earlier. In addition, the correction methods may include the use of recovery coefficients (94) to correct the data for the limited spatial resolution of the imaging system. However, these techniques require a priori information about object size and shape, and generally are only used for simple target geometries (e.g., spheres).

Other investigators have developed methods that incorporate physical models into the quantification process itself. Several methods have been described in the literature combining SPECT and MRI (96) and PET and MRI (97,98) for brain imaging, and SPECT and CT for cardiac and oncologic imaging (17,53,66). Several approaches were developed specifically for quantitation using planar imaging (61,99–102) or for bone marrow dosimetry using SPECT/CT (103). One technique called "template projection" (37,95) can be used to quantify radioactivity in planar images. Regions-of-interest (ROIs) for the target (e.g., tumor) and background regions are defined on the high-resolution CT image, and are then used to define "templates," which represent idealized radionuclide-containing objects (e.g., tumor

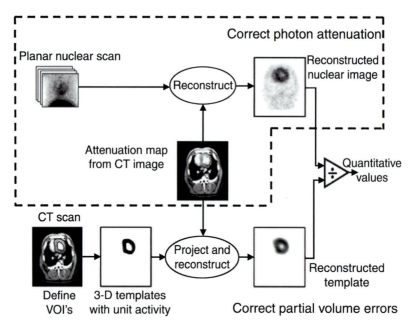

Figure 7 Tomographic reconstruction of the radionuclide data with X-ray-based attenuation correction (enclosed within the *dashed line*) incorporates the attenuation map obtained with transmission data from X-ray or radionuclide source. Compensation for partial volume errors can be performed by defining the anatomic extent of the myocardium with a correlated CT scan. *Abbreviation*: CT, computed tomography. *Source*: From Ref. 17.

vs. background) with unit radionuclide concentration. Planar imaging of these ideal radionuclide objects is modeled by mathematically projecting the templates onto the plane of the radionuclide detector using the known geometrical transformation between the CT coordinate system and the planar radionuclide image provided inherently by the dual-modality imaging system. This process generates "projected templates" that are analogous to conventional ROIs, in that they delineate a target region over which events in the radionuclide image are integrated. Like conventional ROIs, the projected templates specify the geometry of the target and background regions on the projected planar radionuclide images; however, the projected templates are defined on the high-resolution CT images rather than on the low-resolution radionuclide images. The projected templates are nonuniform and contain information about physical effects (photon attenuation, detector response, scatter radiation) included in the projector model. Da Silva et al. have applied this method to quantify the regional myocardial concentration of ^{99m}Tc-sestamibi in a porcine model of myocardial perfusion. In this study, correlated radionuclide and X-ray image data were obtained with a dedicated SPECT/CT system (17). SPECT images were reconstructed using a MLEM algorithm first with no corrections and then with X-ray-based attenuation correction (Fig. 7), from which the regional myocardial content of ^{99m}Tc-sestamibi was quantified in software and calibrated in absolute units of kBq/g. In addition, the SPECT image reconstructed with attenuation correction was compensated for partial volume errors using the template projection method. The in vivo values obtained from SPECT were compared against direct ex vivo measurements of radionuclide content obtained from the excised myocardial tissue, and demonstrated that the correction of partial volume errors significantly

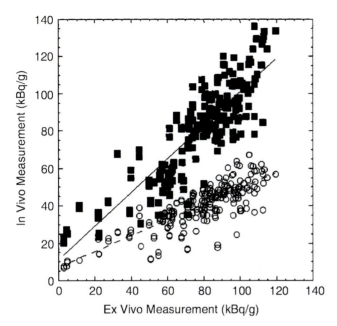

Figure 8 Regional measurements of myocardial ^{99m}Tc-sestamibi uptake in a porcine model of myocardial perfusion showing in vivo measurements from SPECT/CT versus ex vivo measurements obtained directly from excised tissue samples as a gold standard. Measurements obtained with corrections for both photon attenuation and partial volume errors (*solid squares*) were significantly more accurate than measurements obtained from SPECT using attenuation correction alone (*open circles*). When no correction was applied to the SPECT data, the in vivo measurements from SPECT underestimated the direct ex vivo measurements by approximately 90% (not shown). *Abbreviation*: SPECT/CT, single-photon emission computed tomography/computed tomography. *Source*: From Ref. 17.

improved the accuracy of the quantitative values in comparison to those obtained with no correction or with attenuation correction alone (Fig. 8).

Koral et al. have developed a technique that uses CT/SPECT image fusion with conjugate view imaging (104) to quantify the uptake of single-photon radionuclides in vivo. In this analysis, a patient-specific attenuation map is derived from the correlated CT images to correct the radionuclide data for photon attenuation. Volumes of interest delineating the extent of the tumor are defined anatomically on the CT scans, and are superposed on the SPECT data for radionuclide quantification (105–107). This quantification method was developed using images obtained from separate SPECT and CT studies combined by software-based image registration. This quantification technique certainly could also be implemented using a dual-modality SPECT/CT system.

NEED FOR DUAL-MODALITY IMAGING

Even though both PET/CT and SPECT/CT systems have been accepted commercially, the clinical role of these systems is still being debated. Nonetheless, many practitioners now see important roles for dual-modality imaging for diagnostic studies and management of patients with cancer, heart disease, and other disorders. These capabilities arise from several inherent features. The anatomical and functional information from dual-modality imaging is complementary and not redundant (108). The availability of correlated functional and anatomical images improves the detection of disease by highlighting areas of increased radiopharmaceutical uptake on the anatomical image; similarly, regions that look abnormal in the anatomical image can draw attention to areas of disease where radiopharmaceutical uptake may be low. Dual-modality imaging thereby provides information that cannot be easily discerned from structural or functional imaging alone.

The dual-modality system facilitates comparing and correlating the structural and functional information by acquiring and presenting both sets of data with the patient in a consistent configuration and during a single study. Furthermore, this can be done in a faster and more robust way than attempting to register the images by software after they are acquired on separate imaging systems. The correlated CT and radionuclide images can then be viewed during the diagnostic interpretation either side-by-side or as fused images with the SPECT data presented as a color overlay on the gray-scale CT image. In addition to aiding diagnosis and staging, the dual-modality image data can be used to guide radiation treatment planning by providing important information for defining the target volume, as well as for indicating healthy tissues where irradiation should be avoided. Similar roles are played when the dual-modality data is used to guide surgical approach, biopsy, or other interventional procedures.

The complementary nature of dual-modality imaging also is demonstrated by the use of a CT-derived attenuation map and other a priori patient-specific information from CT that can be used to improve the visual quality and the quantitative accuracy of the correlated radionuclide imaging. This occurs in myocardial perfusion imaging of ^{99m}Tc-sestamibi or ^{201}Tl-thallium where SPECT/CT can provide an attenuation map to compensate the radionuclide image for soft-tissue attenuation associated with false-positive errors. The use of a modern CT scanners-derived attenuation map also is practical and cost-efficient because it is generated in a few seconds, resulting in more efficient use of time than with devices that use external radionuclide sources to record the transmission image. The resulting CT-derived attenuation map is of significantly higher quality and potentially more accurate than those acquired using scanning line sources. As dual-modality imaging systems become available with faster and higher spatial-resolution CT systems, it may be possible to perform CT angiography of peripheral or coronary vessels that can be superimposed on a color map of tissue perfusion or metabolism obtained with SPECT. Dual-modality imaging therefore has the potential to provide diagnostic information that is difficult to achieve when the same image data are acquired in separate studies using conventional means.

CHALLENGES OF DUAL-MODALITY IMAGING

Several challenges remain that may represent inherent limitations of dual-modality imaging. All currently used dual-modality imaging systems record the emission and transmission data using separate detectors rather than a single detector. In addition, the X-ray and radionuclide imaging chains are separated by tens of centimeters to facilitate mechanical clearance and to avoid contamination of the radionuclide data by scatter radiation produced by the X-ray scan.

Potential problems may occur when the patient moves either voluntarily or involuntarily between or during the anatomical and functional image acquisitions, when the patient adjusts his or her position while lying on the patient bed, or due to respiration, cardiac motion, peristalsis, and bladder-filling, all of which can lead to motion-blurring or misregistration errors between the radionuclide and CT image acquisitions. Traditionally, CT data are acquired following a breath-hold, whereas SPECT data are acquired over several minutes. Breathing protocols can lead, however, to misregistration artifacts due to anatomical displacement of the diaphragm and chest wall during the SPECT/CT study. For example, if the position of the diaphragm is displaced in the CT scan, which is then used as an attenuation map for the radionuclide data, this displacement can lead to an under- or overestimate of radionuclide uptake in the reconstructed emission data (60). These difficulties can be ameliorated by asking the patient to breathe shallowly for both the CT and SPECT studies. As SPECT/CT technology advances, it is likely that image registration methods will be refined to correct misregistration errors that can occur during the image acquisition process.

The use of iodinated contrast media for the CT study presents a potential challenge when the CT scan is used as an attenuation map for correction of SPECT data. As the X-ray energy of the CT will be different than the photon energy of the radiopharmaceutical emission data, the CT data must be transformed to units of linear attenuation coefficient at the photon energy of the radiopharmaceutical emission data. This is performed by assuming that the CT image contrast is contributed by a mixture of air, soft tissue, and bone. The presence of iodinated contrast complicates this process because two regions that have the same image contrast may have different compositions, such as bone and soft tissue in one case, and iodine contrast and soft tissue in another situation. These artifacts are most severe in cases where the contrast media is concentrated, for example, in abdominal imaging after the patient swallows a bolus of oral contrast. In this case, the higher densities contributed by the oral contrast media can lead to an overestimation of radionuclide uptake, as has been observed for PET/CT (87). Some investigators have proposed using an image segmentation postprocessing technique for PET/CT in which image regions corresponding to iodinated contrast are segmented using image processing from those contributed by bone (109). Other strategies being considered for PET/CT (32) include the acquisition of both precontrast and postcontrast CT scans to minimize possible artifacts contributed by the presence of contrast media. It is likely that similar strategies will have to be adopted for SPECT/CT as they become available with high-end CT subsystems and are used with iodinated contrast media for acquisition of the CT data.

Use of dual-modality systems demands that technologists be cross-trained in both modalities, and that physicians have expertise in interpreting both anatomical and functional image data. Optimal use of SPECT/CT requires a team approach involving physicians from nuclear medicine and radiology, as well as oncology, cardiology, surgery, radiation therapy, and other clinical disciplines.

FUTURE OF DUAL-MODALITY IMAGING

Over the past 10 to 15 years, dual-modality imaging has progressed steadily from experimental studies of technical feasibility to the development of clinical prototypes to its present role as a diagnostic tool that is experiencing widening availability and clinical utilization. Dual-modality imaging has advanced rapidly, primarily by incorporating the latest technological advances of SPECT and CT, and combined MRI/MRSI (110,111), PET/CT (31,32), and SPECT/CT (21,22) systems are currently available for routine clinical use. PET/CT is now becoming well established with virtually all PET scanners now being sold as PET/CT combinations.

The adoption of SPECT/CT has not been as dramatic but has demonstrated clear clinical advantages for both cardiovascular and oncologic imaging. The trend toward dual-modality imaging will likely continue and expand its availability and clinical application.

Recently introduced SPECT/CT systems that include a state-of-the-art dual-headed scintillation camera and multislice helical CT scanners now use CT technology that can acquire the anatomical data as well as transmission data for attenuation correction within a few seconds after the patient is positioned on the patient bed. The addition of advanced CT capability allows anatomical images to be acquired after the patient is administered with contrast media and facilitates CT angiography. With these capabilities, the next-generation SPECT/CT systems could produce high-resolution structural images of the cardiac chambers (112,113) and of coronary and peripheral vasculature (55,114,115) that can be correlated with myocardial perfusion and other functional assessments using radionuclide imaging. In addition, the use of contrast media could enable one to discern the anatomical size of target regions in a way that facilitates correction for partial volume errors in the scintigraphic data (17). This has the potential to improve the quantitative assessment of cancer and cardiovascular disease in comparison to studies acquired with SPECT or CT alone.

Finally, it obviously will be important to balance the cost and performance of the CT versus the SPECT components of the dual-modality imaging system. This is especially important with a dual-modality system incorporating an advanced CT scanner that can complete a study in a few seconds, and then will be idle for several minutes during the radionuclide imaging session. If additional increases in scan speed are needed, these must be implemented using more efficient collimator geometries, which at present limit the speed at which the SPECT data can be acquired. It is also likely that the cost of the CT subsystem will continue to be reduced with technological advances, which certainly will add to the overall cost-effectiveness and clinical practicality of the dual-modality imaging system.

CONCLUDING REMARKS

Dual-modality imaging has been available clinically since 2000, and as such is a relatively recent development in the fields of diagnostic radiology and nuclear medicine. The commercial emergence of both PET/CT and SPECT/CT has been rapid and has benefited significantly from recent technological advances. Over the past year, SPECT/CT has gained increased interest and has been used for clinical applications in myocardial perfusion and oncologic imaging. Newer high-resolution SPECT/CT and PET/CT systems are also becoming available for small animal imaging and are needed for molecular imaging, biological research, and pharmaceutical development in small animal models of human biology and disease.

At present, dual-modality imaging is primarily used for structural–functional correlations. Dual-modality imaging also has important ramifications for radionuclide quantification by facilitating corrections for photon attenuation, scattered radiation, and partial volume errors. The use and understanding of dual-modality imaging for radionuclide quantification is just starting to emerge. The importance of this new technology will only be understood and manifest over the ensuing years as PET/CT, SPECT/CT, and other forms of dual-modality imaging become available and are utilized for clinical studies of humans as well as biological investigations involving animal models of biology and disease.

ACKNOWLEDGMENTS

The authors appreciate the assistance of Mr. Nathan Hermony and Mr. Aharon Peretz (Haifa, Israel), and of Mr. Bruce Levin (San Francisco, California, U.S.A.) in reconstructing the early history of the SPECT/CT development at Elscint Ltd., Elgems, Ltd., and General Electric Healthcare, Inc. in Haifa, Israel.

REFERENCES

1. Larsson A, Johansson L, Sundstrom T, Riklund Ahlstrom K. A method for attenuation and scatter correction of brain SPECT based on computed tomography images. Nucl Med Commun 2003; 24:411–420.
2. Huang SC, Hoffman EJ, Phelps ME, Kuhl DE. Quantitation in positron emission computed tomography. 1. Effects of inaccurate attenuation correction. J Comput Assist Tomogr 1979; 3:804–814.
3. Hoffman EJ, Huang SC, Phelps ME. Quantitation in positron emission computed tomography. I. Effect of object size. J Comput Assist Tomogr 1979; 3:299–308.
4. Tsui B, Frey E, Zhao X, Lalush D, Johnston R, McCartney H. The importance and implementation of accurate 3d compensation methods for quantitative SPECT. Phys Med Biol 1994; 39:509–530.
5. Tsui BMW, Zhao X, Frey E, McCartney WH. Quantitative single-photon emission computed tomography: basics and clinical considerations. Semin Nucl Med 1994; 24:38–65.
6. Rosenthal MS, Cullom J, Hawkins W, Moore SC, Tsui BMW, Yester M. Quantitative SPECT imaging. A review and recommendations by the focus committee of the society of nuclear medicine computer and instrumentation council. J Nucl Med 1995; 36: 1489–1513.
7. Zaidi H, Hasegawa BH. Determination of the attenuation map in emission tomography. J Nucl Med 2003; 44(2):291–315.
8. Shreve PD. Adding structure to function. J Nucl Med 2000; 41:1380–1382.
9. Schillaci O, Simonetti G. Fusion imaging in nuclear medicine—applications of dual-modality systems in oncology. Cancer Biother Radiopharm 2004; 19:1–10.
10. Aizer-Dannon A, Bar-Am A, Ron IG, Flusser G, Evan-Sapir E. Fused functional-anatomic images of metastatic cancer of cervix obtained by a combined gamma camera and an X-ray tube hybrid system with an illustrative case and review of the F-18-fluorodeoxyglucose literature. Gynecol Oncol 2003; 90:453–457.
11. Pfannenberg AC, Eschmann SM, Horger M, et al. Benefit of anatomical-functional image fusion in the diagnostic work-up of neuroendocrine neoplasms. Eur J Nucl Med Mol Imaging 2003; 30:835–843.
12. Maintz JB, Viergever MA. A survey of medical image registration. Med Image Anal 1998; 2(1):1–36.
13. Hutton BF, Braun M. Software for image registration: algorithms, accuracy, efficacy. Semin Nucl Med 2003; 33:180–192.
14. Slomka PJ. Software approach to merging molecular with anatomic information. J Nucl Med 2004; 45(suppl 1):36S–45S.
15. Pellizzari C, Chen GTY, Spelbring DR, Weichselbaum RR, Chen CT. Accurate three-dimensional registration of CT, PET, and/or MR images of the brain. J Comput Assist Tomogr 1989; 13:20–26.
16. Pietrzyk U, Herholz K, Fink G, et al. An interactive technique for three-dimensional image registration: validation for PET, SPECT, MRI and CT brain studies. J Nucl Med 1994; 35:2011–2018.

17. Da Silva AJ, Tang HR, Wong KH, Wu MC, Dae MW, Hasegawa BH. Absolute quantification of regional myocardial uptake of 99mTC-sestamibi with SPECT: experimental validation in a porcine model. J Nucl Med 2001; 42:772–779.
18. Blankespoor SC, Wu X, Kalki K, et al. Attenuation correction of SPECT using X-ray CT on an emission-transmission CT system: myocardial perfusion assessment. IEEE Trans Nucl Sci 1996; 43:2263–2274.
19. Hasegawa BH, Gingold EL, Reilly SM, Liew SC, Cann CE. Description of a simultaneous emission-transmission CT system. Proc SPIE 1990; 1231:50–60.
20. Hasegawa BH, Tang HR, Da Silva AJ, Wong KH, Iwata K, Wu AM. Dual-modality imaging. Nucl Instr Meth Phys Res A 2000; 471:140–144.
21. Hasegawa BH, Iwata K, Wong KH, et al. Dual-modality imaging of function and physiology. Acad Radiol 2002; 9(11):1305–1321.
22. Hasegawa BH, Wong KH, Iwata K, et al. Dual-modality imaging of cancer with SPECT/CT. Technol Cancer Res Treat 2002; 1:449–458.
23. Lang TF, Hasegawa BH, Liew SC, et al. Description of a prototype emission–transmission CT imaging system. J Nucl Med 1992; 33:1881–1887.
24. Beyer T, Kinahan PE, Townsend DW, Sashin D. The use of x-ray CT for attenuation correction of PET data. IEEE Nuclear Science Symposium & Medical Imaging Conference, 1994.
25. Beyer T, Townsend DW, Brun T, et al. A combined PET/CT scanner for clinical oncology. J Nucl Med 2000; 41:1369–1379.
26. Townsend D, Kinahan P, Beyer T. Attenuation correction for a combined 3D PET/CT scanner. Physica Medica 1996; 12(suppl 1):43–48.
27. Townsend DW, Beyer T, Kinahan PE, et al. The SMART scanner: a combined PET/CT tomograph for clinical oncology. IEEE Nuclear Science Symposium and Medical Imaging Conference, 1998.
28. Townsend DW. A combined PET/CT scanner: the choices. J Nucl Med 2001; 42:533–534.
29. Townsend DW, Cherry SR. Combining anatomy and function: the path to true image fusion. Eur Radiol 2001; 11:1968–1974.
30. Townsend DW, Beyer T. A combined PET/CT scanner: the path to true image fusion. Br J Radiol 2002; 75:S24–S30.
31. Townsend DW, Beyer T, Blodgett TM. PET/CT scanners: a hardware approach to image fusion. Semin Nucl Med 2003; 33(3):193–204.
32. Townsend DW, Carney JP, Yap JT, Hall NC. PET/CT today and tomorrow. J Nucl Med 2004; 45(suppl 1):4S–14S.
33. Wahl RL. Why nearly all PET of abdominal and pelvic cancers will be performed as PET/CT. J Nucl Med 2004; 45:82S–95S.
34. Goerres GW, von Schulthess GK, Steinert HC. Why most PET of lung and head-and-neck cancer will be PET/CT. J Nucl Med 2004; 45:66S–71S.
35. Even-Sapir E, Keidar Z, Sachs J, et al. The new technology of combined transmission and emission tomography in evaluation of endocrine neoplasms. J Nucl Med 2001; 42:998–1004.
36. Israel O, Keidar Z, Iosilevsky G, Bettman L, Sachs J, Frenkel A. The fusion of anatomic and physiologic imaging in the management of patients with cancer. Semin Nucl Med 2001; 31:191–205.
37. Tang HR. A combined x-ray CT-scintillation camera system for measuring radionuclide uptake in tumors. In Bioengineering Graduate Group. University of California, 1998.
38. Kinahan PE, Townsend DW, Beyer T, Sashin D. Attenuation correction for a combined 3d PET/CT scanner. Med Phys 1998; 25:2046–2053.
39. Kinahan PE, Hasegawa BH, Beyer T. X-ray-based attenuation correction for position tomography/computed tomography scanners. Semin Nucl Med 2003; 33:166–179.
40. Mayneord WV. The radiology of the human body with radioactive isotopes. Br J Radiol 1952; 25:517–525.

41. Mayneord WV. Radiological research. Br J Radiol 1954; 27:309–317.
42. Mayneord WV, Evans HD, Newberry SP. An instrument for the formation of visual images of ionizing radiations. J Scient Instruments 1955; 32:45–50.
43. Anger HO. Whole-body scanner MARK II. J Nucl Med 1966; 7:331–332.
44. Anger HO, McRae J. Transmission scintiphotography. J Nucl Med 1968; 9:267–269.
45. Cameron JR, Sorenson J. Measurement of bone mineral in vivo: an improved method. Science 1963; 142:230–232.
46. Kuhl DE, Hale J, Eaton WL. Transmission scanning: a useful adjunct to conventional emission scanning for accurately keying isotope deposition to radiographic anatomy. Radiology 1966; 87:278–284.
47. Kuhl DE, Edwards RO, Ricci AR, Yacob RJ, Mich TJ, Alavi A. The Mark IV system for radionuclide computed tomography of the brain. Radiology 1976; 121:405–413.
48. Budinger TF, Gullberg GT. Three-dimensional reconstruction in nuclear medicine emission imaging. IEEE Trans Nucl Sci 1974; NS-21:2–20.
49. Mirshanov DM. Transmission-emission computer tomograph. Tashkent Branch, All-Union Research Surgery Center, USSR. USSR Academy of Medical Science, 1987.
50. Kaplan CH. Transmission/emission registered image (teri) computed tomography scanners. International patent application, 1989.
51. Hasegawa BH, Reilly SM, Gingold EL, Cann CE. Design considerations for a simultaneous emission–transmission CT scanner. Radiology 1989; 173(P):414.
52. Lang TF, Hasegawa BH, Liew SC, et al. A prototype emission–transmission imaging system. IEEE Nuclear Science Symposium and Medical Imaging Conference, Santa Fe, NM: IEEE, 1991.
53. Kalki K, Blankespoor SC, Brown JK, et al. Myocardial perfusion imaging with a combined x-ray CT and SPECT system. J Nucl Med 1997; 38(10):1535–1540.
54. Kalki K, Heanue JA, Blankespoor SC, et al. A combined SPECT and CT medical imaging system. Proc SPIE 1995; 2432:367–375.
55. Foley WD, Karcaaltincaba M. Computed tomography angiography: principles and clinical applications. J Comput Assist Tomogr 2003; 27(suppl 1):S23–S30.
56. Bocher M, Balan A, Krausz Y, et al. Gamma camera-mounted anatomical x-ray tomography: technology, system characteristics, and first images. Eur J Nucl Med 2000; 27:619–627.
57. Hasegawa BH, Zaidi H. Chapter 2: Dual-modality imaging: more than the sum of its components. In: Zaidi H, ed. Quantitative Analysis in Nuclear Medicine Imaging. New York: Springer, 2005:35–81.
58. Cohade C, Osman MM, Leal J, Wahl RL. Direct comparison of F-18-FDG PET and PET/CT in patients with colorectal carcinoma. J Nucl Med 2003; 44:1797–1803.
59. Fricke H, Fricke E, Weise R, Kammeier A, Lindner O, Burchert W. A method to remove artifacts in attenuation-corrected myocardial perfusion SPECT introduced by misalignment between emission scan and CT-derived attenuation maps. J Nucl Med 2004; 45(10):1619–1625.
60. Takahashi Y, Murase K, Higashino H, Mochizuki T, Motomura N. Attenuation correction of myocardial SPECT images with x-ray CT: effects of registration errors between x-ray CT and SPECT. Ann Nucl Med 2002; 16(6):431–435.
61. Blankespoor SC. Attenuation correction of SPECT using x-ray CT for myocardial perfusion assessment. Joint Graduate Group in Bioengineering. Berkeley: University of California, 1996.
62. Koral KF, Dewaraja Y, Li J, et al. Update on hybrid conjugate-view SPECT tumor dosimetry and response in 131i-tositumomab therapy of previously untreated lymphoma patients. J Nucl Med 2003; 44(3):457–464.
63. Keidar Z, Israel O, Krausz Y. SPECT/CT in tumor imaging? Technical aspects and clinical applications. Semin Nucl Med 2003; 33:205–218.
64. Even-Sapir E, Lerman H, Lievshitz G, et al. Lymphoscintigraphy for sentinel node mapping using a hybrid SPECT/CT system. J Nucl Med 2003; 44:1413–1420.

65. Israel O, Mor M, Gaitini D, et al. Combined functional and structural evaluation of cancer patients with a hybrid camera-based PET/CT system using f-18 fdg. J Nucl Med 2002; 43:1129–1136.

66. Tang HR, Da Silva AJ, Matthay KK, et al. Neuroblastoma imaging using a combined CT scanner-scintillation camera and i-131 mibg. J Nucl Med 2001; 42:237–247.

67. Liu Y, Wackers FJ, Natale D, DePuey G, Taillefer R, Anstett F. Validation of a hybrid SPECT/CT system with attenuation correction: a phantom study and multicenter trial. J Nucl Med 2003; 44:209P.

68. Rajasekar D, Datta NR, Gupta RK, Pradhan PK, Ayyagari S. Multimodality image fusion in dose escalation studies in brain tumors. J Appl Clin Med Phys 2003; 4:8–16.

69. Schoder H, Larson SM, Yeung HWD. PET/CT in oncology: integration into clinical management of lymphoma, melanoma, and gastrointestinal malignancies. J Nucl Med 2004; 45:72S–81S.

70. Cook GJR, Ott RJ. Dual-modality imaging. Eur Radiol 2001; 11:1857–1858.

71. Patton JA, Delbeke D, Sandler MP. Image fusion using an integrated, dual-head coincidence camera with x-ray tube-based attenuation maps. J Nucl Med 2000; 41:1364–1368.

72. Chang LT. A method for attenuation correction in radionuclide computed tomography. IEEE Trans Nucl Sci 1978; MS-25:638–643.

73. Harris CC, Greer KL, Jaszczak RJ, Floyd CE, Fearnow EC, Coleman RE. Tc-99m attenuation coefficients in water-filled phantoms determined with gamma cameras. Med Phys 1985; 11:681–685.

74. Malko JA, Van Heertum RL, Gullberg GT, Kowalsky WP. SPECT liver imaging using an iterative attenuation correction algorithm and an external flood source. J Nucl Med 1986; 27:701–705.

75. Thompson CJ, Ranger NT, Evans AC. Simultaneous transmission and emission scans in positron emission tomography. IEEE Trans Nucl Sci 1989; 36:1011–1016.

76. Tsui BMW, Hu HB, Gilland DR, Gullberg GT. Implementation of simultaneous attenuation and detector response correction in SPECT. IEEE Trans Nucl Sci 1988; NS-35:778–783.

77. Tsui BMW, Gullberg GT, Edgerton ER, et al. Correction of nonuniform attenuation in cardiac SPECT imaging. J Nucl Med 1989; 30:497–507.

78. Blankespoor S, Wu X, Kalki K, Brown JK, Cann CE, Hasegawa BH. Attenuation correction of SPECT using x-ray CT on an emission-transmission CT system. IEEE Nuclear Science Symposium and Medical Imaging Conference, 1995.

79. LaCroix KJ. An investigation of the use of single-energy x-ray CT images for attenuation compensation in SPECT. University of North Carolina at Chapel Hill, 1994.

80. Perring S, Summers Q, Fleming J, Nassim M, Holgate S. A new method of quantification of the pulmonary regional distribution of aerosols using combined CT and SPECT and its application to nedocromil sodium administered by metered dose inhaler. Br J Radiol 1994; 67:46–53.

81. Gullberg GT, Huesman RH, Malko JA, Pelc NJ, Budinger TF. An attenuated projector-backprojector for iterative SPECT reconstruction. Phys Med Biol 1985; 30:799–816.

82. Ficaro EP, Wackers FJ. Should SPECT attenuation correction be more widely employed in routine clinical practice? Eur J Nucl Med 2002; 29:409–412.

83. Fricke E, Fricke H, Weise R, et al. Attenuation correction of myocardial SPECT perfusion images with low-dose CT: evaluation of the method by comparison with perfusion PET. J Nucl Med 2005; 46(5):736–744.

84. Romer W, Fiedler E, Pavel M, et al. Attenuation correction of SPECT images based on separately performed CT: effect on the measurement of regional uptake values. Nuklearmedizin 2005; 44(1):20–28.

85. Seo Y, Wong KH, Sun M, Franc BL, Hawkins RA, Hasegawa BH. Correction of photon attenuation and collimator response for a body-contouring SPECT/CT imaging system. J Nucl Med 2005; 46(5):868–877.

86. Ardekani BA, Braun M, Hutton BF, Kanno I, Iida H. A fully automatic multimodality image registration algorithm. J Comput Assist Tomogr 1995; 19(4):615–623.

87. Antoch G, Freudenberg LS, Beyer T, Bockisch A, Debatin JF. To enhance or not to enhance: [F-18]-FDG and CT contrast agents in dual-modality [F-18]-FDG PET/CT. J Nucl Med 2004; 45:56S–65S.

88. Antoch G, Jentzen W, Freudenberg LS, et al. Effect of oral contrast agents on computed tomography-based positron emission tomography attenuation correction in dual-modality positron emission tomography/computed tomography imaging. Invest Radiol 2003; 38(12):784–789.

89. Bockisch A, Beyer T, Antoch G, et al. Positron emission tomography/computed tomography–imaging protocols, artifacts, and pitfalls. Mol Imaging Biol 2004; 6(4):188–199.

90. Beyer T, Antoch G, Bockisch A, Stattaus J. Optimized intravenous contrast administration for diagnostic whole-body 1 PET/CT. J Nucl Med 2005; 46(3):429–435.

91. Carney JP. Attenuation correction and image fusion [abstr]. Med Phys 2004; 31:1695.

92. Townsend DW, Yap JT, Carney JP, Hall NC. Developments in PET/CT: from concept to practice [abstr]. Med Phys 2004; 31:1694.

93. Mukai T, Links JM, Douglass KH, Wagner HN. Scatter correction in SPECT using non-uniform attenuation data. Phys Med Biol 1988; 33:1129–1140.

94. Kessler RM, Ellis JR, Eden M. Analysis of emission tomographic scan data: limitations imposed by resolution and background. J Comput Assist Tomogr 1984; 8(3):514–522.

95. Tang HR, Brown JK, Hasegawa BH. Use of x-ray CT-defined regions of interest for the determination of SPECT recovery coefficients. IEEE Trans Nucl Sci 1997; 44:1594–1599.

96. Koole M, Laere KV, de Walle RV, et al. MRI guided segmentation and quantification of SPECT images of the basal ganglia: a phantom study. Comput Med Imaging Graph 2001; 25(2):165–172.

97. Rousset OG, Ma Y, Evans AC. Correction for partial volume effects in PET: principle and validation. J Nucl Med 1998; 39(5):904–911.

98. Meltzer CC, Kinahan PE, Greer PJ, et al. Comparative evaluation of MR-based partial-volume correction schemes for PET. J Nucl Med 1999; 40(12):2053–2065.

99. Koral KF, Li J, Dewaraja Y, et al. I-131 anti-B1 therapy/tracer uptake ratio using a new procedure for fusion of tracer images to computed tomography images. Clin Cancer Res 1999; 5(suppl 10):3004s–3009s.

100. Koral KF, Dewaraja Y, Clarke LA, et al. Tumor-absorbed-dose estimates versus response in tositumomab therapy of previously untreated patients with follicular non-Hodgkin's lymphoma: preliminary report. Cancer Biother Radiopharm 2000; 15(4):347–355.

101. Dewaraja YK, Wilderman SJ, Ljungberg M, Koral KF, Zasadny K, Kaminiski MS. Accurate dosimetry in 131I radionuclide therapy using patient-specific, 3-dimensional methods for SPECT reconstruction and absorbed dose calculation. J Nucl Med 2005; 46(5):840–849.

102. Minarik D, Sjogreen K, Ljungberg M. A new method to obtain transmission images for planar whole-body activity quantification. Cancer Biother Radiopharm 2005; 20(1): 72–76.

103. Boucek JA, Turner JH. Validation of prospective whole-body bone marrow dosimetry by SPECT/CT multimodality imaging in (131)I-anti-CD20 rituximab radioimmunotherapy of non-hodgkin's lymphoma. Eur J Nucl Med Mol Imaging 2005; 32(4): 458–469.

104. Koral KF, Zasadny KR, Kessler ML, et al. CT-SPECT fusion plus conjugate views for determining dosimetry in iodine-131-monoclonal antibody therapy of lymphoma patients. J Nucl Med 1994; 35(10):1714–1720.

105. Koral KF, Lin S, Fessler JA, Kaminski MS, Wahl RL. Preliminary results from intensity-based CT-SPECT fusion in I-131 anti-B1 monoclonal-antibody therapy of lymphoma. Cancer 1997; 80:2538–2544.

106. Koral KF, Dewaraja Y, Li J, et al. Initial results for hybrid SPECT—conjugate-view tumor dosimetry in 131I-anti-B1 antibody therapy of previously untreated patients with lymphoma. J Nucl Med 2000; 41(9):1579–1586.

107. Koral KF, Francis IR, Kroll S, Zasadny KR, Kaminski MS, Wahl RL. Volume reduction versus radiation dose for tumors in previously untreated lymphoma patients who received iodine-131 tositumomab therapy. Conjugate views compared with a hybrid method. Cancer 2002; 94(suppl 4):1258–1263.

108. Paulino AC, Thorstad WL, Fox T. Role of fusion in radiotherapy treatment planning. Semin Nucl Med 2003; 33:238–243.

109. Carney JP, Beyer T, Brasse D, Yap JT, Townsend DW. Clinical PET/CT using oral CT contrast agents [abstr]. J Nucl Med 2002; 44:272P.

110. Nelson SJ, Vigneron DB, Dillon WP. Serial evaluation of patients with brain tumors using volume MRI and 3D 1H MRSI. NMR Biomed 1999; 12(3):123–138.

111. Kurhanewicz J, Swanson MG, Nelson SJ, Vigneron DB. Combined magnetic resonance imaging and spectroscopic imaging approach to molecular imaging of prostate cancer. J Magn Reson Imaging 2002; 16(4):451–463.

112. Boxt LM, Lipton MJ, Kwong RY, Rybicki F, Clouse ME. Computed tomography for assessment of cardiac chambers, valves, myocardium and pericardium. Cardiol Clin 2003; 21(4):561–585.

113. Chan FP. Cardiac multidetector-row computed tomography: principles and applications. Semin Roentgenol 2003; 38(4):294–302.

114. Nieman K, Rensing B, Munne A, van Geuns RJ, Pattynama P, de Feyter P. Three-dimensional coronary anatomy in contrast-enhanced multislice computed tomography. Prev Cardiol 2002; 5(2):79–83.

115. Hasegawa BH, Tang HR, Da Silva AJ, Iwata K, Wu AM, Wong KH. Implementation and applications of a combined CT/SPECT system. Conference Record of the 1999 Nuclear Science Symposium and Medical Imaging Conference 1999; 3:1373–1377.

2

^{99m}Tc-Sestamibi SPECT/CT for Parathyroid Imaging

Yodphat Krausz

Department of Medical Biophysics and Nuclear Medicine,
Hadassah—Hebrew University Medical Center, Jerusalem, Israel

INTRODUCTION

During the past 10 years, the widespread use of screening serum calcium measurements has led to an early identification of subtle presentation of primary hyperparathyroidism, which has resulted in an increased prevalence of the disease with a larger number of parathyroidectomies performed. Two recent leading articles recommend early surgical intervention in primary hyperparathyroidism (1,2).

The debate regarding the necessity for preoperative localization of a parathyroid adenoma (PTA) has been ongoing since the Consensus Statement was formulated at the NIH Conference on the management of hyperparathyroidism in 1991. Bilateral neck exploration without prior localization was recommended, with preoperative localization reserved for patients with persistent or recurrent hyperparathyroidism after initial surgery (3). A surgical cure rate of 95% to 98% has been achieved following bilateral neck exploration (4), with a relatively low morbidity of less than 5%. However, a solitary PTA accounts for 85% of cases of hyperparathyroidism, omitting the need for invasive bilateral neck exploration. Furthermore, prior localization and limited neck dissection are associated with decreased risk of hypoparathyroidism and recurrent laryngeal nerve injury, as well as shortening of surgery time and hospitalization.

The limited surgical procedures include unilateral neck exploration with assessment of the ipsilateral parathyroid gland (5–7), minimally invasive parathyroidectomy (8,9), endoscopic surgery (10), radioguided surgery (11), and video-assisted thoracic surgery for resection of ectopic mediastinal parathyroid glands (12). Radioguided parathyroidectomy, using the intraoperative gamma probe technique, is also of value in deep-seated neck or ectopic PTAs (13,14).

These surgical procedures are feasible as a consequence of preoperative localization of the PTA, using various imaging modalities, including ^{99m}Tc-methoxyisobutylisonitrile (MIBI) scintigraphy. Even MIBI/single-photon emission computed tomography (SPECT), however, may not provide detailed anatomical information. Other modalities have been therefore suggested, including imaging with a hybrid system combining computed tomography (CT) and gamma camera (15). The present chapter summarizes the contribution of hybrid imaging using MIBI/SPECT/CT for preoperative localization of the PTA and its impact on the surgical procedure.

ANATOMY

The upper parathyroid glands are normally located posterior to the upper two-thirds of the thyroid glands, and posterior to the recurrent laryngeal nerves, cranially to the inferior thyroid artery. The lower parathyroid glands show a more variable location, posterior or lateral to the lower pole of the thyroid lobe.

Ectopic glands occur in up to 20% of cases. They may vary in position along the thyro-thymic tract, from the mandibular level to the lower mediastinum. Ectopic superior parathyroid glands are rare and are usually in a posterior location. They can occur at or above the superior pole of the thyroid, below the inferior thyroid artery, or posterior and medially in the retropharyngeal, retrotracheal, or retroeso-phageal space (16). A deep-seated PTA of superior origin in the retrotracheal/esophageal or paratracheal/paraesophageal space accounts for up to 7% of all cases (13). Ectopic locations of the inferior glands are more common, their position is more variable, and they can occur anywhere from the superior thyroid pole, along the carotid sheath, within the thyroid gland, or they may descend together with the thymus to the anterior mediastinum as far as the pericardium (16).

SURGERY

Bilateral neck exploration has been the traditional surgical approach in patients with primary hyperparathyroidism. It includes exploration of both sides of the neck via a collar incision of 3 cm to 6 cm. The abnormal parathyroid tissue is removed, with or without biopsy of one normal-appearing parathyroid gland. This procedure, considered the "gold standard" for surgical management of primary hyperparathyroidism, is associated with a high success rate (95–98%) and low morbidity when performed by an experienced parathyroid surgeon (17).

With the introduction of improved localization techniques, the era of limited neck exploration has been inaugurated and focused parathyroidectomy has been widely adopted by endocrine surgeons.

Unilateral neck exploration was found suitable for most patients with hyper-parathyroidism, since 85% to 90% of patients will have single-gland disease. Denham and Norman have reviewed the English language literature over the last 10 years, including patients who had undergone preoperative imaging and unilat-eral neck exploration. Of 6331 patients diagnosed with primary hyperparathyroid-ism, 87% had a solitary PTA, with an average sensitivity and specificity for MIBI imaging of 91% and 99%, respectively. Their findings suggested that 78% of all patients with sporadic primary hyperparathyroidism were suitable for unilateral surgical exploration (18). Bergenfelz et al. compared unilateral and bilateral neck exploration in a prospective, randomized, controlled trial in 91 patients. They documented a lower incidence of biochemical and severe symptomatic hypocalce-mia in the early postoperative period in patients undergoing a unilateral proce-dure, compared with patients undergoing bilateral exploration (5). Furthermore, unilateral parathyroid exploration was not found to be associated with an increased incidence of persistent or recurrent hypocalcemia (19). Patients with multigland disease, prior neck surgery, and/or concomitant multinodular goiter (MNG) may, however, be unsuitable for this procedure (13).

Minimally invasive parathyroidectomy (*MIP*) is performed via a small incision in the lateral aspect of the neck, with exploration of the ipsilateral gland following extirpation of the adenoma. This procedure has fewer complications when compared with bilateral exploration, and shows a reduction in operating time as well as a substantial shortening in the postoperative hospitalization period (20).

Endoscopic (following carbon dioxide insufflation) *or video-assisted parathyroidectomy* requires general anesthesia and prolonged operating time when compared with MIP but allows access to both sides of the neck, better visualization of the recurrent laryngeal nerve, and also has a better cosmetic outcome. *Video-assisted thoracoscopic surgery* has been suggested as the first-line treatment in cases of mediastinal adenoma (21), with sparing of median sternotomy that had been previously associated with a 12% incidence in surgical complications.

Radioguided parathyroidectomy, using a gamma probe following MIBI injection two hours prior to surgery, is also associated with reduced morbidity and a decrease in operative time and hospital stay (22,23). Using a simplified approach, intraoperative-probe guidance can be considered as a completion of accurate preoperative scintigraphic imaging (24). This technique is feasible when preoperative scintigraphy identifies a single focus of increased radiotracer uptake. These findings imply a high probability of a solitary PTA, significant MIBI uptake in the adenoma, no coexisting MIBI-avid thyroid nodules, and no history of familial hyperparathyroidism. This procedure is indicated in patients with no history of previous neck irradiation (25,26). An in vitro probe assessment that does not meet the above criteria should direct the surgeon to continue the exploration, obtain intraoperative parathyroid hormone (PTH) measurement, and perform frozen section analysis of the removed tissue.

A recent survey conducted by the members of the International Association of Endocrine Surgeons indicated that in 2000, MIBI-based MIP has been adopted by more than 50% of surgeons worldwide, predominantly the focused approach with a small incision, followed by a video-assisted technique and true endoscopic technique with gas insufflation. PTH assay and probe-guidance were used to ensure completeness of resection (27).

These procedures may be done as outpatient surgery, even with local anesthesia, following MIBI SPECT localization (28). Inabnet et al. documented a successful outcome with a complication rate of less than 1% in 220 of 230 asymptomatic patients (96%) who underwent limited neck exploration under local anesthesia, following successful diagnosis and localization of a solitary adenoma on preoperative imaging. PTH levels were monitored during surgery and the mean operating time was 30 minutes (29). The availability of rapid PTH assays has circumvented the need for visualization of all glands. Failure to achieve a 50% drop in baseline PTH should lead to further exploration. Assays for intraoperative PTH measurements may not be required when there is concordance between preoperative MIBI scan and ultrasound (US).

IMAGING

The surgical approach to hyperparathyroidism has dramatically changed with the introduction of improved preoperative localization modalities that direct treatment algorithms for the surgical management of primary hyperparathyroidism.

A variety of techniques have been suggested for localization of abnormal parathyroid gland(s) prior to surgery, including high-resolution US, CT, magnetic resonance imaging (MRI), selective venous sampling, and different scintigraphic techniques. Functional imaging methods have been used in combination with anatomical imaging, the most sensitive being the scintigraphic techniques and high-resolution US.

Ultrasonography

US plays an important role in preoperative localization of a PTA, especially when the adenoma is negative on parathyroid scintigraphy. US can then localize an additional 14% of enlarged parathyroid glands, further facilitating the surgical procedure while avoiding bilateral exploration (30). The sensitivity of US identification of PTA ranges between 70% and 83% (6,31), but falls to 36% in patients who have had prior failed surgical exploration (32). This modality is affected by multiglandular disease, when a PTA is located close to the trachea or deep in the neck in the para- and retropharyngeal space (26), or when the adenoma is associated with nodular disease of the thyroid gland (thyroid adenoma or MNG) (15).

A comparison study showed that the sensitivity of US is inferior to that of MIBI—77% versus 88%, respectively (33). On the other hand, the combined approach of MIBI and US is considered the imaging modality of choice for noninvasive detection of a PTA localized in the neck (7,23,25,34,35). A sensitivity of 83% increased to 94% when US was combined with MIBI/pertechnetate subtraction technique, the latter having an 85% sensitivity when used alone (35); complementary MIBI and US achieved a 96% sensitivity for solitary PTA, enabling successful limited neck exploration or MIP in 91 of 143 patients, and appropriate bilateral surgery in the others (26). Furthermore, when both MIBI and US suggest the same abnormal parathyroid gland, focused parathyroidectomy can be used with a predicted success rate of about 95% (36).

CT and MRI

CT and MRI are used in the preoperative evaluation of hyperparathyroid patients, mainly in patients with recurrent or persistent hyperparathyroidism.

CT is superior to US in localization of PTA in the retrotracheal, retroesophageal, and mediastinal spaces, but it performs poorly for ectopic lesions located in the lower neck at the level of the shoulders and for lesions close to or within the thyroid gland. An overall sensitivity of CT for preoperative identification of hyperplastic parathyroid glands has been previously reported to range between 40% and 81% (35,37), having the lowest sensitivity among the various modalities in localization of ectopic parathyroid glands (37). However, a recent prospective comparison of MIBI/pertechnetate subtraction scintigraphy and helical CT showed sensitivity for scintigraphy and CT of 88% and 90%, respectively, with 33% of the patients having an ectopic adenoma (38).

On MR examination, enlarged parathyroid glands display considerably increased intensity on T2-weighted and proton density images, with reported sensitivities ranging between 50% and 84% (39–41). Paired evaluations in the same patients have shown that the sensitivity of MRI is equivalent to that of parathyroid scintigraphy (42), and MRI is suggested only when parathyroid scintigraphy is negative, equivocal, or suggestive of an ectopic gland (42,43). A comparison study showed that Tc99m-MIBI/I123 subtraction is more useful than the delayed imaging of MIBI, MRI, and US, with a sensitivity of 71%, 50%, 50%, and 71%, respectively, for the detection of a PTA (40). Furthermore, Ishibashi et al., in a blinded, comparative study, found superior sensitivity of scintigraphy when compared with CT or MRI in detecting ectopic adenomas. The calculated sensitivity and specificity of MIBI imaging were 70% and 88%, respectively, when compared with the sensitivity and specificity of CT of 40% and 88%, and of MRI 60% and 88%, respectively (37).

It has been suggested that CT or MRI should complement positive MIBI mediastinal parathyroid localization. Complementary MIBI and MRI have achieved a sensitivity and positive predictive value of 94% and 98% (42). A combined approach of CT and US also has been described (8), with both imaging modalities providing, however, only morphologic information and CT serving as a surgical road map following US (44).

MIBI Parathyroid Scintigraphy

After an initial experience with MIBI for myocardial perfusion studies, Coakley et al. incidentally observed significant uptake of this tracer in the abnormal parathyroid glands of patients with primary hyperparathyroidism (45), with differential retention compared to thyroid tissue. The single-tracer, dual-phase scintigraphy, originally described by Taillefer et al. (46), is based on the differential washout rate of MIBI from the thyroid and the parathyroid tissue. A dual-tracer subtraction protocol or dual-phase technique may be used, the former showing a higher sensitivity (47), with improvement by MIBI/SPECT (48–50).

Planar imaging of the neck and thorax is recorded starting 15 minutes and then 2 to 3 hours after the intravenous injection of 555 MBq Tc99m-MIBI, using a large field-of-view gamma camera equipped with a parallel-hole collimator. When the PTA is no more intense than the thyroid gland or no differential washout is observed on delayed image, a second radiopharmaceutical is administered that accumulates specifically in the thyroid gland and not in the parathyroid tissue, such as ^{123}I or $^{99m}TcO4^-$ pertechnetate. A distinct focus of increased or separate MIBI uptake relative to the thyroid gland, either on the early or late image or on both, or a focal uptake in the mediastinum is considered positive for abnormal parathyroid tissue.

Tc99m-MIBI scintigraphy was found to play a major role in the preoperative localization of a PTA, with a sensitivity ranging from 85% to 95% (24). A meta-analysis covering 10 years in the English literature reveals that while the MIBI scanning had an average sensitivity and specificity of 90.7% and 99%, respectively (18), 52 studies included in a nonstatistical meta-analysis reported the sensitivity to range from 39% to above 90% (51). Scintigraphy has a high diagnostic accuracy in the presence of intense tracer uptake and retention in a single lesion on delayed images indicative of a high probability of a solitary PTA. Its accuracy decreases, however, in the presence of concomitant MIBI-avid thyroid nodules, a history of familial hyperparathyroidism, multiple endocrine neoplasia, or previous neck surgery or irradiation (13,26).

MIBI/SPECT

The contribution of MIBI/SPECT to enhanced sensitivity prior to initial surgery has been controversial over the years (52). Although only a marginal improvement in the overall detection rate of PTAs is reported with SPECT, most authors now favor a wider application of this imaging modality. Sfakianakis et al. found a 96% sensitivity using MIBI/SPECT and volume-rendered reprojection in both early and delayed stages (53). Perez-Monte recommended the use of early rather than a delayed SPECT, associated with sensitivity for localization of 91% vs. 74%, respectively (54). Billotey et al. showed an increase in sensitivity from 86% of dual-phase planar scintigraphy to 91% by early SPECT (55). The sensitivity of 87% of early planar scintigraphy increased to 95% by the supplementary use of delayed SPECT and a 3-D display (volume-rendered reprojection for visualization). Their contribution was evident mainly for small solitary adenomas of less than 1 g (50). Early SPECT provided in-depth information in 9 out of 30 neck adenomas and in 13 of 15 ectopic adenomas, whose weight ranged from 0.3 to 1.2 g (56).

SPECT offers the advantage of better discrimination of focal MIBI retention in thyroid nodules and parathyroid tissue, but is most helpful when guiding the surgeon for possible sites of ectopic PTAs, with improved sensitivity of 96% for early SPECT as compared to 79% for planar imaging in patients with MNG and in ectopic adenomas (57). These findings led the same group to claim that early MIBI/SPECT can serve as the only localizing study for focused parathyroidectomy (58). They encouraged its use as the only modality to be used in the morning prior to surgery, both to select patients for minimally invasive radioguided surgery and to provide accurate 3-D information on deeply seated or ectopic adenomas. This approach lowered the costs of preoperative localization and intraoperative validation to a single study. Furthermore, the intraoperative gamma probe technique facilitates the surgical planning and reduces surgical trauma and complications. Alternatively, based on 338 patients who subsequently underwent neck exploration, Civelek et al. suggested delayed MIBI/SPECT as the only imaging study that should be performed, with an overall sensitivity of 87%, accuracy of 94%, and a positive predictive value of 86% (48). The use of early SPECT, however, is preferable when the thyroid gland serves as an anatomical landmark and prior to the rapid washout that may occur in a PTA (59).

MIBI/SPECT/CT Image Fusion

Localization of PTAs in the neck, using planar MIBI scintigraphy, is facilitated by visualization of the thyroid gland in the early phase of the study. In the majority of patients, this technique, with occasional SPECT providing in-depth information, is sufficient for localization of the adenoma in the neck. Coregistered anatomical mapping, however, may help localize deep-seated or ectopic adenomas, and may facilitate the surgical re-exploration in patients with distorted anatomy after previous neck surgery or irradiation.

Fusion of functional and anatomical data may follow separate acquisition of the two modalities, using a specific software coregistration package, or sequential acquisition on a dual-modality imaging system.

MIBI/SPECT and CT image coregistration, following overlapping of separately acquired techniques, improved the localization of an ectopic mediastinal PTA located in the upper mediastinum, with removal by limited median sternotomy (60). The thoracoscopic approach diagnosed a 9 mm ectopic PTA located anterior to the aorta at the level of the carina, using an image fusion software package for MIBI and CT when findings in isolation were inconclusive (61). Profanter et al. studied six consecutive patients with primary hyperparathyroidism, using the CT/MIBI image fusion following separate acquisition of the CT and the MIBI scans. The patient's head and neck were fixed with a head holder and a vacuum cushion, and radiographic and scintigraphic markers were mounted at the head holder and the patient.

Fusion of CT and MIBI images, performed by overlaying the radiographic markers using commercial software and workstation, correctly predicted the localization of solitary adenomas in five patients (62). Using the same technique in a prospective study of 24 patients, these authors further described the incremental role of MIBI/CT image fusion compared with MIBI SPECT alone, with a sensitivity of 93% vs. 31%, respectively (63). However, these authors found SPECT data to be inferior to the SPECT sensitivity documented in the literature.

An additional fusion technique using Tc99m-MIBI and Tc99m-albumin, the latter as an indicator of the intravascular space, enabled visualization of an ectopic PTA localized to the aorto-pulmonary window (24).

Fusion of MRI and SPECT MIBI data was evaluated in 17 patients with primary hyperparathyroidism, using a modified version of the Express 5.0 software. MRI detected 10 (71%) and scintigraphy localized 12 (86%) of 14 adenomas. Image fusion improved the anatomical assignment of 13 scintigraphic foci in five patients, and was helpful in the final interpretation of inconclusive MR-findings in two patients (41).

Software-based fusion of independently acquired nuclear medicine and morphological data is subject to poor alignment and time-consuming processing. In contrast, a device combining a dual-head variable angle gamma camera with a low-dose CT provides hybrid images following sequential acquisition of both modalities on the same device. The CT images obtained are of inferior resolution when compared with the state-of-the-art high-resolution CT, and cannot be used for independent anatomic lesion detection and characterization. These images, however, may serve as a road map for anatomical localization of the increased MIBI uptake seen on SPECT.

SPECT/CT of the neck and chest is acquired, as a rule, after the early planar imaging, using a dual-head variable-angle gamma camera and a low-power X-ray CT transmission system, mounted on the same slip-ring gantry (Millennium VG & Hawkeye, General Electric Medical Systems). Transmission data are reconstructed using filtered back-projection (FBP) to produce cross-sectional attenuation images. Following transmission, the SPECT component is acquired in 120 projections, 3° angle step, in a 128×128 matrix and acquisition time of 20 seconds per frame. Reconstruction is performed by FBP or iteratively using the ordered subsets expectation maximization (OSEM) technique. The resultant emission images, obtained in transaxial, sagittal, and coronal planes, are inherently registered to the anatomical maps, using the workstation software with generation of fused images of the overlying transmission (CT) and emission (SPECT) data.

Several case reports and small series of patients have been described, using the hybrid device (64,65). SPECT/CT imaging was the only procedure that precisely localized two ectopic adenomas, in the anterior and posterior mediastinum, respectively (66), and led to precise localization of mediastinal parathyroid glands in four additional patients (67), with facilitation of surgery. In a retrospective study of 36 patients who underwent focused parathyroidectomy, Krausz et al. evaluated the incremental value of SPECT/CT in localization of PTAs and its impact on neck exploration. Referral criteria for SPECT/CT included the visualization of an ill-defined focus in the neck or an ectopic site on planar scintigraphy (56). The hybrid images provided precise topographic localization of the adenoma with respect to proximity to adjacent structures. SPECT/CT improved the lesion localization in 14 of 33 patients with positive MIBI scintigraphy (42%). It precisely indicated the proximity of the PTA to the thyroid, trachea, and esophagus in 4 of 23 patients (17%) with cervical PTA (Fig. 1), and it identified the relationship of the lesion to the

(A)

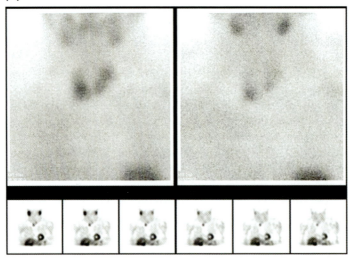

(B)

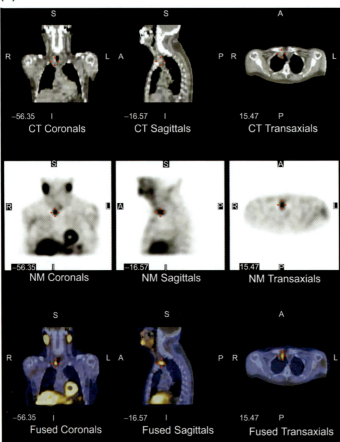

Figure 1 (*Caption on facing page*)

(C)

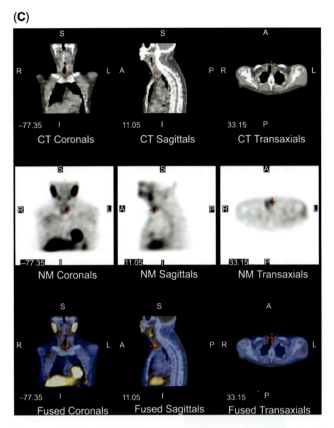

Figure 1   Hybrid MIBI SPECT/CT imaging in a patient with a deep-seated PTA in the neck, in the presence of a thyroid nodule. (**A**) Early (*upper left*) and delayed (*upper right*) anterior planar MIBI images show uptake and retention in the right lower neck, and very mild retention overlapping the upper half of the left thyroid lobe. Series of coronal sections of MIBI SPECT (*lower row*) show the posterior location of a right PTA, whereas a thyroid nodule is located anteriorly in the upper half of the left thyroid gland. (**B**) SPECT/CT (CT, *upper row*; SPECT, *center row*; fusion, *lower row*) shows the precise localization of the PTA in the right posterior paratracheal space, at the level of the SSN. (**C**) SPECT/CT (CT, *upper row*; SPECT, *center row*; fusion, *lower row*) shows the paratracheal anterior location of the left thyroid nodule. These findings guided the surgical exploration, and a 2.0 cm diameter, 1.1 g PTA was excised, with a left colloid nodule, 1.6 cm in diameter. *Abbreviations*: PTA, parathyroid adenoma; MIBI, methoxyisobutylisonitrile; SPECT, single-photon emission computed tomography; CT, computed tomography.

trachea, esophagus, thymus, spine, or sternum in all 10 patients with ectopic PTAs (Fig 2). In the 23 patients with a PTA confined to the neck, SPECT/CT played a limited role prior to initial surgery. The anatomical mapping, however, improved the surgical management in deep-seated adenomas and in special clinical settings, such as in the presence of distorted neck anatomy after initial neck exploration, and with nonvisualization of the thyroid gland after thyroidectomy. SPECT/CT was also of additional value in differentiating PTAs from thyroid nodules in patients with associated MNG, despite the MIBI-avidity of both lesions, and in guiding their further exploration (Fig. 1). In 10 patients with ectopic adenomas, fused images determined the topographical localization of all 10 lesions with respect to proximity to trachea, esophagus, or spine; directed the surgical search; and spared these

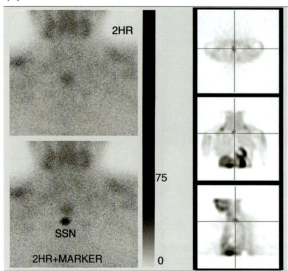

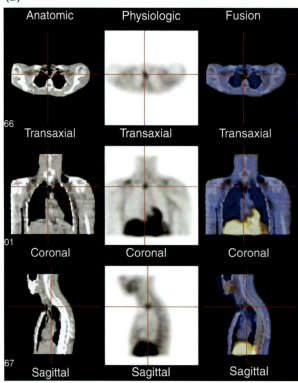

Figure 2 Hybrid MIBI SPECT/CT imaging, after failed initial surgery, in a patient with an ectopic PTA deeply seated in the neck. (**A**) Delayed (*upper left*) anterior planar MIBI image shows faint uptake and retention in a mid-neck, above the suprasternal notch (Marker: *lower left*). MIBI/ SPECT sections (*right*), transaxial (*upper*), coronal (*mid*), and sagittal (*bottom*), show the posterior location of the PTA. (**B**) SPECT/CT (CT, *left column*; SPECT, *center column*; fusion, *right column*) shows the precise localization of the PTA in the right pre-esophageal space. The latter findings guided the surgical re-exploration, and a 2.5 cm diameter, 2.0 g PTA was excised. *Abbreviations*: PTA, parathyroid adenoma; MIBI, methoxyisobutylisonitrile; SPECT, single-photon emission computed tomography; CT, computed tomography.

patients unnecessary re-exploration. Six lesions were paratracheal, anterior in location at the level of the suprasternal notch or manubrium, with sparing of sternotomy following precise SPECT/CT localization. Four PTAs were posterior in location: a deeply situated PTA adjacent to the thymus, a pre-esophageal lesion (Fig. 2), and two PTAs detected in the retrotracheal plane. After failed initial surgery, SPECT/CT improved the lesion localization of four adenomas, both in the neck and in an ectopic site, and facilitated the surgical resection in patients (67%) who underwent re-exploration (56).

The findings were also reported in a preliminary study including 23 patients with hyperparathyroidism. SPECT/CT precisely localized 88% of patients, with results further confirmed at surgery, as compared to a precise localization rate of 53% for both planar MIBI scintigraphy and cervical ultrasound (68). Similar results were found in a study on 37 patients showing for SPECT/CT a sensitivity of 91%, specificity 95%, positive predictive value of 82%, and negative predictive value of 97%, as compared to 85%, 93%, 76%, and 95%, respectively, for MIBI/SPECT and 68%, 90%, 65%, and 91%, respectively, for CT. The authors conclude that SPECT/CT allowed for precise localization of a PTA in the majority of cases and has the advantage of reducing the radiation exposure and time of examination to the patient, which made the use of external markers unnecessary and decreased the costs of diagnostic procedures (69).

CONCLUSION

Localization of parathyroid adenomas in the neck, using MIBI scintigraphy, is facilitated by visualization of the thyroid gland in the early phase of the study and is improved with occasional SPECT/MIBI. SPECT/CT, however, contributes to precise PTA localization and to planning of the surgical strategy in patients with deep-seated adenomas in the neck, or in those with distorted cervical anatomy following neck surgery or with concomitant MNG or ectopic adenomas. The major role of SPECT/CT is in patients with an ectopic PTA and in patients after prior failed surgery. The 3-D presentation of an adenoma on SPECT/CT in the presence of known anatomical structures provides an optimized surgical road map towards easy accessibility of the PTA, a shorter invasive procedure, and a higher success rate.

SPECT/CT need not necessarily be applied to patients with a well-defined cervical PTA on planar MIBI scintigraphy prior to initial surgery, in the absence of associated medical conditions of the neck. Combined functional and morphologic information provided by SPECT/CT appears to be a promising prerequisite tool for minimally invasive surgery in patients with primary hyperparathyroidism with a more complicated clinical background or in a difficult anatomic location.

REFERENCES

1. Utiger RD. Treatment of primary hyperparathyroidism. N Engl J Med 1999; 341:1301–1302.
2. Toft AD. Surgery for primary hyperparathyroidism–sooner rather than later. Lancet 2000; 355:1478–1479.
3. NIH Conference: diagnosis and management of asymptomatic primary hyperparathyroidism consensus development conference statement. Consensus Development Conference, NIH. Ann Intern Med 1991; 114:593–597.
4. Clark OH. How should patients with primary hyperparathyroidism be treated? J Clin Endocrinol Metab 2003; 88:3011–3014.

5. Bergenfelz A, Lindblom P, Tibblin S, Westerdahl J. Unilateral versus bilateral neck exploration for primary hyperparathyroidism: a prospective randomized controlled trial. Ann Surg 2002; 236:543–551.

6. Gofrit ON, Lebensart PD, Pikarsky A, Lackstein D, Gross DJ, Shiloni E. High-resolution ultrasonography: highly sensitive, specific technique for preoprative localization of parathyroid adenoma in the absence of multinodular thyroid disease. World J Surg 1997; 21: 287–290.

7. Krausz Y, Lebensart PD, Klein M, et al. Preoperative localization of parathyroid adenoma in patients with concomitant thyroid nodular disease. World J Surg 2000; 24:1573–1578.

8. van Vroonhoven TJ, van Dalen A. Successful minimally invasive surgery in primary hyperparathyroidism after combined preoperative ultrasound and computed tomography imaging. J Intern Med 1998; 243:581–587.

9. Chen H. Surgery for primary hyperparathyroidism: what is the best approach? Ann Surg 2002; 236:552–553.

10. Ohshima A, Simizu S, Okido M, Shimada K, Kuroki S, Tanaka M. Endoscopic neck surgery: current status for thyroid and parathyroid diseases. Biomed Pharmacother 2002; 56(suppl 1):48S–52S.

11. Norman J, Chheda H. Minimally invasive parathyroidectomy facilitated by intraoperative nuclear mapping. Surgery 1997; 122:998–1003.

12. Medrano C, Hazelrigg SR, Landreneau RJ, Boley TM, Shawgo T, Grasch A. Thoracoscopic resection of ectopic parathyroid glands. Ann Thorac Surg 2000; 69:221–223.

13. Casara D, Rubello D, Cauzzo C, Pelizzo MR. 99mTc-MIBI Radioguided minimally invasive parathyroidectomy: experience with patients with normal thyroids and nodular goiters. Thyroid 2002; 12:53–61.

14. Rubello D, Casara D, Pagetta C, Piotto A, Pelizzo MR, Shapiro B. Determinant role of Tc-99m MIBI SPECT in the localization of a retrotracheal parathyroid adenoma successfully treated by radioguided surgery. Clin Nucl Med 2002; 27:711–715.

15. Coakley AJ. Nuclear Medicine and parathyroid surgery; a change in practice. Nucl Med Commun 2003; 24:111–113.

16. Loevner LA. Imaging of the parathyroid glands. Semin Ultrasound CT MRI 1996; 17:563–575.

17. Chen H, Zeiger MA, Gordon TA, Udelsman R. Parathyroidectomy in Maryland: effects of an endocrine center. Surgery 1996; 120:948–952.

18. Denham DW, Norman J. Cost-effectiveness of preoperative sestamibi scan for primary hyperparathyroidism is dependent solely upon the surgeon's choice of operative procedure. J Am Coll Surg 1998; 186:293–305.

19. Sidhu S, Neill AK, Russell CF. Long-term outcome of unilateral parathyroid exploration for primary hyperparathyroidism due to presumed solitary adenoma. World J Surg 2003; 27:339–342.

20. Udelsman R. Six hundred fifty-six consecutive explorations for primary hyperparathyroidism. Ann Surg 2002; 235:665–670.

21. Amar L, Guignat L, Tissier F, et al. Video-assisted thoracoscopic surgery as a first-line treatment for mediastinal parathyroid adenomas: strategic value of imaging. Eur J Endocrinol 2004; 150:141–147.

22. Goldstein RE, Blevins L, Delbeke D, Martin WH. Effect of minimally invasive radioguided parathyroidectomy on efficacy, length of stay, and costs in the management of primary hyperparathyroidism. Ann Surg 2000; 231:732–742.

23. Rubello D, Casara D, Pelizzo MR. Symposium on parathyroid localization. Optimization of preoperative procedures. Nucl Med Commun 2003; 24:133–140.

24. Mariani G, Gulec SA, Rubello D, et al. Preoperative localization and radioguided parathyroid surgery. J Nucl Med 2003; 44:1443–1458.

25. Casara D, Rubello D, Piotto A, Pelizzo MR. 99mTc-MIBI Radioguided minimally invasive parathyroid surgery planned on the basis of a preoperative combined

99mTc-pertechnetate/99mTc-MIBI and ultrasound imaging protocol. Eur J Nucl Med 2000; 27:1300–1304.

26. Casara D, Rubello D, Pelizzo MR, Shapiro B. Clinical role of 99mTcO4/MIBI scan, ultrasound and intraoperative gamma probe in the performance of unilateral and minimally invasive surgery in primary hyperparathyroidism. Eur J Nucl Med 2001; 28: 1351–1359.

27. Sackett WR, Barraclough B, Reeve TS, Delbridge LW. Worldwide trends in the surgical treatment of primary hyperparathyroidism in the era of minimally invasive parathyroidectomy. Arch Surg 2002; 137:1055–1059.

28. Chen H, Sokoll LJ, Udelsman R. Outpatient minimally invasive parathyroidectomy: a combination of sestamibi/SPECT localization, cervical block anesthesia, and intraoperative parathyroid hormone assay. Surgery 1999; 126:1016–1021.

29. Inabnet WB, Fulla Y, Richard B, Bonnichon P, Icard P, Chapuis Y. Unilateral neck exploration under local anesthesia: the approach of choice for asymptomatic primary hyperparathyroidism. Surgery 1999; 126:1004–1009.

30. Quiros RM, Alioto J, Wilhelm SM, Ali A, Prinz RA. An algorithm to maximize use of minimally invasive parathyroidectomy. Arch Surg 2004; 139:501–506.

31. Ammori BJ, Madan M, Gopichandran TD, et al. Ultrasound guided unilateral neck exploration for sporadic primary hyperparathyroidism: is it worthwhile? Ann R Coll Surg Eng 1998; 80:433–437.

32. Miller DL, Doppman JL, Shawker TH, et al. Localization of parathyroid adenomas in patients who have undergone surgery. Part I. Noninvasive imaging methods. Radiology 1987; 162:133–137.

33. Haber RS, Kim CK, Inabnet WB. Ultrasonography for preoperative localization of enlarged parathyroid glands in primary hyperparathyroidism: comparison with (99m)technetium sestamibi scintigraphy. Clin Endocrinol (Oxf) 2002; 57:241–249.

34. De Feo ML, Colagrande S, Biagini C, et al. Parathyroid glands: combination of 99mTc-MIBI scintigraphy and US for demonstration of parathyroid glands and nodules. Radiology 2000; 214:393–402.

35. Lumachi F, Ermani M, Basso S, Zucchetta P, Borsato N, Favia G. Localization of parathyroid tumours in the minimally invasive era: which technique should be chosen? Population-based analysis of 253 patients undergoing parathyroidectomy and factors affecting parathyroid gland detection. Endocr Relat Cancer 2001; 8:63–69.

36. Arici C, Cheah WK, Ituarte PH, et al. Can localization studies be used to direct focused parathyroid operations? Surgery 2001; 129:720–729.

37. Ishibashi M, Nishida H, Hiromatsu Y, et al. Localisation of ectopic parathyroid glands using technetium-99m sestamibi imaging: comparison with magnetic resonance and computed tomographic imaging. Eur J Nucl Med 1997; 24:197–201.

38. Lumachi F, Tregnaghi A, Zucchetta P, et al. Technetium 99m sestamibi scintigraphy and helical CT together in patients with primary hyperparathyroidism: a prospective clinical study. Br J Radiol 2004; 77:100–103.

39. Lee VS, Spritzer CE, Coleman RE, Wilkinson RH Jr, Coogan AC, Leight GS Jr. The complementary roles of fast spin-echo MR imaging and double-phase 99mTc-sestamibi scintigraphy for localization of hyperfunctioning parathyroid glands. Am J Roentgenol 1996; 167:1555–1562.

40. Wakamatsu H, Noguchi S, Yamashita H, et al. Parathyroid scintigraphy with 99mTc-MIBI and 123I subtraction: a comparison with magnetic resonance imaging and ultrasonography. Nucl Med Commun 2003; 24:755–762.

41. Ruf J, Hanninen EL, Steinmuller T, et al. Preoperative localization of parathyroid glands. Use of MRI, scintigraphy, and image fusion. Nuklearmedizin 2004; 43:85–90.

42. Gotway MB, Reddy GP, Webb WR, Morita ET, Clark OH, Higgins CB. Comparison between MR imaging and 99mTc-MIBI scintigraphy in the evaluation of recurrent or persistent hyperparathyroidism. Radiology 2001; 218:783–790.

43. Fayet P, Hoeffel C, Fulla Y, et al. Techetium-99m-sestamibi, magnetic resonance imaging and venous blood sampling in persistent and recurrent hyperparathyroidism. Br J Radiol 1997; 70:459–464.

44. van Dalen A, Smit CP, van Vroonhoven TJ, Burger H, de Lange EE. Minimally invasive surgery for solitary parathyroid adenomas in patients with primary hyperparathyroidism: role of US with supplemental CT. Radiology 2001; 220:631–639.

45. Coakley AJ, Kettle AG, Wells CP, O'Doherty MJ, Collins RE. 99Tcm sestamibi–a new agent for parathyroid imaging. Nucl Med Commun 1989; 10:791–794.

46. Taillefer R, Boucher Y, Potvin C, Lambert R. Detection and localization of parathyroid adenomas in patients with hyperparathyroidism using a single radionuclide imaging procedure with technetium-99m-sestamibi (double-phase study). J Nucl Med 1992; 33: 1801–1807.

47. Leslie WD, Dupont JO, Bybel B, Riese KT. Parathyroid 99mTc-sestamibi scintigraphy: dual-tracer subtraction is superior to double-phase washout. Eur J Nucl Med Mol Imaging 2002; 29:1566–1570.

48. Civelek AC, Ozalp E, Donovan P, Udelsman R. Prospective evaluation of delayed technetium-99m sestamibi SPECT scintigraphy for preoperative localization of primary hyperparathyroidism. Surgery 2002; 131:149–157.

49. Neumann DR, Esselstyn CB Jr, Madera AM. Sestamibi/iodine subtraction single photon emission computed tomography in reoperative secondary hyperparathyroidism. Surgery 2000; 128:22–28.

50. Moka D, Voth E, Dietlein M, Larena-Avellaneda A, Schicha H. Technetium 99m-MIBI/ SPECT: A highly sensitive diagnostic tool for localization of parathyroid adenomas. Surgery 2000; 128:29–35.

51. Gotthardt M, Lohmann B, Behr TM, et al. Clinical value of parathyroid scintigraphy with technetium-99m methoxyisobutylisonitrile: discrepancies in clinical data and a systematic metaanalysis of the literature. World J Surg 2004; 28:100–107.

52. Neumann DR, Esselstyn CB Jr, Kim EY, Go RT, Obuchowski NA, Rice TW. Preliminary experience with double-phase SPECT using Tc-99m sestamibi in patients with hyperparathyroidism. Clin Nucl Med 1997; 22:217–221.

53. Sfakianakis GN, Irvin GL III, Foss J, et al. Efficient parathyroidectomy guided by SPECT-MIBI and hormonal measurements. J Nucl Med 1996; 37:798–804.

54. Perez-Monte JE, Brown ML, Shah AN, et al. Parathyroid adenomas: accurate detection and localization with Tc-99m-sestamibi SPECT. Radiology 1996; 201:85–91.

55. Billotey C, Sarfati E, Aurengo A, et al. Advantages of SPECT in technetium-99m-sestamibi parathyroid scintigraphy. J Nucl Med 1996; 37:1773–1778.

56. Krausz Y, Bettman L, Guralnik L, et al. Tc99m-MIBI SPECT/CT in primary hyperparathyroidism. World J Surg 2006; 30:76–83.

57. Lorberboym M, Minski I, Macadziob S, Nikolov G, Schachter P. Incremental diagnostic value of preoperative ^{99m}Tc-MIBI SPECT in patients with a parathyroid adenoma. J Nucl Med 2003; 44:904–908.

58. Schachter PP, Issa N, Shimonov M, Czerniak A, Lorberboym M. Early, postinjection MIBI/SPECT as the only preoperative localizing study for minimally invasive parathyroidectomy. Arch Surg 2004; 139:433–437.

59. Krausz Y, Shiloni E, Bocher M, Agranovicz S, Manos B, Chisin R. Diagnostic dilemmas in parathyroid scintigraphy. Clin Nucl Med 2001; 26:997–1001.

60. Rubello D, Casara D, Fiore D, et al. An ectopic mediastinal parathyroid adenoma accurately located by a single-day imaging protocol of Tc-99m pertechnetate-MIBI subtraction scintigraphy and MIBI/SPECT-computed tomographic image fusion. Clin Nucl Med 2002; 27:186–190.

61. Ng P, Lenzo NP, McCarthy MC, Thompson I, Leedman PJ. Ectopic parathyroid adenoma localized with sestamibi SPECT and image-fused computed tomography. Med J Aust 2003; 179:485–487.

62. Profanter C, Prommegger R, Gabriel M, et al. Computed axial tomography-MIBI image fusion for preoperative localization in primary hyperparathyroidism. Am J Surg 2004; 187:383–387.
63. Profanter C, Wetscher GJ, Gabriel M, et al. CT-MIBI image fusion: a new preoperative localization technique for primary, recurrent, and persistent hyperparathyroidism. Surgery 2004; 135:157–162.
64. Bocher M, Balan A, Krausz Y, et al. Gamma camera-mounted anatomical X-ray tomography: technology, system characteristics and first images. Eur J Nucl Med 2000; 27:619–627.
65. Patton JA, Delbeke D, Sandler MP. Image fusion using integrated, dual-head coincidence camera with X-ray tube-based attenuation maps. J Nucl Med 2000; 41:1364–1368.
66. Even-Sapir E, Keidar Z, Sachs J, et al. The new technology of combined transmission and emission tomography in evaluation of endocrine neoplasms. J Nucl Med 2001; 42:998–1004.
67. Kaczirek K, Prager G, Kienast O, et al. Combined transmission and (99m)Tc-sestamibi emission tomography for localization of mediastinal parathyroid glands. Nuklearmedizin 2003; 42:220–223.
68. Buhl T, Mollerup C, Mortensen J. Precise preoperative localization of parathyroid adenomas with combined 99mTc-MIBI SPECT/Hawkeye (low-dose CT) scanning [abstr]. J Nucl Med 2004; 45(suppl):16.
69. Martin P, Alcan I, Berges L, et al. Contribution of 99mTc-MIBI SPECT/CT fusion imaging in hyperparathyroidism: first experience with the new hybrid SPECT/CT GE Hawkeye [abstr]. Eur J Nucl Med Mol Imaging 2004; 31(suppl 2):246.

3

^{131}Iodine SPECT/CT for Thyroid Cancer Imaging

William H. Martin and Dominique Delbeke
Vanderbilt University Medical Center, Nashville, Tennessee, U.S.A.

INTRODUCTION

Thyroid carcinoma accounts for 90% of all endocrine tumors and 1.5% of all malignancies, with approximately 19,000 new cases occurring annually in the United States. Thyroid cancer mortality is 1200 deaths per year, resulting in a relatively high prevalence of disease, with over 200,000 patients living in the United States having undergone thyroidectomy for thyroid cancer and requiring regular assessment (1). In recent years, the number of radiopharmaceuticals used in the detection and ongoing evaluation of thyroid carcinoma has expanded beyond ^{131}Iodine (^{131}I) to include ^{123}Iodine, ^{201}Thallium (Tl), ^{99m}Tc-sestamibi, ^{99m}Tc-tetrofosmin, ^{99m}Tc-furifosmin, ^{99m}Tc-(V)-DMSA, ^{123}I/^{131}I-metaiodobenzylguanidine (MIBG), somatostatin receptor (SSR) analogs, ^{18}F-Fluoro-deoxyglucose (FDG), and radiolabeled monoclonal antibodies (2). The utility of these various radiopharmaceuticals is dependent on the histologic type of thyroid cancer present. Thyroid malignancies include mostly differentiated thyroid carcinomas (DTC) (80%) as well as medullary thyroid carcinomas (MTC) (7%), lymphomas (5%), and undifferentiated anaplastic carcinomas (less than 5%). These tumors present specific challenges in imaging. Carcinoma metastatic to the thyroid is not uncommon at autopsy, but it is usually detected clinically as an incidental finding by ultrasonography (US), computed tomography (CT), magnetic resonance imaging (MRI), or even positron emission tomography (PET) with FDG, in a patient who is being treated for a non-thyroid malignancy (3).

DIFFERENTIATED THYROID CARCINOMA

Seventy to eighty percent of DTC are of mixed papillary/follicular histology and the remainder are follicular. The behavior of the two tumor types differs with papillary thyroid cancer typically metastasizing to locoregional nodes and the lungs, and follicular malignancies disseminating hematogenously to the bones. DTC usually maintains the capacity to trap and organify iodine and to synthesize and release thyroglobulin (Tg). These characteristics of DTC allow postthyroidectomy treatment of iodine-avid disease with high-dose ^{131}I and the monitoring of therapy using serum Tg measurements and radioiodine scintigraphy (RIS). Dedifferentiation occurs, however, to a variable extent with both types of DTC with loss of the sodium iodide symporter (NIS) and/or loss of Tg expression, thus presenting challenges for further imaging and monitoring of these patients. For example, the Hurthle cell variant of follicular cell carcinoma is not often radioiodine-avid but is virtually always FDG-avid (4). Other less differentiated thyroid malignancies have characteristics (such as calcitonin expression or increased glucose metabolism) that permit specific imaging and posttherapy monitoring as well.

Traditional Methods of Follow-up for Patients with Differentiated Thyroid Carcinomas

The traditional methods of follow-up for patients with DTC are whole-body RIS and serum Tg monitoring. [131]I uptake by functioning thyroid carcinomas and their metastases is usually less than 10% that of normal thyroid tissue (5). Therefore, distant and nodal metastases may not be visualized by RIS prior to ablation of the postsurgical thyroid remnant. Optimal ^{131}I uptake by neoplastic tissue is also dependent on thyroid stimulating hormone (TSH) stimulation. Adequate endogenous TSH levels of greater than $30\,\mu U/mL$ can be attained 10 to 14 days after the discontinuation of exogenous triiodothyronine (T_3) or one to four weeks after the discontinuance of thyroxine (T_4) therapy (6–8). Recombinant human TSH (rhTSH) is now a method of stimulating radioiodine uptake (and Tg release) that can be used in patients maintained on thyroid hormone therapy (9–11). In a multicenter trial of 226 patients, findings observed on diagnostic RIS acquired after rhTSH administration, while the patients continued exogenous thyroxine treatment, were congruent with or superior to thyroid hormone withdrawal scans in 93% of patients. There was no significant difference in detection rate for metastases. When coupled with rhTSH-stimulated Tg determinations, the detection rate for disease or tissue limited to the thyroid bed was 93% and for metastatic disease was 100% (10). These results have been confirmed in a report of 289 patients, which found no difference in the diagnostic accuracy between 128 patients who underwent rhTSH-stimulated ^{131}I imaging and the remainder population who underwent RIS following thyroxine-withdrawal imaging (11). RIS is, therefore, performed under conditions of TSH stimulation, either endogenous via thyroid hormone withdrawal or following administration of rhTSH.

Circulating Tg, a complex iodinated glycoprotein synthesized by thyroid cells but no other tissues, should be undetectable in the absence of functioning thyroid tissue. Measurement of Tg serum values is used as a tumor marker in the postoperative follow-up of patients with DTC after ^{131}I ablation of the normal remnant. An elevated serum Tg determination ($>2\,ng/mL$) in such patients is a highly sensitive and specific indicator of residual or metastatic thyroid carcinoma (12,13). Most patients with elevated serum Tg but no other evidence of disease are frequently found to have metastatic disease at follow-up (5,10,11,14). Tg determinations done after thyroid hormone withdrawal are, in most reports, more sensitive for the detection of metastases than measurements in patients on suppressive therapy (15,16). Only 80% of patients with metastases will demonstrate an elevated serum Tg level while on thyroid hormone therapy (10), but Tg is elevated in virtually 100% of patients with residual thyroid tissue or metastatic DTC given rhTSH or after thyroid hormone withdrawal (10). Although some authors have recommended performing RIS only in patients who exhibit elevation in serum Tg (17), others, including the National Cancer Center Network, have advocated the use of RIS in combination with TSH-stimulated serum Tg measurements, due to the occurrence of recurrent disease in the absence of TSH-stimulated Tg elevation in as many as 18% of patients (11,18–20).

Whole-body RIS is first performed 6 to 10 weeks after thyroidectomy and 2 weeks after the discontinuation of exogenous T_3 therapy or following rhTSH administration. Although the specificity of ^{131}I scanning is 95%, it is important to differentiate the normal physiologic radiotracer activity in the salivary glands, nose, gastric mucosa, urinary bladder, liver, bowel, and lactating breast from abnormal uptake in sites of metastatic disease. Rare false-positive scans have been described due to the presence of radiotracer in body secretions (urine, saliva, lactation, and perspiration), and in benign processes such as pathologic transudates, Zenker's or Meckel's diverticulum, Warthin's tumor, scrotal hydrocele, bronchogenic tumor, gastric tumor, struma ovarii, and other nonthyroidal pathologies (21). More accurate localization of malignant lesions and differentiation between metastases and physiologic accumulation of radioiodine is essential to identify patients who require further therapy. It is in this context that ^{131}I-SPECT (single-photon emission computed tomography)/CT fusion imaging has the potential to contribute to patient management (22,23).

The sensitivity of ^{131}I scintigraphy for the detection of persistent or metastatic thyroid carcinoma is reported to be 50% to 75%, dependent in part on the diagnostic dose of ^{131}I used for imaging (13,15). A meta-analysis of seven reports indicates that 75% of patients with recurrent DTC will have positive findings with RIS (24). Imaging at three to seven days after the administration of a therapeutic dose of 100 mCi to 200 mCi of ^{131}I may increase the detection rate of metastatic lesions by up to 45%. In fact, diffuse miliary involvement of the lungs is usually detected only on posttreatment RIS (25). The combination of ^{131}I imaging and serum Tg determination augments the detection of metastatic disease to 85% to 100% (12,13,16,26). A schedule of follow-up examinations at 6- to 12-month intervals is recommended until the serum Tg is undetectable and RIS demonstrates no abnormal uptake (27). Follow-up should thereafter continue over the years, although at less frequent intervals, because up to 50% of DTC recurrences occur at more than five years after initial treatment, with one-third of these events being diagnosed after 10 years and more (18).

Stunning is the phenomenon by which the initial diagnostic dose of ^{131}I, as a routine 3 mCi to 5 mCi, reduces trapping of the subsequently administered therapeutic dose. The frequency of this effect and its clinical significance is controversial (28). Quantitative dosimetric studies seem to confirm a 30% to 50% reduction in the therapeutic radioiodine uptake as compared to the diagnostic dose (29,30). A prospective study demonstrated a highly significant reduction in tracer uptake following a first dose of ^{131}I in 171 consecutive patients with benign thyrotoxic disease, which significantly correlated to the absorbed energy dose delivered by the initial treatment (31). To avoid these potential adverse consequences of stunning, many institutions now perform ^{123}I scintigraphy using doses between 1 and 5 mCi (32). In most reports there is little, if any, difference in the sensitivity for detection of thyroid remnant and metastases using ^{123}I versus posttreatment high-dose ^{131}I imaging, and SPECT acquisitions with or without fusion with CT can be performed with diagnostic ^{123}I imaging as well, when appropriate (33–35).

Other Radiopharmaceuticals to Follow-Up Patients with Differentiated Thyroid Carcinomas

Owing to the somewhat limited sensitivity of RIS for metastatic DTC, and because RIS is a cumbersome procedure requiring thyroid hormone withdrawal or rhTSH administration and multiple visits to the nuclear medicine facility, other DTC-avid radiopharmaceuticals have been sought for functional assessment of thyroid malignancies (2). Agents such as ^{201}Tl, ^{99m}Tc-sestamibi, ^{99m}Tc-tetrofosmin, SSR analogs, and FDG may potentially detect non-iodine-avid tumor deposits, including less differentiated varieties such as Hurthle cell carcinoma (36), to image euthyroid patients on thyroxine replacement, and patients who have an expanded iodine pool related to recent iodinated contrast administration. These alternative radiopharmaceuticals should not be used, however, instead of RIS, unless the patient is known from earlier studies to be ^{131}I-negative. [201]Tl and ^{99m}Tc-agents are more convenient for the patient. They do not require withdrawal from thyroid hormone, and are not dependent on TSH levels or iodine trapping. They do not, however, accumulate well in normal thyroid tissue and hence are not as sensitive as ^{131}I or ^{123}I in detecting the presence of postoperative thyroid remnant (36). In addition, false-positive findings may be seen with nonthyroidal tumors, vascular structures, and salivary gland uptake.

FDG-PET imaging appears to be particularly helpful in patients with elevated Tg levels but negative RIS. In a review of 14 studies, FDG-PET demonstrated a consistently high sensitivity for detection of recurrent thyroid neoplasm in patients with elevated serum Tg and negative RIS, in direct proportion to the degree of elevation in serum Tg, probably representing a measure of tumor volume (37). Few initial reports emphasize the potential value of hybrid PET/CT imaging using FDG or [124]I for improved diagnostic assessment of thyroid cancer (38,39).

Prognosis and Treatment Options for Patients with Recurrent Differentiated Thyroid Carcinomas

Although the prognosis for patients with DTC is as a rule very good, with long-term survival rates of 84% to 94%, 40-year recurrence rates are approximately 35%. Mortality rises in patients with recurrent disease, particularly when located within the soft tissues of the neck and with distant metastases (27). Long-term survival in patients with distant recurrences decreases to approximately 50%. Because one-third of these recurrences occur more than 10 years after initial treatment, patients with DTC are at risk for recurrence and death for decades. With the aim of decreasing the recurrence rate and its attendant morbidity and mortality, most patients with DTC currently undergo total thyroidectomy, often with lymph node dissection, followed by postsurgical radioiodine ablation of the remnant thyroid, even in those patients with "low risk" disease (40). In a series of 1510 patients without distant metastases at the time of initial treatment, remnant ablation was reported to reduce cervical recurrences, distant recurrences, and mortality (27).

Treatment of patients found to have local or distant recurrences depends on the site and the extent of disease. Surgical excision of the amenable-to-surgery lesions and ^{131}I administration remain the mainstay. External beam radiation therapy (XRT) may be administered to areas of unresectable gross residual disease. Patients found to have recurrent disease first require tumor localization. Determination of iodine-avidity and an estimate of tumor burden are easily accomplished by whole-body RIS. Patients with functioning pulmonary metastases, for instance, can survive for years following treatment with repeated doses of [131]I. Following ongoing radio-iodine treatment, patients with iodine-avid lung metastases had an overall 10-year survival rate of 84% for an average follow-up of 12.7 years. The prognosis for patients with micronodular disease was better than for those with positive findings on chest X-ray (25). Patients with nonfunctioning metastases have a worse prognosis. However, localization requires correlation with conventional radiologic studies, such as CT, MR, and US. Because surgical extirpation is the treatment of choice for cervical nodal metastases and for resectable distant macroscopic metastatic deposits, accurate localization is critical. Resection of superior mediastinal metastases, for instance, may be curative in over 90% of patients (41). Patients who are able to undergo resection of cerebral metastases have an improved survival rate as compared to those undergoing external XRT or ^{131}I treatment, and often die of extracranial disease (42,43). Surgical removal of metastatic lesions in the lungs and bones, when combined with ^{131}I and/or XRT, can result in prolonged survival (78% at five years) (43,44). The addition of ^{131}I administration or XRT to the palliative embolization of symptomatic skeletal metastases is reported to prolong the duration of the disease-free interval by a factor of 2 (45). Multimodality treatment is, in fact, the rule in the majority of patients with metastatic DTC.

LOCALIZATION OF RECURRENT AND METASTATIC DISEASE WITH ^{131}I-SPECT/CT

Although whole-body RIS provides excellent functional data regarding the extent and iodine avidity of disease in patients with DTC, precise anatomic localization is often difficult due to the lack of anatomic landmarks and the low resolution and increased image noise when performing planar and SPECT scintigraphy using [131]I. These problems are confounded by physiologic radioiodine uptake in the salivary glands, genitourinary system, gastrointestinal system, and liver. As discussed previously, differentiating tumor uptake from sites of physiologic tracer activity or contamination requires accurate localization of ^{131}I-avid sites and is critically important for optimal further treatment planning. Correlation of planar whole-body imaging findings with tomographic modalities such as CT, MR, and even ultrasound, can be unfulfilling. Owing to the suboptimal physical characteristics of ^{131}I, SPECT is not routinely used, even following high-dose treatment doses. The ability to perform SPECT with ^{123}I imaging provides an advantage over ^{131}I, although there are no studies comparing SPECT findings of these two agents. Coregistration of SPECT and CT (or MR) data sets can be performed by using external fiduciary markers or by using an integrated SPECT/CT imaging system wherein both functional and anatomic image data sets are acquired with the patient in exactly the same position. Data has accumulated to indicate that the acquisition of SPECT/CT images using ^{131}I favorably impacts clinical decision-making in patients with DTC (23,46–51).

In 1995, investigators from Memorial Sloan-Kettering Cancer Center reported the use of image registration of SPECT and CT using an external fiduciary band and a 3-D surface-fitting algorithm to accurately localize metastases to the liver, lungs, and vertebral bodies in a patient with metastatic thyroid carcinoma (49). Because this degree of localization could not be obtained from the planar or SPECT images alone, the fusion images were credited with considerably altering the management of the patient. Subsequently, Perault et al. (48) was able to obtain reliable and reproducible coregistration of CT and thoraco-abdominal SPECT images using dual isotope tomoscintigraphy with ^{131}I and ^{99m}Tc-HDP in 13 patients with thyroid carcinoma. Fused images permitted localization and characterization of 10 sites in eight patients, six of whom were unsuspected by CT alone. Most recently, Yamamoto et al. (47) used fiducial markers to coregister CT and SPECT images acquired seven days after the administration of a therapeutic ^{131}I dose of 100 to 200 mCi to 17 patients with DTC. Fusion images were considered of incremental value in 88% (15/17) of patients by detecting occult bone ($n = 5$), nodal ($n = 4$), and soft-tissue ($n = 1$) metastases not detected by CT, and also by precisely localizing planar findings to the skeleton in three, and by differentiating physiologic radioiodine uptake from tumor in two patients. Although the ability to coregister SPECT and CT data sets is theoretically available at most medical centers without the purchase of additional hardware, practical issues of misregistration have stimulated the development of integrated SPECT/CT imaging systems.

Even-Sapir et al. (23) used an integrated SPECT/CT imaging system (Millennium VG Hawkeye^{TM}, General Electric Medical Systems) to study 27 patients with known or suspected neuroendocrine tumors, four of which were studied using ^{131}I for DTC with distant metastases. In two of the four DTC patients, fused images impacted clinical care of the patients by accurately localizing abnormal radioiodine uptake to bone metastases in the coracoid and sternum. Other investigators (47) have also reported on the utility of fusion SPECT/CT Hawkeye images, in one case for the differentiation of physiologic sites of radioiodine deposition in the respiratory tract from tracer uptake in lung metastases.

Tharp et al. at Vanderbilt University and Rambam Medical Center (46) reported the use of the same integrated SPECT/CT imaging system in 71 patients with DTC. The fusion images had an incremental diagnostic value in 57% (41/71) of patients. SPECT/CT fusion images allowed for the precise characterization of equivocal neck lesions identified on planar imaging in 14/17 patients and changed the assessment of the lesion location in five patients as compared to the planar studies. Overall there was an additional diagnostic value in 31% (19/61) of patients undergoing SPECT/CT of the neck. Of 36 patients undergoing SPECT/CT imaging for evaluation of extracervical abnormalities seen on planar imaging, precise localization of malignant lesions to the skeleton was possible in 12, and to the lungs versus the mediastinum in five patients. Equivocal lesions were clarified as benign in nine patients.

SPECT/CT imaging had an incremental diagnostic value in 65% (11/17) of patients who underwent diagnostic whole-body RIS after administration of 3 mCi to 5 mCi, and in 58% (15/26) of patients after high-dose therapeutic radioiodine administration for suspected or known recurrence.

Findings on SPECT/CT changed the therapeutic approach in 41% (7/17) of the patients who had a diagnostic study: the dose of ^{131}I for treatment was increased in two patients with lymph node metastases, surgery was indicated and the surgical approach planned in three patients (including two with localization of involved lymph nodes prior to neck dissection and one with a soft tissue metastasis and skeletal involvement of the pelvis), and unnecessary therapy was avoided in two patients with physiological tracer uptake.

The value of SPECT/CT in patients with inconclusive uptake foci on planar whole-body scintigraphy was further assessed in a study including 25 patients with DTC. SPECT/CT improved the anatomical assignment of 44% of suspicious lesions. Changes in characterization of abnormal uptake foci were therapeutically relevant in 25% of patients (52).

A preliminary communication by Kienast et al. (53) reported on the clinical impact of SPECT/CT in 32 patients with DTC, indicating that fused images led to a change in management in 31% of patients. Weckesser et al. (54) compared planar whole-body images and SPECT/CT images in 28 patients with DTC evaluated after the first or second fraction of radioiodine therapy. SPECT/CT of the neck and upper chest was performed in 22 patients and of other regions in six patients. The authors conclude that the SPECT/CT images provided additional diagnostic information in 43% (12/28) of patients, confirming lymph node involvement in five patients, changing the interpretation from negative to positive in two patients, from positive to negative in two patients, and from ambiguous to definite lymph node involvement in two additional cases. The correct anatomical localization modified the interpretation in two patients: in one patient, a suspected skeletal metastasis could be attributed to abnormal iodine accumulation in a tumor of the ovary and, in the other patient, suspected mediastinal involvement was correctly attributed to physiological intraluminal esophageal activity.

Lehmkuhl et al. (55) and Costa et al. (56) reported similar results in studies including 25 and 10 patients with thyroid cancer examined following postablative therapy. Both studies concluded that SPECT/CT images improved the diagnostic interpretation and had a significant clinical impact. As in the studies previously described, the additional diagnostic information was either for the characterization of ^{131}I-avid foci as physiologic or malignant or for the precise localization of malignant foci, especially to lymph nodes, lungs, and skeleton.

Krausz et al. (57) showed that SPECT/CT improved the interpretation accuracy of foci of increased tracer uptake in 70% of 67 patients with DTC mainly for diagnosis of cervical metastatic lymphadenopathy, the differential diagnosis of pelvic soft-tissue versus bone involvement, and the exclusion of suspected malignancy in sites of physiologic tracer uptake such as the bowel or thymus. This optimized characterization and localization of suspicious lesions further affected clinical management of 22 patients (41%), leading to referral to previously unplanned radiotherapy or surgery and sparing unnecessary radioiodine treatment.

A patient treated with ^{131}I for mandibular invasion diagnosed at surgery, in whom SPECT/CT fusion images allowed for precise localization of a focus of uptake in the femur, further documented as a skeletal metastasis, is shown in Figure 1. In another patient, the posttherapy (150 mCi) [131]I planar images demonstrated increased tracer uptake at the base of the neck and an additional focus of intense ^{131}I uptake in the lower body. The SPECT/CT fusion images allowed localization of the suspicious foci to the thyroid bed in the neck and to the right ovary in the pelvis (Fig. 2). A follow-up ultrasound demonstrated a 2-cm partially cystic and solid mass consistent with a dermoid cyst. A patient previously treated for metastatic thyroid cancer received ^{123}I whole-body assessment for follow-up. A focus of increased tracer activity was localized by SPECT/CT to a region of physiologic biodistribution of radioiodine, thus excluding the presence of new metastases (Fig. 3).

OTHER TYPES OF THYROID CANCERS

MTC represents approximately 7% of all thyroid cancers, with nearly 50% of patients presenting with metastatic cervical adenopathy at diagnosis (58,59). The 10-year survival rate after surgery is 86%, even with persistent hypercalcitoninemia (60).

Five-year survival decreases from 94% in patients with metastatic lymphadeno-pathy to only 41% in those with extranodal disease (61). MTC does not, as a rule, concentrate radioiodine, so RIS is not useful, and ^{131}I treatment results in no improvement in recurrence rate or survival (62). Recurrent or persistent disease is diagnosed by elevated serum calcitonin and carcinoembryonic antigen (CEA) levels (58). The optimal modality for localization of metastases continues to evolve. An improved prognosis in up to 40% of patients with persistent postoperative disease has been reported following repeat microdissection of residual metastatic cervical nodes (59). Selective venous catheterization for serum calcitonin determinations is sensitive and specific for localization of the tumor (63,64) but is invasive and technically demanding. CT and MRI, although providing images with a good spatial resolution for surgical planning, are similar to or slightly inferior to radio-nuclide scanning for the detection and localization of MTC recurrences (65,66). [201]Tl, ^{99m}Tc-sestamibi, and [99m]Tc-(V)-DMSA may be highly sensitive and specific if basal calcitonin levels are above 1000 pg/mL (67–69). In a similar fashion, FDG PET is evolving into a primary modality for detection of MTC (70) with a sensitivity of 73% in 20 consecutive patients with MTC and elevated calcitonin levels (71,72), and 94% in a cohort of 36 MTC patients (73). Metastatic MTC

(A)

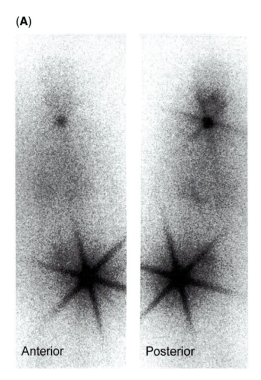

Anterior   Posterior

Figure 1 A 60-year-old female with papillary thyroid carcinoma was treated with 260 mCi of ^{131}I for mandibular involvement found during thyroidectomy. (**A**) Planar whole-body images show increased ^{131}I uptake in the thyroid bed as well as at the level of the left thigh. (**B**) SPECT/CT images of the region of the thighs demonstrate that the focus of uptake is localized to the left proximal femur indicating a skeletal metastasis. (**C**) Plain X-rays of the left femur were normal. Follow-up whole-body ^{131}I scintigraphy was normal, indicating a good response to therapy. *Abbreviations*: CT, computed tomography; SPECT, single-photon emission computed tomography. *Source*: From Ref. 46.

(B)

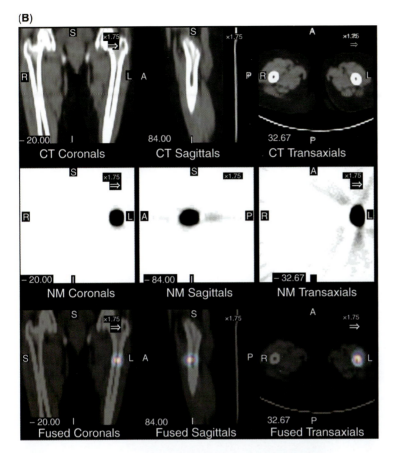

(C)

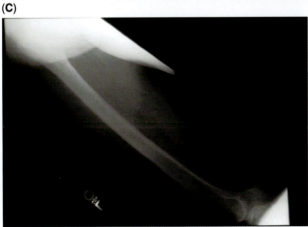

Figure 1 (*Caption on facing page*)

can be also visualized using ^{111}In-labeled SSR scintigraphy, with false-positive foci due to chronic inflammation (65,74). Similar to other neuroectodermal tumors, 40% to 50% of MTC will take up ^{123}I/^{131}I-MIBG (75,76). Early detection of recurrent MTC and localization of metastases are important because microdissection offers the chance for long-term remission or cure, improvement in symptomatology, improved prognosis, and potentially prolonged survival. No single diagnostic modality is able to reliably demonstrate the full extent of disease in these patients, but the hybrid imaging using each of the above-mentioned tracers in combination with diagnostic CT may prove in the future to play an important role in management of patients with MTC.

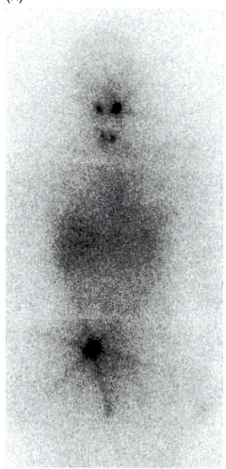

Figure 2 A 28-year-old female underwent thyroidectomy for a 2-cm follicular variant of papillary thyroid carcinoma with evidence of invasion of vessels and capsule. She was treated with 150 mCi of ^{131}I and whole-body ^{131}I scintigraphy was performed 10 days later. (A) Planar whole-body images show the presence of increased ^{131}I uptake in the thyroid bed and an additional focus of intense activity in the lower part of the torso. (B) SPECT/CT images of the neck, transaxial slices, and MIP demonstrate that the focus of increased tracer activity is localized to the thyroid bed, consistent with residual thyroid tissue. (C) SPECT/CT images of the abdomino-pelvic region, transaxial slices, and MIP localize the intense abnormal focus of radioiodine uptake to the right ovary. A previously suspected skeletal metastasis was excluded. Follow-up pelvic ultrasound demonstrated the presence of a 2-cm partially cystic and solid mass consistent with a dermoid cyst. *Abbreviations*: CT, computed tomography; MIP, maximum intensity projection; SPECT, single-photon emission computed tomography.

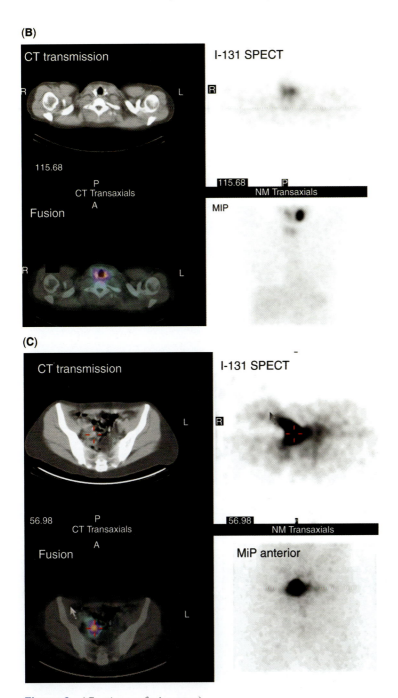

Figure 2 (*Caption on facing page*)

Lymphoma of the thyroid constitutes up to 5% of all thyroid malignancies and only 2% of all extranodal lymphomas. ^{99m}Tc and ^{131}I scintigraphy are of little utility in differentiating lymphomatous involvement from thyroiditis. Detection of lymphomatous involvement of the thyroid by US, CT, and MRI are comparable (77,78). Preliminary data suggests that ^{201}Tl and ^{99m}Tc-sestamibi may be useful in the detection

(A)

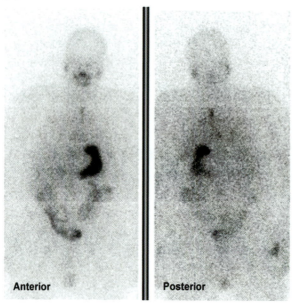

(B)

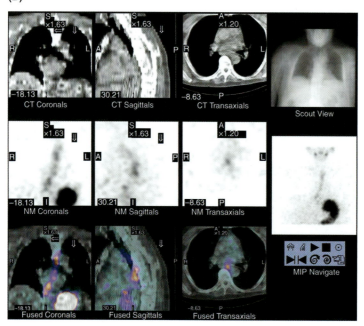

Figure 3   A 67-year-old male underwent thyroidectomy for papillary thyroid carcinoma in 1992. Twelve years later, he underwent left modified radical neck dissection for a recurrence to the neck and mediastinum, followed by ^{131}I-radioiodine therapy (206 mCi). He was referred for a follow-up diagnostic whole-body ^{123}I-study. (**A**) Planar whole-body scintigraphy indicated the presence of increased ^{123}I uptake in the mediastinum, in addition to physiologic uptake of ^{123}I in the stomach and bowel. (**B**) SPECT/CT images of the chest demonstrated that the area of uptake has a linear shape and is localized along the course of the esophagus, therefore consistent with an additional area of physiological tracer activity. *Abbreviations*: CT, computed tomography; SPECT, single-photon emission computed tomography.

of thyroid lymphoma as well as its response to therapy (79,67). [67]Gallium (^{67}Ga) scanning, although positive, is not the recommended imaging procedure (80). FDG uptake may be increased in lymphoma, PET being of value in defining extrathyroidal disease. Biopsy is, however, required for tissue sampling and should be guided by hybrid imaging.

Undifferentiated or anaplastic carcinoma of the thyroid also accounts for less than 5% of thyroid malignancies (81). [201]Tl or ^{99m}Tc-sestamibi may be useful in evaluating recurrent anaplastic carcinoma, especially if CT or US is equivocal (82,83). Although the median survival is less than six months, response to aggressive multimodality radiotherapy and chemotherapy in patients without distant metastases may result in a few enduring responses of over two years, provided that extent and localization of disease are clearly determined (84), potentially using SPECT/CT.

SUMMARY AND FUTURE DIRECTIONS

There is adequate clinical data available now to suggest that the use of ^{131}I SPECT/CT fusion imaging, especially after the administration of a relatively high treatment dose, will provide clinically relevant diagnostic information of incremental value in a high proportion of patients with recurrent or metastatic DTC. It is most useful in clarifying the location of foci of increased uptake seen on planar images and differentiating physiologic radioiodine activity from uptake in malignant sites, thus permitting the avoidance of additional, at times invasive, diagnostic procedures. The improved diagnostic interpretation provided by the availability of combined SPECT/CT images has been reported to change management decisions in at least one-third of patients with DTC. In view of the findings reported by many investigators using SPECT/CT with a variety of radiopharmaceuticals in patients with neuroendocrine neoplasms (85), combined analysis of SPECT/CT images with those of concurrent contrast-enhanced high-end CT may provide the highest degree of accuracy.

At present, published data regarding integrated SPECT/CT imaging have been obtained with the only system available commercially (Millenium Hawkeye, GEMS, Milwaukee, Wisconsin) since the late 1990s. This imaging technology provides low-dose noncontrast-enhanced CT images that are of limited diagnostic value. For surgical candidates, there is a need for a high-resolution diagnostic CT scan with oral and intravenous contrast administration in settings that make the evaluation of extent of disease and of the relationship of lesions to vascular structures clinically significant. In these patients, SPECT/low-dose CT data can and have been used to further select the appropriate field of view for performing or reviewing a high-quality diagnostic CT study that matches up with the area of increased metabolism localized by fused images. It should, however, be cautioned that a diagnostic CT with contrast is contraindicated in the follow-up of patients being considered for radioidine therapy within two months after the SPECT/CT study.

Because of the incremental diagnostic value of integrated SPECT/CT imaging for various pathological processes and its impact on the management of patients, manufacturers have now developed integrated SPECT/CT systems with high-performance multidetector CT scanners. These new technological developments and potential future clinical applications appear to be similar to the ones related to integrated PET/CT imaging, a well established technology in oncology.

Data are currently accumulating regarding the value of integrated PET/CT imaging for thyroid cancers that are not ^{131}I-avid. It is assumed that studies using various radiopharmaceuticals in different histologic types of malignancy, in large patient populations, will, in the future, establish the role of simultaneous functional and structural assessment, and its impact on patient management and the prognosis of thyroid cancer.

REFERENCES

1. Cancer facts and figures, 2000. New York: American Cancer Society, 2000.
2. Casara D, Rubello D, Saladini G. Role of scintigraphy with tumor-seeking agents in the diagnosis and preoperative staging of malignant thyroid nodules. Biomed Pharmacother 2000; 54:334–336.
3. Takashima S, Takayama F, Wang JC, et al. Radiologic assessment of metastases to the thyroid gland. J Comput Assist Tomogr 2000; 24:539–545.
4. Lowe VJ, Mullan BP, Hay ID, McIver B, Kasperbauer JL. 18F-FDG PET of patients with Hurthle cell carcinoma. J Nucl Med 2003; 44:1402–1406.
5. Mazzaferri EL. Radioiodine and other treatments and outcomes. In: Braverman LE, Utiger RD, eds. Werner & Ingbar's The Thyroid: A Fundamental and Clinical Text. 8th ed. Philadelphia: Lippincott Williams & Wilkins, 2000:904–929.
6. Goldman JM, Line BR, Aamodt RL, Robbins J. Influence of triiodothyronine withdrawal time on ^{131}I uptake post-thyroidectomy for thyroid cancer. J Clin Endocrinol Metab 1980; 50:734–739.
7. Serhal DI, Nasrallah MP, Arafah BM. Rapid rise in serum thyrotropin concentrations after thyroidectomy or withdrawal of suppressive thyroxine therapy in preparation for radioactive iodine administration to patients with differentiated thyroid cancer. J Clin Endocrinol Metab 2004; 89:3285–3289.
8. Grigsby PW, Siegel BA, Bekker BA, Clutter WE, Moley JF. Preparation of patients with thyroid cancer for 131I scintigraphy or therapy by 1–3 weeks of thyroxine discontinuation. J Nucl Med 2004; 45(4):567–570.
9. Ladenson PW, Braverman LE, Mazzaferri EL, et al. Comparison of administration of recombinant human thyrotropin with withdrawal of thyroid hormone for radioactive iodine scanning in patients with thyroid carcinoma. N Engl J Med 1997; 337:888–896.
10. Haugen BR, Pacini F, Reiners C, et al. A comparison of recombinant human thyrotropin and thyroid hormone withdrawal for the detection of thyroid remnant or cancer. J Clin Endocrinol Metab 1999; 84:3877–3885.
11. Robbins RJ, Tuttle RM, Sharaf RN, et al. Preparation by recombinant human thyrotropin or thyroid hormone withdrawal are comparable for the detection of residual differentiated thyroid carcinoma. J Clin Endocrinol Metab 2001; 86:619–625.
12. van Sorge-van Botel RA, van Eck-Smit BL, Goslings BM. Comparison of serum thyroglobulin, 131-I and 201-Tl scintigraphy in the postoperative follow-up of differentiated thyroid cancer. Nucl Med Commun 1993; 14:365–372.
13. Lubin E, Mechlis-Frish S, Zatz S. Serum thyroglobulin and iodine-131 whole-body scan in the diagnosis and assessment of treatment for metastatic differentiated thyroid carcinoma. J Nucl Med 1994; 35:257–262.
14. Black EG, Sheppard MC. Serum thyroglobulin measurements in thyroid cancer: evaluation of "false" positive results. Clin Endocrinol 1991; 35:519–520.
15. Ashcraft MW, Van Herle AJ. The comparative value of serum thyroglobulin measurements and iodine-131 total body scans in the follow-up study of patients with treated differentiated thyroid cancer. Am J Med 1981; 71:806–814.
16. Schneider AB, Line BR, Goldman JM, Robbins J. Sequential serum thyroglobulin determinations, I-131 scans, I-131 uptakes after triiodothyronine withdrawal in patients with thyroid cancer. J Clin Endocrinol Metab 1981; 53:1199–1206.
17. Cailleux AF, Baudin E, Travagli JP, Ricard M, Schlumberger M. Is diagnostic iodine-131 scanning useful after total thyroid ablation for differentiated thyroid cancer? J Clin Endocrinol Metab 2000; 85:175–178.

18. Mazzaferri EL, Kloos RT. Using recombinant human TSH in the management of well-differentiated thyroid cancer: current strategies and future directions. Thyroid 2000; 10: 767–778.
19. Westbury C, Vini L, Fisher C, Harmer C. Recurrent differentiated thyroid cancer without elevation of serum thyroglobulin. Thyroid 2000; 10:171–176.
20. Torlontano M, Attard M, Crocetti U, et al. Follow-up of low risk patients with papillary thyroid cancer: role of neck ultrasonography in detecting lymph node metastases. J Clin Endocrinol Metab 89:3402–3407.
21. Shapiro B, Rufini V, Jarwan A, et al. Artifacts, anatomical and physiological variants, and unrelated disease that might cause false-positive whole-body 131-I scans in patients with thyroid cancer. Semin Nucl Med 2000; 30:115–132.
22. Martin WH, Patton JA, Delbeke D, Sandler MP. Improved localization of endocrine sites using an integrated CT-SPECT fused imaging system. J Nucl Med 2000; 41:9P.
23. Even-Sapir E, Keidar Z, Sachs J, et al. The new technology of combined transmission and emission tomography in evaluation of endocrine neoplasms. J Nucl Med 2001; 42: 998–1004.
24. Maxon HR, Smith HS. Radioiodine-131 in the diagnosis and treatment of metastatic well-differentiate thyroid cancer. Endocrinol Metab Clin North Am 1990; 19: 958–718.
25. Hindie E, Melliere D, Lange F, et al. Functioning pulmonary metastases of thyroid cancer: does radioiodine influence the prognosis? Eur J Nucl Med Mol Imaging 2003; 30: 974–981.
26. Ikekubo K, Hino M, Ito H, et al. The early detection of metastatic differentiated thyroid cancer using 131-I total body scan and treatment with 131-I. Jpn J Nucl Med 1991; 28: 247–259.
27. Mazzaferri EL, Kloos RT. Current approaches to primary therapy for papillary and follicular thyroid cancer. J Clin Endocrinol Metab 2001; 86:1447–1463.
28. Morris LF, Waxman AD, Braunstein GD. The nonimpact of thyroid stunning: remnant ablation rates in 131I-scanned and nonscanned individuals. J Clin Endocrinol Metab 2001; 86:3501–3511.
29. Leger FA, Izembart M, Dagousset F, et al. Decreased uptake of therapeutic doses of iodine-131 after 185-MBq iodine-131 diagnostic imaging for thyroid remnants in differentiated thyroid carcinoma. Eur J Nucl Med 1998; 25:242–246.
30. Huie D, Medvedec M, Dodig D, et al. Radioiodine uptake in thyroid cancer patients after diagnostic application of low-dose 131I. Nucl Med Commun 1996; 17:839–842.
31. Sabri O, Zimny M, Schreckenberger M, Meyer-Oelmann A, Reinartz P, Buell U. Does thyroid stunning exist? A model with benign thyroid disease. Eur J Nucl Med 2000; 27: 1591–1597.
32. Haugen BR, Lin EC. Isotope imaging for metastatic thyroid cancer. Endocrinol Metab Clin North Am 2001; 30:469–492.
33. Mandel SJ, Shankar LK, Benard F, Yamamoto A, Alavi A. Superiority of iodine-123 compared with iodine-131 scanning for thyroid remnants in patients with differentiated thyroid cancer. Clin Nucl Med 2001; 26:6–9.
34. Park H-M, Park Y-H, Zhou X-H. Detection of thyroid remnant/metastasis without stunning: an ongoing dilemma. Thyroid 1997; 7:277–280.
35. Yaakob W, Gordon L, Spicer KM, Nitke SJ. The usefulness of iodine-123 whole-body scans in evaluating thyroid carcinoma and metastases. J Nucl Med Technol 1999; 27: 279–281.
36. Ramanna L, Waxman A, Braunstein G. Thallium-201 scintigraphy in differentiated thyroid cancer: comparison with radioiodine scintigraphy and serum thyroglobulin determinations. J Nucl Med 1991; 32:441–445.
37. Grunwald F, Kalicke T, Feine U, et al. Fluorine-18 fluorodeoxyglucose positron emission tomography in thyroid cancer: results of a multicentre study. Eur J Nucl Med 1999; 26:1547–1552.

38. Balon HR, Fink-Bennet TD, Stoffer SS. Technetium-99m-sestamibi uptake by recurrent Hürthle cell carcinoma of the thyroid. J Nucl Med 1992; 33:1393–1395.

39. Zimmer LA, McCook B, Meltzer C, et al. Combined positron emission tomography/ computed tomography imaging of recurrent thyroid cancer. Otolaryngol Head Neck Surg 2003; 128:178–184.

40. Hundahl SA, Cady B, Cunningham MP, et al. Initial results from a prospective cohort study of 5583 cases of thyroid carcinoma treated in the US during 1996. Cancer 2000; 8:1012–1021.

41. Khoo ML, Freeman JL. Transcervical superior mediastinal lymphadenectomy in the management of papillary thyroid carcinoma. Head Neck 2003; 25:10–14.

42. McWilliams RR, Giannini C, Hay ID, et al. Management of brain metastases from thyroid carcinoma: a study of 16 pathologically confirmed cases over 25 years. Cancer 2003; 98:356–362.

43. Pak H, Gourgiotis L, Chang W, et al. Role of metastasectomy in the management of thyroid carcinoma: The NIH experience. J Surg Oncol 2003; 82:10–18.

44. Stojadinovic A, Shoup M, Ghossein RA, et al. The role of operations for distantly metastatic well-differentiated thyroid carcinoma. Surgery 2002; 131:636–643.

45. Eustatia-Rutten CF, Romijn JA, Guijt MJ, et al. Outcome of palliative embolization of bone metastases in differentiated thyroid carcinoma. J Clin Endocrinol Metab 2003; 88:3184–3189.

46. Tharp K, Israel O, Hausmann J, et al. Impact of I-131 SPECT/CT images obtained with an integrated system in the follow-up of patients with thyroid carcinoma. Eur J Nucl Med Mol Imaging 2004; 31(10):1435–1442.

47. Yamamoto Y, Nishiyama Y, Monden T, et al. Clinical usefulness of fusion of I-131 SPECT and CT images in patients with differentiated thyroid carcinoma. J Nucl Med 2003; 44:1905–1910.

48. Perault C, Schvartz C, Wampach H, Liehn JC, Delisle MJ. Thoracic and abdominal SPECT-CT image fusion without external markers in endocrine carcinomas. J Nucl Med 1997; 38(8):1234–1242.

49. Scott AM, Macapinlac H, Zhang J, et al. Image registration of SPECT and CT images using an external fiduciary three-dimensional surface fitting in metastatic thyroid cancer. J Nucl Med 1995; 36:100–103.

50. Schillaci O, Danieli R, Manni C, Simonetti G. Is SPECT/CT with a hybrid camera useful to improve scintigraphic imaging interpretation? Nucl Med Commun 2004; 25(7): 705–710.

51. Kienast O, Hofmann M, Ozer S, et al. Retention of iodine-131 in respiratory tract in a patient with papillary thyroid carcinoma after radionuclide therapy: a rare false-positive finding. Thyroid 2003; 13:509–510.

52. Ruf J, Lehmkuhl L, Bertram H, et al. Impact of SPECT and integrated low-dose CT after radioiodine therapy on the management of patients with thyroid carcinoma. Nucl Med Commun 2004; 25:1177–1182.

53. Kienast O, Dobrozemsky G, Ozer S, et al. Clinical impact of image fusion by means of XCT/SPECT: Vienna experience with 111In-labeled somatostatin analogues, 123I-MIBG and 131I-WBS. Eur J Nucl Med 2003.

54. Weckesser M, Kies P, Franzius A, Brunegraf A, Schober O. Iodine -131 SPECT/CT in patients with differentiated thyroid cancer after remnant ablation. J Nucl Med 2004; 45(suppl):349P.

55. Lehmkuhl L. Bertram H, Sandrock D, Amthauer H, Munz DL, Felix R. Value of SPECT and integrated low-dose CT in patients with thyroid carcinoma after radioiodine therapy. Eur J Nucl Med Mol Imaging 2004; 31 (suppl 2):S356.

56. Costa GLM, Albuquerke A, Vieira F, Rovira E, Lima J. Value of SPECT/CT image fusion in the assessment of endocrine tumors with 131I scintigraphy. Eur J Nucl Med Mol Imaging 2004; 31(suppl 2):S357.

57. Krausz Y, Klein M, Uziely B, et al. Impact of SPECT/CT on assessment of I131-avid sites in differentiated thyroid cancer. J Nucl Med 2004; 31(suppl 2):S357.

58. Ball DW, Baylin SB, de Bustros AC. Medullary thyroid carcinoma. In: Braverman LE, Utiger RD, eds. Werner & Ingbar's The Thyroid: A Fundamental and Clinical Text. 8th ed. Philadelphia: Lippincott Williams & Wilkins, 2000:930–943.

59. Gagel RF, Robinson MF, Donovan DT, Alford BR. Medullary thyroid carcinoma: recent progress. J Clin Endocrinol Metab 1993; 76:809–814.

60. van Heerden JA, Grant CS, Gharib H, et al. Long term course of patients with persistent hypercalcitoninemia after apparent curative primary surgery for medullary thyroid carcinoma. Ann Surg 1990; 212:395–400.

61. Ellenhorn JDI, Shah JP, Brennan MF. Impact of therapeutic regional lymph node dissection for medullary carcinoma of the thyroid gland. Surgery 1993; 114:1078.

62. Saad MF, Guido JJ, Samaan NA. Radioacive iodine in the treatment of medullary carcinoma of the thyroid. J Clin Endocrinol Metab 1983; 57:124–128.

63. Norton JA, Doppman JL, Brennan MF. Localization and resection of clinically inapparent carcinoma of the thyroid. Surgery 1980; 87:616–622.

64. Medina-Franco H, Herrera MF, Lopez G, et al. Persistent hypercalcitoninemia in patients with medullary thyroid cancer: a therapeutic approach based on selective venous sampling for calcitonin. Rev Invest Clin 2001; 53:212–217.

65. Dorr U, Wurstlin S, Frank-Raue K, et al. Somatostatin receptor scintigraphy and magnetic resonance imaging in recurrent medullary thyroid carcinoma: a comparative study. Horm Metab Res 1993; 27(suppl):48–55.

66. Wang Q, Takashima S, Fukuda H, Takayama F, Kobayashi S, Sone S. Detection of medullary thyroid carcinoma and regional lymph node metastases by magnetic resonance imaging. Arch Otolaryngol Head Neck Surg 1999; 125:842–848.

67. Arslan N, Ilgan S, Yuksel D, et al. Comparison of In-111 octreotide and Tc-99m (V) DMSA scintigraphy in the detection of medullary thyroid tumor foci in patients with elevated levels of tumor markers after surgery. Clin Nucl Med 2001; 26:683–688.

68. Hoefnagel CA, Delprat CC, Marcuse HR, de Vijlder JJM. Role of thallium-201 totalbody scintigraphy in follow-up of thyroid carcinoma. J Nucl Med 1986; 27:1854–1857.

69. Talpos GB, Jackson CE, Froelich JW, Kambouris AA, Block MA, Tashnian AH Jr. Localization of residual medullary thyroid cancer by thallium/technetium scintigraphy. Surgery 1985; 98:1189–1196.

70. Conti FS, Durski JM, Bacqai F, Grafton ST, Singer PA. Imaging of locally recurrent and metastatic thyroid cancer with positron emission tomography. Thyroid 1999; 9:797–804.

71. Gasparoni P, Rubello D, Ferlin G. Potential role of fluorine-18-deoxyglucose (FDG) positron emission tomography (PET) in the staging of primitive and recurrent medullary thyroid carcinoma. J Endocrinol Invest 1997; 20:527–530.

72. Brandt-Mainz K, Muller SP, Gorges R, Saller B, Bockisch A. The value of fluorine-18 fluorodeoxyglucose PET in patients with medullary thyroid cancer. Eur J Nucl Med 2000; 27:490–496.

73. Esik O, Szavcsur P, Szakall S Jr, et al. Angiography effectively supports the diagnosis of hepatic metastases in medullary thyroid carcinoma. Cancer 2001; 91:2084–2095.

74. Kwekkeboom DJ, Reubi JC, Lamberts SW, et al. The potential value of somatostatin receptor scintigraphy in medullary thyroid carcinoma. J Clin Endocrinol Metab 1993; 76:1413–1417.

75. Troncone L, Rufini V, Montemaggi P, Danza FM, Lasorella A, Mastrangelo R. The diagnostic and therapeutic utility of radioiodinated metaiodobenzylguanidine (MIBG): 5 years of experience. Eur J Nucl Med 1990; 16:325–335.

76. Clarke SE. [131I]metaiodobenzylguanidine therapy in medullary thyroid cancer: guy's hospital experience. J Nucl Biol Med 1991; 35:323–326.

77. Shibata T, Noma S, Nakano Y, Konishi J. Primary thyroid lymphoma: MR appearance. J Comp Assist Tomogr 1991; 15:629–633.

78. Takashima S, Morimoto S, Ikezoe J, et al. Primary thyroid lymphoma: comparison of CT and US assessment. Radiology 1989; 171:439–443.

79. Scott AM, Kostakoglu L, O'Brien JP, Straus DJ, Abdel-Dayem HM, Larson SM. Comparison of Technetium-99m-MIBI and thallium-201-chloride uptake in primary thyroid lymphoma. J Nucl Med 1991; 33:1396–1398.

80. O, Miyakawa M, Shirota H, et al. Comparison of Tl-201 and Ga-67-citrate scintigraphy in the diagnosis of thyroid tumor [Concise communication]. J Nucl Med 1982; 23:225–228.

81. Nel CJC, van Heerden JA, Goellner JR, et al. Anaplastic carcinoma of the thyroid: a clinicopathologic study of 82 cases. Mayo Clin Proc 1985; 60:51–58.

82. Iida Y, Hidaka A, Hatabu H, Kasagi, Konishi J. Follow-up study of postoperative patients with thyroid cancer by thallium-201 scintigraphy and serum thyroglobulin measurement. J Nucl Med 1991; 32:2098–2100.

83. Montes TC, Munoz C, Rivero JI, Mota JA, Pustilnik N, Garcia F. Uptake of Tc-99m sestamibi and Tc-99m MDP in anaplastic carcinoma of the thyroid (nondiagnostic CT and ultrasound scans). Clin Nucl Med 1999; 24:355–356.

84. Nilsson O, Lideberg J, Zedenius J, et al. Anaplastic giant cell carcinoma of the thyroid gland: treatment and survival over a 25-year period. World J Surg 1998; 22:725–730.

85. Pfannenberg AC, Eschmann SM, Horger M, et al. Benefit of anatomical–functional image fusion in the diagnostic work-up of neuroendocrine neoplasms. Eur J Nucl Med Mol Imaging 2003; 30:835–843.

4

^{123}Iodine MIBG SPECT/CT for Tumor Imaging

Carl-Martin Kirsch
Department of Nuclear Medicine, Saarland University Medical Center, Homburg/Saar, Germany

Ora Israel
Rambam Health Care Campus, B. Rapaport School of Medicine, Technion—Israel Institute of Technology, Haifa, Israel

INTRODUCTION

The theme of the "Bauhaus" movement, "form follows function," may also apply to medical imaging. In many instances, organ dysfunction precedes anatomical changes. Imaging methods aimed primarily at anatomy such as ultrasound, computed tomography (CT), and magnetic resonance imaging (MRI) display morphological changes with high spatial and contrast resolution. Nuclear medicine imaging, targeted at the assessment of organ and tissue function, provides early information about pathophysiological changes, albeit with reduced spatial resolution.

The greater the disease-specificity of a tracer uptake, the lesser the nonspecific uptake found in surrounding tissues. As a consequence, there is often a paucity of background anatomic information, rendering the accurate localization of suspicious or abnormal foci of increased activity, and the assessment of their precise characteristics, difficult. Such is the case with imaging using metaiodobenzylguanidine (MIBG) as a radiotracer labeled with either ^{131}I or ^{123}I.

RADIOLABELED MIBG

MIBG was synthesized and introduced into clinical practice in the early 1980s (1,2). The synthesis of MIBG evolved from efforts to develop antihypertensive drugs based upon the coupling of the adrenergic neuron blockers bretylium and guanethidine to synthesize benzyl-guanethidine derivatives. The benzyl group in the new compound provided a mechanism to introduce radioactive halogens, particularly the radioactive isotopes ^{131}I or ^{123}I. Radioiodinated MIBG is taken up by chromaffin catecholamine storage granules because of its structural similarity to norepinephrine (Fig. 1).

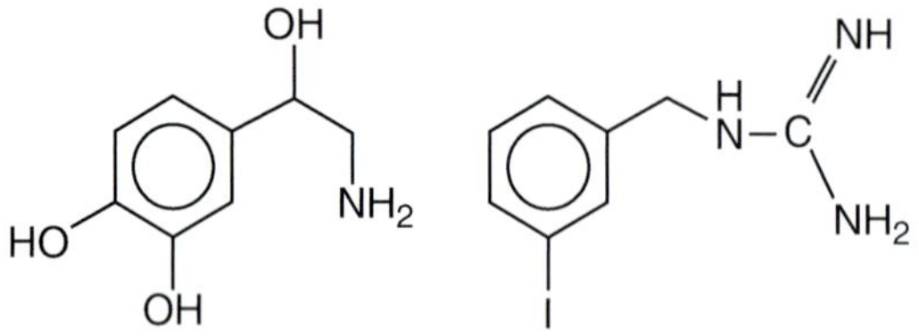

Figure 1 Chemical structure of norepinephrine (*left*) and MIBG (*right*). *Abbreviation*: MIBG, metaiodobenzylguanidine.

As an analogue of norepinephrine, MIBG is taken up by adrenergic nerve endings primarily via the sodium dependent uptake-1 pathway. MIBG is transported into storage vesicles that are found in all organs with adrenergic innervation, as well as in tumors arising from adrenergic tissue, such as pheochromocytoma, neuroblastoma, and paraganglioma. The intravesicular MIBG uptake leads to a temporary accumulation of the radiolabeled tracer since, unlike norepinephrine, MIBG is not subject to further metabolism by monaminoxidase and catechol-methyl-transferase. This mechanism allows for scintigraphy of the tracer accumulated in normal organs and disease processes.

[123]I AND ^{131}I MIBG SCINTIGRAPHY

As a radiotracer, MIBG has been used initially as ^{131}I-MIBG, but more recently, ^{123}I-MIBG has become the primary agent for diagnostic imaging. Physiologic uptake of radiolabeled MIBG is seen, as a rule, in the normal salivary glands, thyroid, lungs, heart, liver, and bowel, as well as in its excretory pathways in the kidneys and urinary bladder. Faint physiologic uptake may also be observed in normal functioning adrenal glands.

Unbound extraneural MIBG is excreted through the kidneys by glomerular filtration. Approximately 50% of the injected radioactivity is cleared within the first 24 hours. A relatively constant uptake level of the radiopharmaceutical in organs with pronounced adrenergic innervations like the heart or the above-mentioned tumors is reached within four to six hours after injection. Delayed imaging at 24 hours postinjection may benefit from whole-body clearance of the unbound tracer, and may further improve the detectability rate of the specific tumor uptake.

A number of drugs can interfere with the accumulation of MIBG (3–5). These include reuptake inhibitors such as tricyclic antidepressants and cocaine (which inhibit the synaptic uptake-1 mechanism), reserpine (inhibiting the active transport into the vesicles), MAO-inhibitors (which compete with MIBG uptake into vesicles), sympathomimetics such as amphetamine, and to a lesser extent, reserpine (by increased depletion of the vesicle content of both norepinephrine and MIBG) and calcium antagonists (by affecting the neural Ca^{2+} channels).

Of the two radioiodine isotopes that have been used to label MIBG, [123]Iodine is at present the principal tracer for diagnostic purposes. [131]Iodine labeling of MIBG for diagnostic applications is at large outdated, but this radiopharmaceutical is used to treat metastatic malignant pheochromocytoma and other tumors arising from the adrenergic tissue, having as a prerequisite demonstrated tracer uptake on a diagnostic study using ^{123}I MIBG.

(B)

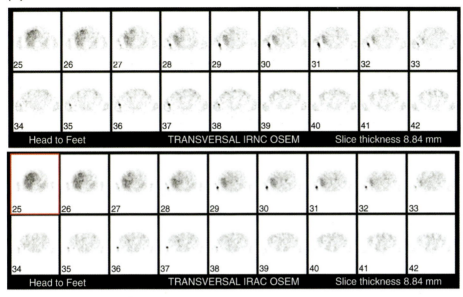

(C)

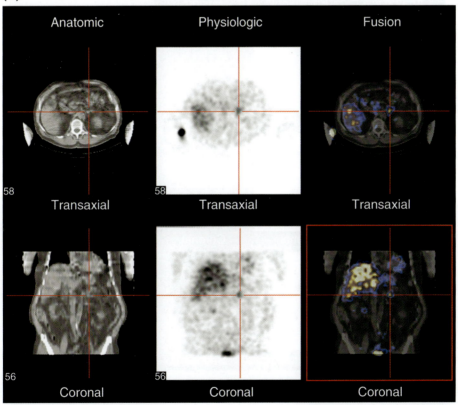

Figure 3 (*Caption on facing page*)

(A)

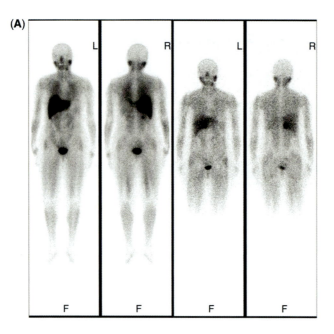

(B)

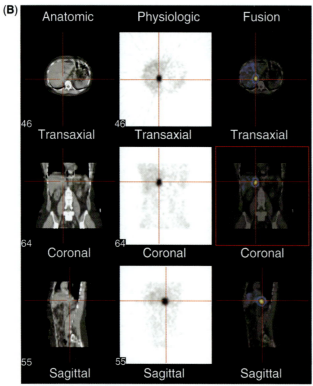

Figure 4   (*Caption on facing page*)

(B)

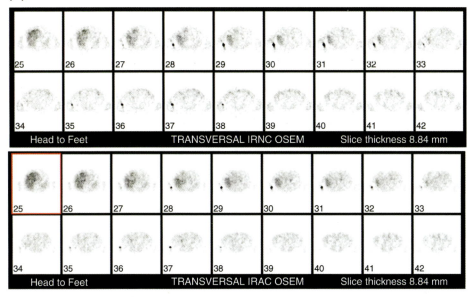

(C)

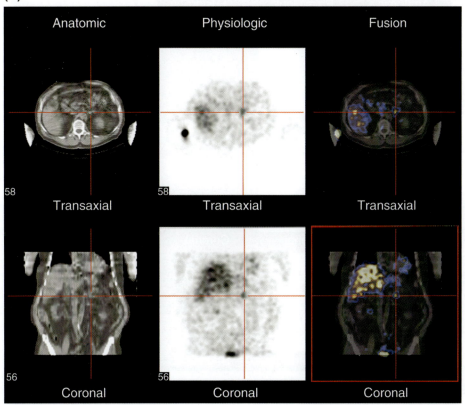

Figure 3 (*Caption on facing page*)

(A)

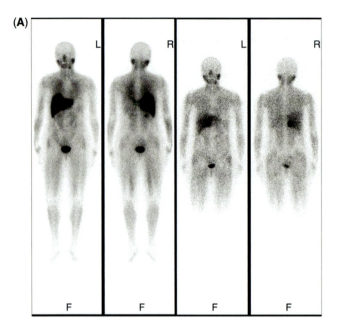

(B)

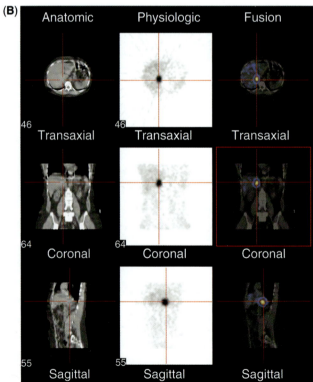

Figure 4   (*Caption on facing page*)

there have been attempts to coregister nuclear medicine images with conventional imaging modalities such as CT and/or MRI (18,20,21). Particular care for precise coregistration has to be taken, mainly in the thorax and abdomen, due to the displacement of thoracic and abdominal structures with time. This has limited the use of these coregistration methods in the torso. In addition, coregistration of separately performed functional and structural studies requires cumbersome technique with fiducial markers, and are processed using complicated mathematical algorithms. These "software" solutions have been used mainly for research applications and have been less successful in routine clinical applications.

MIBG SPECT/CT

Most of the limitations of coregistration have been overcome by the development of the hybrid imaging technology that combines CT and SPECT acquisition in a single device (22–24). The addition of CT transmission images accomplishes two tasks. It provides an attenuation map to correct for tissue absorption and scatter of photons emitted from foci of increased tracer accumulation. In addition, it provides a cross-sectional image to assist in precise localization of sites showing increased MIBG uptake. SPECT/CT appears to be of main value for assessment of scintigraphy performed with tracers such as ^{123}I-MIBG, an agent that provides images with a high target-to-nontarget ratio. The high tumor-specific uptake of radiolabeled MIBG is responsible for the low tracer activity in surrounding tissues.

The availability of X-ray attenuation-corrected images results in a different distribution pattern of the radiotracer. The analysis of ^{123}I-MIBG SPECT/CT studies has to consider a few previously unknown scintigraphic patterns related to the use of hybrid imaging. Assessment of both non-attenuated and attenuation-corrected image sets may be of value in selected cases.

SPECT/CT may be of value both in defining the pathological nature of a suspicious focus, as well as excluding the presence of disease. Physiological tracer uptake in sites of the normal MIBG biodistribution may be of low intensity but are nevertheless difficult to differentiate from pathology related to early, small tumor-load stages of disease, or to residual viable tumor after treatment. Precise SPECT/CT localization of a low-intensity uptake focus in areas with congruent structural disease-related changes may be of help for the final characterization of equivocal scintigraphic MIBG findings (Fig. 3). It may also allow the detection of previously unexpected disease-related lesions and therefore precisely define the whole extent of disease (Figs. 5–7).

Figure 4   (*Facing page*) Fifty-year-old woman was referred for further assessment following several hypertensive crises, elevated urinary catecholamine metabolites, and a suspicion of a right-sided pheochromocytoma on CT. (**A**) ^{123}I-MIBG whole-body scintigraphy performed at 6 hours after injection (*left*) shows increased tracer uptake in the right lumbar region, obscured in part by physiologic activity in the liver. Scintigraphy performed at 24 hours post-injection (*right*) shows clearance of nonspecific activity and improved visualization of the suspected lesion in the right lumbar region. (**B**) SPECT/CT indicates the presence of a right adrenal pheochromocytoma, clearly delineated from the surrounding anatomical structures. *Abbreviations*: CT, computed tomography; MIBG, metaiodobenzylguanidine; SPECT, single-photon emission computed tomography.

(A)

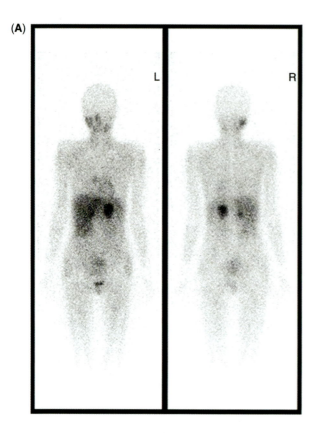

(B)

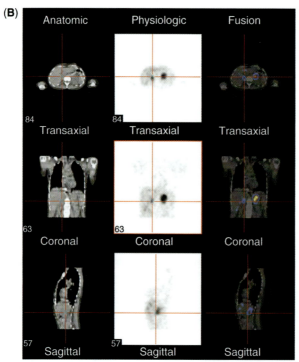

Figure 5 (*Caption on facing page*)

(C)

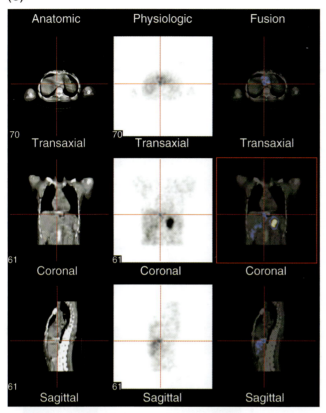

Figure 5 Thirty four-year-old woman after thyroidectomy for medullary carcinoma and genetic analysis and diagnosis of a MEN II syndrome with bilateral pheochromocytoma was referred for further assessment of rising serum calcitonin. (**A**) (*Facing page*) ^{123}I-MIBG whole-body scintigraphy shows markedly increased tracer uptake in the left adrenal and an additional low-intensity lesion in the right lumbar region. (**B**) (*Facing page*) SPECT/CT confirmed the presence of lesions in both adrenal glands. (**C**) Reveals the presence of previously unknown liver metastases. The diagnosis of bilateral pheochromocytoma and medullary thyroid carcinoma metastatic to the liver was confirmed at surgery. *Abbreviations*: CT, computed tomography; MIBG, metaiodobenzylguanidine; SPECT, single-photon emission computed tomography.

SPECT/CT may also be of value in delineating and separating foci of disease with increased ^{123}I-MIBG uptake located in close proximity to sites of physiologic tracer activity in normal tissues (Fig. 4). On SPECT alone, sites of physiological activity may mask lesions showing increased MIBG uptake that are of clinical significance.

Even-Sapir et al. described the first clinical use of MIBG SPECT/CT in three patients, being part of a larger patient population with endocrine neoplasms (25). Fused images were of value in excluding a clinically suspected pheochromocytoma in one patient, by precisely localizing a suspicious focus of increased tracer uptake in the lumbar region to physiologic excretion of MIBG in the ureter. The role of ^{123}I-MIBG was further assessed in 11 patients, included in a larger study population of neuroendocrine tumors (26). Although this later study does not provide a separate analysis of MIBG SPECT/CT studies, it shows that hybrid imaging had the highest accuracy (99%) in classifying malignant lesions in this patient population. The specificity of SPECT/CT was also significantly higher as compared to SPECT or CT as stand-alone modalities. A change in clinical management was performed in 28% of patients based on the results of SPECT/CT imaging. This included sparing unnecessary invasive diagnostic procedures (as well as unnecessary surgery), changes in the planned surgical approach, and providing

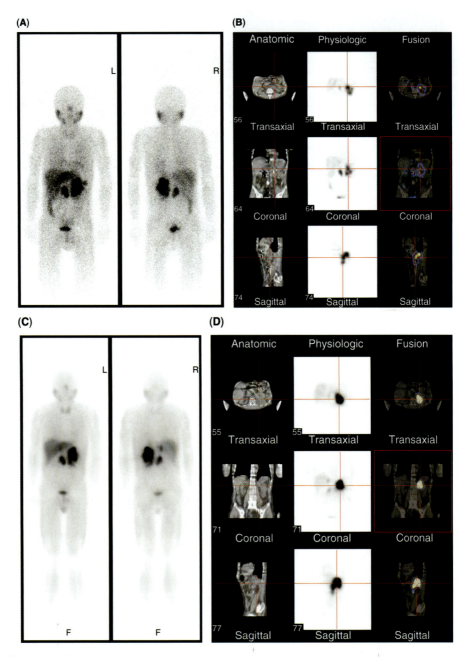

Figure 6   (*Caption on facing page*)

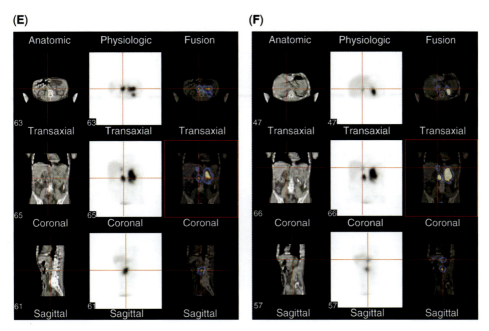

Figure 6 Sixty-year-old male patient, one year after an incompletely resected recurrent pheochromocytoma, was referred for assessment of residual active disease. (**A**) ^{123}I-MIBG wholebody scintigraphy at 24 hours after injection shows significant MIBG uptake in the residual unresected tumor. (**B**) SPECT/CT shows the whole extent and precise anatomic location of the mass. The clinical team decided to treat the patient with 5.5 GBq ^{131}I-MIBG. (**C**) Whole-body scintigraphy performed at six days after the administration of the therapeutic dose shows extensive accumulation of ^{131}I-MIBG in residual viable tumor masses in the upper abdomen. (**D**) SPECT/CT localizes this tumor to the left adrenal bed, extending dorsally to the paravertebral region, (**E**) to a metastatic right paraaortic lymph node, and (**F**) to an additional mass in the right paravertebral region. *Note*: The higher target-to-nontarget ratio of the posttherapy scan may allow for the identification of additional sites of disease, precisely localized by SPECT/CT. *Abbreviations*: CT, computed tomography; MIBG, metaiodobenzylguanidine; SPECT, single-photon emission computed tomography.

previously unconsidered medical and radiopeptide treatment. In a preliminary report on MIBG SPECT/CT performed on 45 patients, hybrid imaging changed the further management in all patients, due to a better interpretation of scintigraphic findings (27). The authors found that in addition to the objective measure of performance of SPECT/CT, providing fused images to the referring physicians resulted in a higher acceptance and understanding of functional nuclear medicine data in the medical community.

A report on the use of SPECT/CT in 31 patients undergoing ^{123}I-MIBG scintigraphy showed that in 74% of the study population, fused images excluded the presence of malignancy in sites representing pathways of physiologic biodistribution of the tracer, including the gastrointestinal tract and the renal pelvis (28). Furthermore, in 6% of patients, SPECT/CT allowed for the precise characterization of equivocal scintigraphic findings as adrenal lesions. Hybrid imaging enabled the differentiation of primary tumors and metastatic sites. An additional preliminary report in 10 patients (including 7 children) assessed the contribution of fused images using

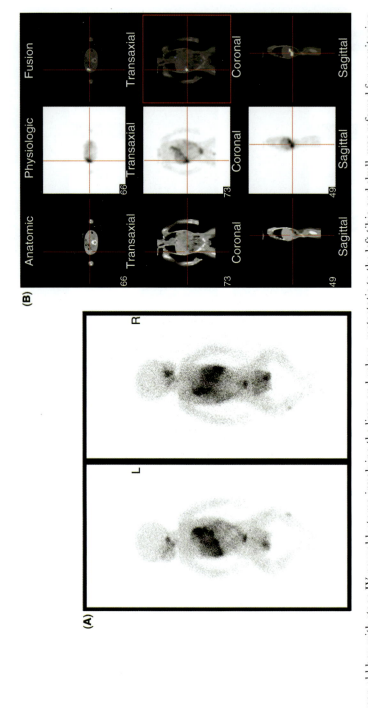

Figure 7 A one-year-old boy with stage IV neuroblastoma involving the liver and spleen, metastatic to the left tibia and skull, was referred for monitoring response to treatment and assessment of residual viable tumor after six cycles of chemotherapy and surgical removal of the bone lesions. (**A**) ^{123}I-MIBG whole-body scintigraphy shows increased uptake in the metastasis in the left tibia and in residual viable primary tumor in the right abdomen. (**B**) SPECT/CT identifies an additional liver metastasis in the caudal portion of the right hepatic lobe, previously missed on assessment of planar and SPECT images. *Abbreviations*: CT, computed tomography; MIBG, metaiodobenzylguanidine; SPECT, single-photon emission computed tomography.

low-dose CT as a bridge between scintigraphic findings and diagnostic conventional imaging (29). The authors conclude that in cases of equivocal imaging studies SPECT/CT may be used as a guiding tool, leading to an increase in the diagnostic accuracy and interpretation certainty in 57% of patients.

SPECT/CT moves nuclear medicine procedures closer to one of the ultimate goals of functional imaging: namely, the quantitative measurement of radionuclide concentration. This appears to be of particular significance for MIBG, which, as stated above, when labeled with ^{131}I, is being used for treatment of advanced stage neuroendocrine tumors and neuroblastoma. The ability to precisely determine the percent of injected dose of MIBG taken up by the tumor and the effective half-life of the drug, using tracer doses of either the ^{123}I or ^{131}I labeled agent, may allow for better treatment-tailoring in the individual patient, and may potentially lead to higher chances for response to this therapeutic modality.

CONCLUSION

Based upon the data of initial research available at present, it is not premature to predict that in the future all or at least the majority of MIBG tumor imaging procedures will include SPECT/CT fusion images (30). The advantages of anatomical mapping will prevail over the additional time needed and additional radiation delivered for the transmission study. The next generation of SPECT/CT devices, equipped with a diagnostic quality CT component, will reduce the acquisition time of the combined study. The addition of diagnostic CT procedures to the routine study protocols should result in ^{123}I-MIBG SPECT/CT studies becoming the procedure of choice for the evaluation of patients with a clinical suspicion of adrenergic tissue ^{123}I-MIBG-avid tumors, as well as for more advanced clinical applications such as quantitative dosimetry studies preceding therapeutic administration of ^{131}I-MIBG or monitoring response to treatment of patients with tumors of the neural crest.

The hybrid SPECT/CT technology provides a long-desired tool for oncologic nuclear medicine imaging: technology that combines high-contrast functional imaging with anatomic mapping of the physiologic information. Since SPECT/CT imaging also provides information that allows for cross-sectional pixel-by-pixel attenuation correction, it has the potential to address the need for absolute radionuclide quantification of special significance for radiation-absorbed dose estimates before and following ^{131}I-MIBG therapy, as well as for the quantitative assessment of therapeutic response.

REFERENCES

1. Wieland DM, Swanson DP, Brown LE, Beierwaltes WH. Imaging the renal medulla with an I-131 labelled antiadrenergic agent. J Nucl Med 1979; 20:155–158.
2. Wieland DM, Wu J, Brown LE, Manger TJ, Swanson DP, Beierwaltes WH. Radiolabeled adrenergic neuron-blocking agents: adrenomedullary imaging with [131I] iodobenzylguanidine. J Nucl Med 1980; 21:349–353.
3. Courbon F, Brefel-Courbon C, Thalamas C, et al. Cardiac MIBG scintigraphy is a sensitive tool for detecting cardiac sympathetic denervation in Parkinson's disease. Mov Disord 2003; 18:890–897.
4. Hmada K, Hirayama M, Watanabe H, et al. Onset age and severity of motor impairment are associated with reduction of myocardial 123-I-MIBG uptake in Parkinson's disease. J Neurol Neurosurg Psychiat 2003; 74:423–426.

5. Solanki KK, Bomanji J, Moyes J, Mather SJ, Trainer PJ, Britton KE. A pharmacologic guide to medicines which interfere with the biodistribution or radiolabeled meta-iodobenzylguanidine (MIBG). Nucl Med Commun 1992; 13:513–521.

6. 131-I/123-I-metaiodobenzylguanidine (MIBG) scintigraphy. Procedure guidelines for tumour imaging. Eur J Nucl Med 2003; 30:B132 (www.eanm.org/guidelines/).

7. Piepsz A, Hahn K, Roca I, et al. A radiopharmaceutical schedule for imaging in paediatrics. EANM Paediatric Task Group. Eur J Nucl Med 1990; 17:127–129.

8. Datz FL. Gamuts in Nuclear Medicine. 3rd ed. Mosby-Year Book Inc., 1995:39–45.

9. O'Connor MK, ed. The Mayo Clinic Manual of Nuclear Medicine. Mayo Foundation, 1996:251–256.

10. Kline RC, Swanson DP, Wieland DM, et al. Myocardial imaging in man with I-123 meta-iodobenzylguanidine. J Nucl Med 1981; 22:129–132.

11. Hattori N, Schwaiger M. Metaiodobenzylguanidine scintigraphy of the heart: what have we learnt clinically? Eur J Nucl Med 2000; 27:1–6.

12. Schnell O, Kirsch CM, Stemplinger J, Haslbeck M, Standl E. Scintigraphic evidence for cardiac sympathetic dysinnervation in long-term IDDM patients with and without ECG-based autonomic neuropathy. Diabetologia 1995; 38:1345–1352.

13. Samnick S, Bader JB, Müller M, et al. Improved labelling of no-carrier-added 123J-MIBG and preliminary clinical evaluation in patients with ventricular arrhythmias. Nucl Med Commun 1999; 20:537–545.

14. Fujimoto S, Inoue A, Hisatake S, et al. Usefulness of (123) I-metaiodobenzylguanidine myocardial scintigraphy for predicting the effectiveness of beta-blockers in patients with dilated cardiomyopathy from the standpoint of long-term prognosis. Eur J Nucl Med Mol Imaging 2004; 31:1356–1361.

15. Wieland DM, Brown LE, Tobes MC, et al. Imaging the primate adrenal medulla with I-123 and I-131 meta-iodo-benzylguanidine: concise communication. J Nucl Med 1981; 21:349–364.

16. Sisson JC, Frager MS, Valk TW, et al. Scintigraphic localisation of pheochromocytoma. N Engl J Med 1981; 305:12–17.

17. Krebs im Saarland 1998–2000. Published by: Epidemiologisches Krebsregister Saarland, Juli 2004, ISBN 3-980 8880-2-9.

18. Kirsch CM, Hellwig D, Schaefer A. Form follows function: molecular and anatomical information in conventional nuclear medicine. Der Nuklearmediziner 2003; 26:251–258.

19. Kirsch CM. Nuklearmedizin und die multimediale Zukunft [editorial]. Nuklearmedizin 1998; 37:1.

20. Pietrzyk U, Herholz K, Fink G, et al. An interactive technique for three-dimensional image registration: validation for PET, SPECT, MRI and CT brain studies. J Nucl Med 1994; 35:2008–2011.

21. Dey D, Slomka PJ, Hahn LJ, Kloiber R. Automatic three-dimensional multimodality registration using radionuclide transmission CT attenuation maps: a phantom study. J Nucl Med 1999; 40:448–455.

22. Beyer T, Townsend DW, Brun T, et al. A combined PET/CT scanner for clinical oncology. J Nucl Med 2000; 41:1369–1379.

23. Townsend DW, Cherry SR. Combining anatomy and function: the path to true image fusion. Eur Radiol 2001; 11:1968–1974.

24. Bar-Shalom R, Yefremov N, Guralnik L, et al. Clinical performance of PET/CT in evaluation of cancer: additional value of diagnostic imaging and patient management. J Nucl Med 2003; 44:1200–1209.

25. Even-Sapir E, Keidar Z, Sachs J, et al. The new technology of combined transmission and emission tomography in evaluation of endocrine neoplasms. J Nucl Med 2001; 42:998–1004.

26. Pfannenberg AC, Eschmann SM, Horger M, et al. Benefit of anatomical–functional image fusion in the diagnostic work-up of neuroendocrine neoplasms. Eur J Nucl Med 2003; 30:835–843.

27. Keinast O, Dobrozemsky G, Ozer S, et al. Clinical impact of image fusion by means of XCT/SPECT: Vienna experience with 111In-labeled somatostatin analogues, 123I-MIBG and 131I-WBS. Eur J Nucl Med 2003; 30:182.

28. Ozer S, Dobrozemsky G, Kienast O, et al. Value of combined XCT/SPECT technology for avoiding false positive planar 123I-MIBG scintigraphy. Nuklearmedizin 2004; 43:164–170.

29. Klein M, Kopelewitz B, Krausz Y, et al. Contribution of SPECT/CT to bridging between MIBG scan and diagnostic CT. J Nucl Med 2004; 45:1138.

30. Schillaci O. Functional–anatomical image fusion in neuroendocrine tumors. Cancer Biother Radiopharm 2004; 19:129–134.

5

^{111}Indium Octreotide SPECT/CT for Neuroendocrine Tumor Imaging

Stanley J. Goldsmith, Rosna M. Mirtcheva, and Lale Kostakoglu
New York-Presbyterian Hospital/New York Weill Cornell Medical Center, New York, New York, U.S.A.

INTRODUCTION

^{111}In-octreotide scintigraphy has been widely available to image somatostatin receptor positive neuroendocrine tumors for over a decade. Throughout this time, lesion detection and overall clinical accuracy has improved as imaging techniques have improved. Currently, 222 MBq (6 mCi) of ^{111}In-octreotide, also referred to as ^{111}In-DTPA-pentetreotide (OctreoScan®, Mallinckrodt, St. Louis, Missouri) is routinely used for clinical scintigraphy, whereas initial studies were performed with only 111 MBq (3 mCi). Single-photon emission computed tomography (SPECT) imaging using dual-head gamma cameras has become more widely available. Display quality and software options have further improved the quality of the images available for interpretation. This chapter reviews the evolution of ^{111}In-octreotide scintigraphy and the current value of SPECT/computed tomography (CT) fusion as a nuclear medicine imaging technique for the diagnosis and management of patients with somatostatin receptor positive neuroendocrine tumors.

SOMATOSTATIN AND SOMATOSTATIN RECEPTORS

In 1977, Guillemin and Schally were awarded a portion of the Nobel Prize in Medicine and Physiology for the identification of somatostatin and other hypothalamic regulatory peptides. Somatostatin is a 14-amino-acid peptide (with 12 of the amino acids arranged in a ring structure) that inhibits the secretion of growth hormone (or somatotropin; hence, *somato statin*) by the anterior pituitary. Somatostatin was also observed to have an effect on a number of processes throughout the body, including inhibition of smooth muscle contraction and hormonal secretion. These peripheral effects correlated with the expression of specific somatostatin receptors on tissues and cells throughout the body. In humans, five different subtypes of somatostatin receptors have been identified and cloned. Tumors that arose from neuroendocrine tissue had an increased expression of somatostatin receptors, most commonly receptor subtypes 2 and 5 (SSTR2; SSTR5). In order to evaluate the therapeutic potential of this inhibitory peptide, a variety of analogs that had high binding affinity to somatostatin receptors but a longer biologic effect were synthesized as they had greater resistance to the peptidases responsible for somatostatin's short biologic half-life.

Of these analogs, octreotide, an eight-amino-acid cyclic peptide, has been the most extensively studied. It has a biologic half-life measured in hours and is used as an injectable therapeutic agent to inhibit excess secretions from neuroendocrine tumors, particularly in patients with acromegaly and carcinoid syndrome. Octreotide therapy has also been shown to inhibit tumor growth. The availability of octreotide as a therapeutic agent has had a profound effect on the management of patients suffering from these diseases.

The potential use of octreotide as a carrier of a radionuclide for diagnostic imaging or targeting therapy was recognized early. By substituting a tyrosyl moiety in position 3 of the cyclic amino acid ring, the tyrosyl[3]-octreotide could be labeled with Iodine-123, producing an agent that demonstrated that scintigraphy of somatostatin receptor tumors was feasible (1). It was recognized, however, that I-123 was not a satisfactory choice as a radiolabel since it was expensive, short-lived, and required sophisticated personnel to perform labeling at individual clinical sites. This prompted the further development of a method to use the radiometal [111]Indium to be bound to the octreotide molecule, producing ^{111}In-DTPA-pentetreotide, in which the original octreotide eight-amino-acid molecule is covalently bound to DTPA that, in turn, serves to link the radiometal (2).

[111]IN-OCTREOTIDE

Several centers in Europe provided the initial reports of the clinical efficacy of ^{111}In-octreotide for the detection of neuroendocrine tumors (2–4). In 1993, Krenning et al. published the Rotterdam experience in over 1000 patients demonstrating the remarkable clinical utility of ^{111}In-octreotide (^{111}In-DTPA-pentetreotide) as an imaging agent to identify somatostatin receptor expression in tumors, both neuroendocrine and non-neuroendocrine, as well as in granulomas and autoimmune processes based on the accumulation of activated inflammatory cells that contained somatostatin receptors (5). At that time, they demonstrated in a large group of patients the high efficacy of this agent for in vivo scintigraphic detection of pituitary tumors, carcinoid, gastrinoma, islet cell tumors, medullary thyroid carcinoma, pheochromocytoma, and paraganglioma, as well as neuroblastoma, small cell lung carcinoma, Hodgkin's and non-Hodgkin's lymphoma, active sarcoid and tuberculous lesions, hyperfunctioning thyroid glands, and clinically active Graves' ophthalmopathy. Other reports from centers throughout Europe followed (6,7). In 1995 ^{111}In-octreotide (^{111}In-DTPA-pentetreotide) was approved by the Food and Drug Administration (FDA) and became available to nuclear medicine departments throughout the United States.

[111]IN-OCTREOTIDE PLANAR AND SPECT SCINTIGRAPHY

Since the introduction of ^{111}In-octreotide as a diagnostic imaging tool, there have been many reports that scintigraphy using this radiotracer not only identified neuroendocrine tumors, but also that somatostatin receptor scintigraphy (SRS) outperformed conventional imaging techniques for the detection of these difficult-to-detect tumors (5–9). In patients with abdominal carcinoid tumors and in patients with gastrinoma, the sensitivity for detection approached 100%. In 1996, Cardiot et al. prospectively compared ^{111}In-octreotide scintigraphy, conventional imaging techniques, and endoscopic ultrasound to the findings at surgery in 21 consecutive patients with Zollinger-Ellison syndrome (8). ^{111}In-octreotide scintigraphy was the only positive preoperative technique in 32% of the patients. Using both planar spot and SPECT images, but administering only an average of 135 MBq of ^{111}In-octreotide, these investigators identified foci of abnormal tracer uptake in 11 of 19 patients in whom duodenal gastrinoma was found at surgery, none of them located in the pancreas. Used in conjunction with conventional imaging techniques, overall sensitivity for tumor localization was 90%.

Gibril et al. performed a similar study in 80 consecutive patients with Zollinger-Ellison syndrome (9). SRS using ^{111}In-octreotide identified liver metastases in 70% of the patients compared with only 45% by magnetic resonance imaging (MRI), 38% by CT, and 19% by ultrasound. Primary tumors were identified by scintigraphy alone in 58% of the cases. Although suboptimal, this number outperformed ultrasound (9%), CT (31%), MRI (30%), and angiography (28%). These investigators concluded that ^{111}In-octreotide scintigraphy was the single most sensitive preoperative technique for the detection of either primary or metastatic gastrinomas and recommended that "because of its simplicity, it should be the first imaging method used in these patients," even though it was not 100% accurate in their series. The authors further recommended that other imaging techniques should be employed when ^{111}In-octreotide scintigraphy is negative in patients in whom there is other evidence, such as elevated serum markers, for the presence of active disease (9).

Despite these impressive results, it is appropriate to consider the range of sensitivity of the SRS reported in various series. While carcinoid tumors and gastrinomas, for example, are frequently detected, the overall sensitivity varies amongst series from 58% to 90% (5–7). In small series of patients with insulin-producing tumors of the pancreatic islet cells, only 50% of the tumors were detected. Variable expression of somatostatin receptor subtypes, as well as tumor size, were among the main clinical factors determining this variability. Technical factors including differences in the injected radionuclide dose, the instrumentation used for imaging, acquisition protocols, and image display characteristics, as well as differences in observer interpretation skills, also contributed to the variations in sensitivity, specificity, and overall accuracy of SRS in the assessment of neuroendocrine tumors.

In the early European reports, planar and SPECT images were obtained at 4, 24, and occasionally 48 hours after the intravenous injection of 111–222 MBq (3–6 mCi) of ^{111}In-octreotide, using a large field of view gamma camera with a medium energy collimator, and dual energy windows for the photon peaks of [111]In (5–8). In all likelihood, only 111 MBq of ^{111}In-octreotide were used initially in these series, partly because of the high cost of the radiopharmaceutical. By 1994, however, clinical trails in the United States were routinely using 222 MBq of ^{111}In-octreotide. This was also the dose recommended when ^{111}In-octreotide was approved by the FDA (9–12).

For imaging of the region of the head and neck, it was suggested that at 24 hours, 300,000 counts, or up to 15 minutes, acquisition should be performed. Because of the nonspecific uptake of the tracer in the liver, spleen, and kidneys, a higher number of counts (500,000) or a longer acquisition were recommended for imaging of the torso and mainly of the abdomen. In addition, for the same reason, acquisition of separate views of the chest and abdomen were recommended, in order to decrease the potential false-negative results related to small but potentially clinically significant lesions in the field of view. This is illustrated in a recently published review, which demonstrates the variable scintigraphic appearance of thyroid gland activity (13). When the liver activity was included in the imaged field of view, it accounted, as expected, for a great fraction of the total counts, and the thyroid was barely visible, whereas in an acquisition dedicated to the area of the head and neck in the same patient, a greater fraction of the total counts accumulated were derived from uptake in the thyroid gland. This observation can be extrapolated and become relevant to the scintigraphic detectability rate of small tumors, particularly in the abdomen where, in addition to the liver and spleen activity, there is also intense renal uptake due to tracer excretion in the urine. In addition to obscuring tumor sites located in close vicinity to the kidneys, count-based acquisition protocols may not include adequate time for detection of small but tracer-avid lesions. The degree of the detection rate of a tumor is dependent, therefore, on the amount of counts acquired, which is in turn dependent upon the administered dose, the fractional perfusion, and uptake, and also upon the imaging system sensitivity.

The increasingly available whole-body scintigraphy, while convenient to perform and providing aesthetically pleasing studies, contains too little information when performed at a scan speed of 6–8 cm/min in order to limit the patient's total time on the imaging table. Experts in the field have continuously underscored that strict adherence to technique, with the acquisition of sufficient counts in each region of the body, is essential for optimal scintigraphic results (12,14). A better image is acquired by slowing scan speed, even if total patient imaging time is maintained by limiting the total scan area, when this is consistent with the clinical indication.

To address the concerns about the effect of technique on proper and optimal SRS, the Society of Nuclear Medicine has published a "Procedure Guideline for Somatostatin Receptor Scintigraphy with ^{111}In-Pentetreotide" (12). The recommended administered ^{111}In-octreotide activity is 222 MBq (6 mCi) and 5 MBq/kg (0.14 mCi/kg) in children. Planar images should be acquired for 10 to 15 minutes per field of view, using either a 256×256 or a 512×512 word matrix. This finer matrix display is possible because of the greater total counts available with increased injected dose and acquisition time. Whole body scintigraphy should be performed at a speed of 3 cm/min, or a minimum of 30 minutes for the head to upper femur imaging field, to assure acquisition of adequate data.

Scintigraphy (either whole-body or selected regions of interest) is performed as a rule at 4 and 24 hours after injection. Routine practice currently includes performing SPECT of the appropriate regions of interest, preferably at 24 hours (5,13,15,16). The Guideline acknowledges inherent differences amongst various imaging systems but recommends a SPECT protocol including 3° angular sampling, with 20 to 30 seconds per stop.

Digital display stations which are currently increasingly available should be used for image interpretation. Workstations with a digital display system allow continuous manipulations in intensity and provide a variety of black and white and color maps, with wide inherent variations in contrast, as well as other display features that enhance lesion detection. In addition, with sufficient counts accumulated, the availability of a maximum intensity projection (MIP) volume display for SPECT images, as well as the possibility for triangulation, also benefit image interpretation.

Following intravenous injection, ^{111}In-octreotide clears rapidly from the blood. Normal biodistribution includes uptake by the liver, spleen, and kidneys. Splenic uptake is more intense than the liver activity due to the greater density of somatostatin receptor positive lymphocytes. The kidneys show the most intense uptake. As a small molecule, ^{111}In-Octreotide is excreted in the glomerular filtrate and is reabsorbed in the proximal and distal convoluted tubules, where it either binds to intracellular proteins or is digested into smaller insoluble fragments that remain in the renal tubule cells. Intense renal accumulation is the dose-limiting factor in targeted radionuclide therapy with radiolabeled octreotide and other somatostatin analogs. In diagnostic imaging, particularly with planar scintigraphy, renal activity interferes with interpretation of suspicious foci in the head and tail of the pancreas or in the adrenal glands, as well as with lesions in the bowel located in the vicinity to the kidneys.

Another limitation in interpretation is related to the presence of gallbladder uptake. Approximately 10% to 15% of the administered dose is secreted into the bile and will be detected in the gall bladder and in bowel content. This may lead to false-positive interpretations. On the other hand, if these foci of intra-abdominal activity cannot be identified as gallbladder or bowel content, the patient may be misdiagnosed or subjected to repeated imaging after efforts to promote bowel evacuation. These maneuvers make the test inconvenient to the patient, and prolong the instrument and personnel time involved, while nevertheless resulting in inconclusive or incorrect interpretations.

The use of SPECT imaging in general, and MIP volume displays in particular, helps to more rapidly and correctly identify bowel activity. This allows satisfactory assessment of the abdominal and retroperitoneal regions that would be otherwise obscured by kidney activity. Availability of anatomic imaging information (CT) is also helpful in this regard. Nevertheless, precise assignment of the anatomic location of activity may be difficult even with side-by-side comparison with other imaging modalities, specifically CT.

Schillaci et al. compared SRS using planar and SPECT studies with conventional imaging procedures in 149 patients with gastroenteropancreatic (GEP) neuroendocrine tumors and concluded that SPECT imaging is the procedure of choice (15). Of 65 cases with histologically confirmed liver metastases, SPECT identified the lesions in 60 (92% sensitivity) compared with only 38 of 65 (58% sensitivity) for planar imaging and 52 of 65 (80% sensitivity) for conventional imaging including CT and MRI. There were no false-positive SPECT images, resulting in 100% specificity versus 86% for conventional imaging of the liver. On the basis of SPECT, management changes were indicated in 28 of 149 patients (19%), due to identification of additional, previously missed metastatic lesions, or due to excluding suspicious lesions on conventional imaging as false positives. In comparison, planar scintigraphy resulted in management changes in 13 patients (9%). The classification based on SPECT had an error rate of 3% compared with 17% for planar imaging (15).

Briganti et al. performed SPECT, as well as a semiquantitative analysis using a triple-head camera, in 38 patients with suspected pancreatic neuroendocrine neoplasms. Although the injected dose of ^{111}In-octreotide ranged from 111 to 220 MBq, SPECT results were true positive in 18 of 19 patients. SRS was the only positive study in 14 of the 19 patients who had further confirmation of disease. Twenty of the thirty-eight patients had modification of their management plan as a result of the SRS SPECT (16).

With the use of SPECT, the problem of lesions being obscured by overlying structures is eliminated in a large proportion. Triangulation of suspect focal accumulations of activity and subsequent examination of transaxial, coronal, and sagittal planes permit improved placement of the activity accumulation and contribute to greater accuracy in interpretation. However, the area to be examined by SPECT needs to be identified by the nuclear medicine physician based on good knowledge of the clinical problem at hand, regardless of whether a lesion is seen or is absent on planar images. To do this effectively requires familiarity with the clinical expression of neuroendocrine tumors. For example, symptoms suggestive of carcinoid syndrome imply that in addition to the primary, usually intra-abdominal tumor, the patient is likely to have hepatic metastases (17,18). The examination and interpretation is incomplete without careful SPECT examination of the liver with subsequent manipulation of the image display.

Most neuroendocrine tumors, including carcinoids, islet cell tumors, gastrinomas, and pheochromocytomas, arise from cells located in one of the abdominal organs. Carcinoid may arise in the wall of the small intestine, the pancreas, ceacum or appendix, the biliary tract, rectum, or ovary, as well as the bronchial tree. Bronchial carcinoids, if functional, usually secrete peptides hormones such as adrenocorticotrophic hormone (ACTH) (18). Hence, patients suspected of having an ACTH-producing tumor should always include examination of the thorax, with iterative manipulation of the displayed images, as well as SPECT of the abdomen to identify pancreatic or intestinal carcinoid tumors as the alternative source of ACTH production. SPECT imaging of the region of the head and neck should be performed in patients who undergo SRS due to a suspicion of, or knowledge of having paragangliomas, medullary carcinoma of the thyroid, differentiated (papillary-follicular) thyroid carcinoma, or a functioning pituitary tumor.

[111]IN-OCTREOTIDE CO-REGISTRATION AND SPECT/CT

Despite the valuable contribution of ^{111}In-octreotide scintigraphy, including planar and SPECT studies, to the diagnosis and management of patients with known or suspected neuroendocrine tumors, or other processes characterized by the increased expression of somatostatin receptors, the patterns of distribution of ^{111}In-octreotide indicate the need for correlating the functional imaging findings with anatomic imaging. To begin with, many of the tumors of interest are found in the abdomen, and it is often difficult to precisely localize a suspicious lesion and differentiate whether a focus of abnormal uptake is in the pancreas, small bowel, liver, or bone without anatomic correlation. In the region of the liver, it is difficult to distinguish between physiologic gallbladder accumulation versus a lesion in the head of the pancreas, in the right adrenal, or in the small bowel.

The two image sets (SPECT and CT) were acquired separately for side-by-side interpretation. Methods for fusion of separately acquired anatomic imaging (usually CT) and SPECT ^{111}In-octreotide imaging have been developed (Fig. 1). These techniques work quite well for fusion of brain data, as there is no shift of intracranial content from one study to another. In the thorax, there are differences in organ and lesion position depending on respiratory dynamics. However, in general, the mediastinal central structures have limited excursion, and satisfactory fusion can be achieved (Fig. 1). In the abdomen and, to a lesser degree, in the pelvis, there is the potential for significant shift of lesions depending upon patient positioning and variations in stomach, bowel, or bladder distention. This may challenge post-hoc coregistration between the SPECT and CT examinations, even when those are obtained within a close temporal interval, leading to misalignment of suspicious foci.

In 2002, Forster et al. (19) reported the use of a software package to fuse helical CT and SPECT images (performed following the administration of only 111 MBq of ^{111}In-octreotide) of 28 lesions identified in 10 patients, and compared the use of

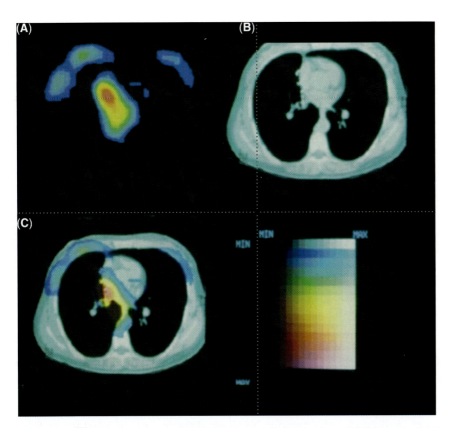

Figure 1 ^{111}In-octreotide scintigraphy (transaxial slices) of an ACTH-producing bronchial carcinoid in a 38-year-old woman with Cushing's syndrome; ectopic ACTH production suspected based upon elevated plasma ACTH and negative brain MRI. The patient had a Cushing's syndrome secondary to ectopic ACTH-producing right main stem bronchial carcinoid 20 years ago and reached complete remission of the Cushing's syndrome following resection. Several months prior to this scan, repeated CT examinations with attention to the site of the previous lesion were interpreted as negative. Images were obtained by the senior author at Memorial Sloan-Kettering Cancer Center in New York in 1994 during a "Compassionate Use Clinical Trial." At that time, no commercial image fusion systems were available. An in-house system was used in selective cases to assign the scintigraphic transaxial slice (**A**) to the corresponding CT slice (**B**) to produce the post-hoc fusion image (**C**) demonstrating recurrence at the site of the previous resection. Based on these images, the patient underwent repeat thoracotomy and resection of tissue in the region of the demonstrated activity although no tumor mass was identified visually at surgery. A 3 mm diameter cluster of carcinoid tissue was identified at histopathological examination. Cushing's syndrome symptomatology cleared within weeks of resection. *Abbreviations*: CT, computed tomography; MRI, magnetic resonance imaging.

external fiducial markers and internal anatomic landmarks (spleen and kidney contour). They demonstrated a shift of a few millimeters in organ location. External markers produced better alignment as compared to internal anatomic landmarks. This suggests that altered patient positioning may account for a significant component of internal organ shift. The authors made no attempt to compare accuracy of interpretation with and without fusion. Rather, the goal was simply to clarify CT findings by superimposing the ^{111}In-octreotide images and to reliably define the location of the functional information. Nevertheless, a figure in the publication (19) demonstrates a carcinoid tumor with a liver metastasis that is unremarkable on non-contrast-enhanced CT and only faintly visible with contrast, but is readily demonstrated with ^{111}In-octreotide on the 5 and 24 hours transaxial SPECT images that have been superimposed on CT.

Subsequently, Amthauer et al. evaluated the use of image fusion in the preoperative staging of patients with GEP neuroendocrine tumors following ^{111}In-Octreotide administration (20). The imaging results were compared to findings at surgery or clinical follow-up. A precise anatomic location was assigned to 87 lesions by two independent readers based on SPECT images alone and the results were compared to the findings after fusion with contrast-enhanced CT obtained within 14 days of the scintigraphic study. Image fusion was possible in 36 of the 38 patients included in the study. The accuracy of successfully assigning the anatomical location by two readers increased from 57% and 61% to 91% and 93%. The authors report an improvement in the diagnosis and localization of liver metastases to a specific segment, rising from 45% and 58% to 98% and 100%. This difference between SPECT stand-alone and coregistration is of clinical importance when wedge resection of hepatic lesions is being considered. In this paper, the improved anatomic localization was relevant for therapeutic decisions in 7 of the 36 patients (19%) (20).

Despite this study having demonstrated the value of fusion of CT images acquired separately as much as 14 days after the SPECT study, one should not conclude that this approach is as good as or better than near-simultaneously acquired SPECT/CT fusion images acquired on the same imaging table. The above-mentioned study evaluated only a narrow indication. It is not appropriate to assess the relative value of CT and ^{111}In-octreotide scintigraphy in the management of patients with neuroendocrine tumors only in this limited context. Evaluation of patients with neuroendocrine tumors involves identification of primary tumors and extrahepatic metastases as well as hepatic metastases.

In 2000, Patton et al. demonstrated the feasibility of imaging with a dual-head gamma camera with an integrated low-power CT (Millenium and Hawkeye, General Electric Medical Systems) (21). This initial report utilized the dual detectors in a coincident mode with [18]FDG as the diagnostic radionuclide. The authors concluded that an additional advantage of the system was its ability to perform attenuation correction and fusion for SPECT procedures as well. For the SPECT/CT procedure, multiple fields of view can be obtained. Following the CT acquisition, SPECT is performed in 3 to 6° angular steps, 45 to 60 seconds each, with a 128×128 matrix. SPECT is reconstructed using the OSEM reconstruction program and images are processed on the accompanying digital processor. Attenuation-corrected emission images are viewed side-by-side with the transmission image and a SPECT/CT fused image in which the emission data is overlaid on the transmission images. Images can be viewed in all three planes: coronal, sagittal and transaxial as functional (SPECT) images alone, anatomical (low-dose CT) images alone, or a combination of functional, anatomic, and fused image for each slice. A variety of color displays can be selected.

Even-Sapir et al. used this hybrid imaging system to evaluate 27 patients with endocrine tumors, 10 of whom had neuroendocrine tumors and were imaged following the injection of 148 MBq ^{111}In-octreotide (22). Eight of the ten patients had abnormal studies. In one patient suspected of having recurrent gastrinoma, the fused images demonstrated abnormal uptake in a normal-sized retroperitoneal lymph node. The lymph node would not have been identified as abnormal on CT alone and in all likelihood it would not have been properly localized on SPECT alone. Another

(A)

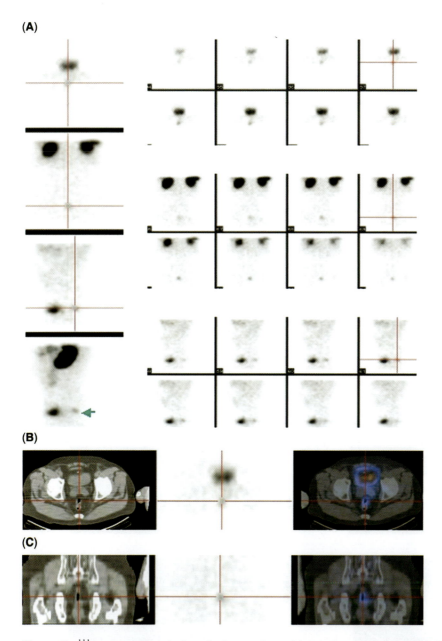

(B)

(C)

Figure 2   ¹¹¹In-octreotide scintigraphy in a patient with rectal carcinoid. (**A**) SPECT display of the abdomino-pelvic region shows a focus of abnormal tracer uptake in the region of rectum. (**B**) Transaxial and (**C**) coronal SPECT/CT slices confirm the localization of lesion to the rectum. Given to the known location of the lesion prior to study, SPECT/CT did not provide additional information (category 1). *Abbreviations*: CT, computed tomography; SPECT, single-photon emission computed tomography.

example identified a second tumor site within the pancreas that had been previously missed by CT alone. The impact of precise localization using a selection of therapeutic options was demonstrated in a case with extensive liver metastases. Fusion scintigraphy in this case excluded extrahepatic foci enabling the patient to undergo hepatic

(A)

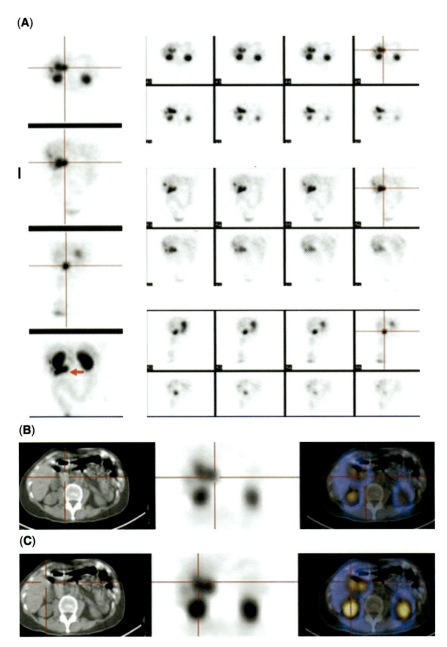

(B)

(C)

Figure 3 ^{111}In-octreotide scintigraphy in a patient with a pancreatic mass. (**A**) SPECT display of the abdominal region shows two foci of increased tracer uptake. On the MIP (*lower left*) and selected enlarged coronal slice, the medial focus appears to be extrahepatic and was interpreted as corresponding to the pancreatic mass. The lateral focus was suspected as representing a hepatic metastasis. (**B**) CT, SPECT, and SPECT/CT confirmed the location of the medial focus to the head of pancreas. (The additional tube-like site of activity appears to represent bowel content). (**C**) CT, SPECT, and SPECT/CT confirmed the location of the lateral focus to the liver, representing a hepatic metastasis. The SPECT/CT findings in this patient did not induce a change in interpretation but greater diagnostic confidence (category 2) in interpretation of the SRS. *Abbreviations*: CT, computed tomography; MIP, maximum intensity project; SPECT, single-photon emission computed tomography.

(A)

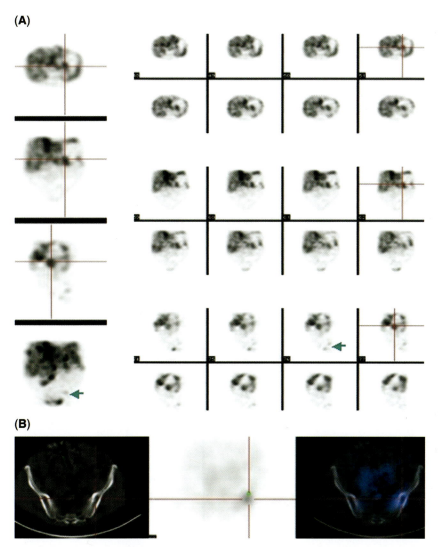

(B)

Figure 4   ^{111}In-octreotide scintigraphy in a patient with a nonfunctioning pancreatic islet cell carcinoma with hepatic metastases known by previous CT. (A) SPECT display of the abdominal region indicates the presence of increased tracer uptake in an epigastric tumor and multiple hepatic metastases seen in all projections. In the MIP image, an additional focus is identified in the left lower quadrant and was considered to represent an intra-abdominal lymph node ("shed" tumor). (B) CT, SPECT, and SPECT/CT images demonstrate that this focus represents a previously unknown and unexpected skeletal metastasis in the left ilium. The SPECT/CT findings led to a change in diagnosis and defined the whole extent of metastatic disease but had no impact on further patient management (Category 3). *Abbreviations*: CT, computed tomography; MIP, maximum intensity projection; SPECT, single-photon emission computed tomography.

chemoembolization. The authors concluded that hybrid imaging improved image interpretation mainly by providing a more precise anatomic localization of lesions detected by SPECT.

Pfannenberg et al. reported an analysis of 43 patients with neuroendocrine tumors who were imaged with a similar SPECT/CT system following the injection of 200 MBq of ^{111}In-octreotide (23). The SPECT/CT results were compared to those of SPECT stand-alone and to high-end CT images, with histopathological findings or clinical and imaging follow-up used as the diagnostic standard. There were four false-negative SPECT examinations with a total of eight undetected hepatic metastases subsequently demonstrated on diagnostic CT. Separate SPECT and CT interpretations were in agreement for 56 of 114 lesions overall (49% concordance). For

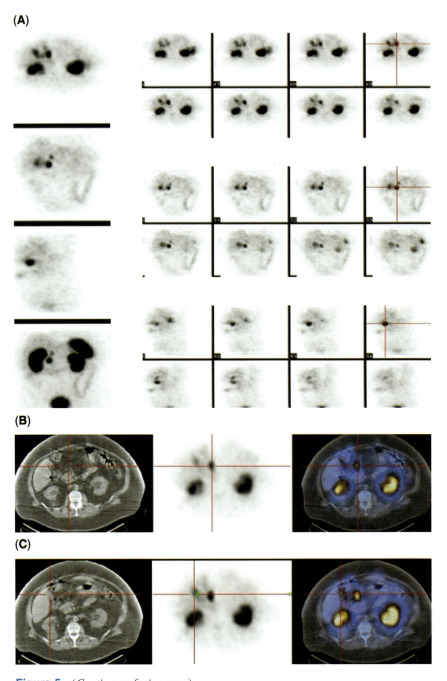

(A)

(B)

(C)

Figure 5   (*Caption on facing page*)

the remaining 58 lesions (51%), consensus readings of the fused SPECT/CT images resulted in a change from the original interpretation of 39 CT and 19 SPECT examinations. Overall, SPECT/CT outperformed SPECT alone and was significantly better than high-end CT alone. The greatest accuracy involved the use of SPECT/CT with side-by-side availability of high-end CT. In fact, in this report, SPECT and side-by-side high-end CT performed slightly better than SPECT/CT.

A preliminary report presented the results of our group with SPECT/CT fusion of ^{111}In-octreotide scintigraphy in 27 patients with suspected or known neuroendocrine tumors, primarily of the GEP type (24). In order to assess the incremental value of SPECT/CT, the studies were initially interpreted as SPECT alone followed by analysis of the combined SPECT/CT study. All studies were scored as to the incremental value of SPECT/CT: category 1, if the fusion images provided no change in the diagnostic impression reported by SPECT images alone (Fig. 2); category 2, if there was no change in the impression but there was marked improvement in the diagnostic confidence following review of the SPECT/CT (Fig. 3); category 3, change in impression based upon review of the SPECT/CT (Fig. 4); category 4, change in impression using SPECT/CT with a potential or actual impact on patient management (Fig. 5). Fifteen of the twenty-seven studies were interpreted as abnormal. SPECT/CT fusion scintigraphy improved the diagnostic confidence in 15 of 27 cases overall and changed the impression (categories 3 and 4) in 5 of the 27 cases (19%) or 5 of the 15 cases (33%) scored as abnormal on planar and SPECT scintigraphy alone. In two of these five patients, the change in impression had an impact on management (8%). No additional value of the combined imaging technique (category 1) was scored in 5 of the 27 cases (18%).

In a large series including 72 patients with neuroendocrine tumors, Krausz et al. (25) evaluated the impact of SPECT/CT on the diagnostic accuracy of SRS and on further clinical management. SRS was positive in 44 of the 72 patients. Fused images provided no additional information in 48 of the 72 patients, including all the 28 studies reported as negative. Hybrid imaging, however, improved the study interpretation in 23 patients (32% of the total study population, 52% of the positive SRS). SPECT/CT allowed for the precise localization of foci of increased ^{111}In-octreotide activity thereby defining the whole extent of disease in 17 patients, it diagnosed previously unsuspected bone metastases in 3 patients, and defined suspicious lesions as sites of physiologic activity unrelated to cancer in 3 additional patients. Of higher clinical significance was the fact that SPECT/CT altered the further management of 10 patients (14%). Results of fused images modified the previously planned surgical approach in six patients, spared unnecessary surgery in two patients with

Figure 5 (*Facing page*) ^{111}In-octreotide scintigraphy in a patient with Zollinger–Ellison syndrome. (**A**) SPECT display of the abdominal region shows two foci of abnormal tracer activity seen on MIP (*lower left*) and selected slices. These foci are considered to be located either within the liver or to represent bowel uptake. (**B**) SPECT/CT identifies the most intense focus of uptake as being located within the bowel, which is considered to represent the primary lesion. (**C**) SPECT/CT demonstrates the second lesion to represent physiologic activity within the gallbladder and excludes the previously suspected hepatic metastasis. Based on SPECT/CT, extent of disease classification was changed and the patient was down-staged. Since on SPECT alone, a false–positive interpretation of hepatic metastasis would have been made, this case was classified as resulting in a significant change in interpretation with management impact (Category 4). *Abbreviations*: CT, computed tomography; MIP, maximum intensity project; SPECT, single-photon emission computed tomography.

newly diagnosed involvement of the skeleton, and led to referral of one patient each to liver transplant and to chemoembolization, rather than to systemic therapy. The authors conclude that SRS-SPECT/CT indicates the functional status of the tumor, its precise localization, and whole extent. Fused images are therefore a useful tool to choose the optimal treatment strategy, mainly in patients with advanced disease.

Fusion images provide much greater accuracy in localization of findings than is possible on functional SPECT imaging alone and much greater specificity than is possible even on diagnostic CT. SPECT/CT plays a role in localizing a lesion in the event that it had been an unexpected finding (Fig. 2). The impact on reader confidence and increased credibility with referring clinicians is an important add-on feature for SPECT/CT, as the fusion image provides a convincing demonstration of the anatomic location of the radiotracer focus, whereas interpretation of the SPECT images is more dependent on the experience and skill of the specific reader (Fig. 3). The concept of incremental confidence is difficult to quantify or to assign a value in trying to define a benefit-to-cost ratio. It is clear that despite the effort to evaluate the impact of combined SPECT/CT in a quantitative manner, given the complexity of image interpretation, this assessment remains a subjective process. In the study by Mirtcheva et al. (24), all SRS were initially interpreted by nuclear medicine residents and subsequently by an attending nuclear medicine physician. The category assigned to each case was a consensus ranking after attending physician review. During the review process, however, another observation was made that was not possible to quantify. Early in the process, the resident interpretation of the SPECT alone was often uncertain. With SPECT/CT fusion image interpretation, the resident was able to formulate an impression more quickly and with greater confidence. These impressions were frequently consistent with the impression of the more experienced attending nuclear medicine physicians. With increased ease of interpretation of the SPECT/CT fusion images, it became increasingly difficult to adhere to the original study format requiring interpretation of the SPECT images alone prior to reviewing the SPECT/CT images that were conveniently available. In a short time interval, there was regular agreement between the resident's and attending nuclear medicine physician's interpretation and the residents developed increased confidence in their initial interpretation based on the SPEC/CT images. In other words, experienced nuclear medicine physicians often render an interpretation but the referring clinician remains unconvinced or uncertain because of the difficulties in visualizing the location of the finding on scintigraphy alone. Moreover, it is difficult to define how different nuclear medicine physicians come to a conclusion with so little visual evidence other than to attribute it to experience. With CT image correlation and the near precise image registration possible with a hybrid imaging system, the interpretation of high signal to background functional images when combined with better anatomic information is less dependent upon individual experience and skill. The evidence shows that using near simultaneously acquired SPECT and even low-power CT images improves the accuracy of side-by-side interpretation with diagnostic CT more than interpreting SPECT and diagnostic CT images side by side.

SRS with SPECT/CT also results in more meaningful communication with referring physicians. The hybrid imaging study interpretation is more credible to the clinician who is able to see the location of the somatostatin-receptor positive tissue. Clinician confidence in the results of a diagnostic imaging procedure is important, particularly when the findings suggest an alteration in patient management from traditional practice.

SUMMARY AND CONCLUSIONS

Despite the favorable impact that ^{111}In-octreotide scintigraphy, particularly SPECT imaging, has had on the diagnosis and management of patients with neuroendocrine tumors, these results improve further when correlated with anatomic imaging data, regardless of whether it is available as a low-power CT in a hybrid SPECT/CT system or whether it is side-by-side comparison of SPECT images with diagnostic quality CT images. Measurements of improvement include greater sensitivity and specificity of diagnosis as well as impact on patient management.

In addition, although there is minimal objective quantitative data, personal experience and discussion with other practitioners confirm the impression that confidence and speed of interpretation improves, interpretative performance of less-experienced readers improves, and the images are more useful and credible to clinicians.

While a case can be made that it would be cost-effective to continue to perform separate anatomical (CT) and functional (SPECT) examinations, it is clear that the side-by-side comparison is time-consuming and tedious—even more so without at least low-power CT images for ease of geographical lesion identification, which makes it difficult to be certain that the slices examined are comparable. In addition, there is the issue of the shift in abdominal contents due to physiologic variations, as well as change in position on different imaging tables. This would suggest that near-simultaneous acquisition of SPECT images and CT images on the same imaging table (hybrid SPECT/CT imaging) is the preferable mode of performing SRS examination. Near-simultaneous acquisition of both CT and SPECT image sets (hybrid SPECT/CT images) represents state-of-the-art for diagnostic ^{111}In-octreotide imaging.

Upgrading the CT capacity to perform diagnostic quality CT images would seem to be ideal from a diagnostic image interpretation point-of-view. The only remaining issue is cost-effectiveness. It seems likely that over time, with increased awareness of the value of ^{111}In-octreotide hybrid SPECT/CT image fusion, this technique will become the standard of care for the evaluation of the patient with known or suspected neuroendocrine tumors.

REFERENCES

1. Krenning EP, Bakker WH, Breeman WA, et al. Localization of endocrine-related tumors with radioiodinated analogue of somatostatin. Lancet 1989; 242–244.
2. Krenning EP, Bakker WH, Kooij PP, et al. Somatostatin receptor scintigraphy with indium-111-DTPA-D-Phe-1-octreotide in men: metabolism, dosimetry and comparison with iodine-123-Tyr-3-octreotide. J Nucl Med 1992; 33:652–658.
3. van Dongen A, Verhoeff NPLG, Bemelman F, van Royen EA. Somatostatin receptor imaging with ^{111}In-pentetreotide for whole body studies and high resolution brain SPECT. Eur J Nucl Med 1992; 19:679.
4. Ivanovec V, Nauck C, Sandrock D, et al. Somatostatin receptor scintigraphy with ^{111}In-pentetreotide in gastro-enteropancreatic endocrine tumors (GEP). Eur J Nucl Med 1992; 19:736.
5. Krenning EP, Kwekkeboom DJ, Lamberts SWJ. Somatostatin receptor scintigraphy with [^{111}In-DTPA-D-Phe^1] and [^{123}I-Tyr^3]-octreotide: the Rotterdam experience with more than 1000 patients. Eur J Nucl Med 1993; 20:716–731.
6. Krenning EP, Kwekkeboom DJ, Oel HY, et al. Somatostatin-receptor scintigraphy in gastroenteropancreatic tumors: an overview of the European results. In: Wiedenmann

B, Kvols LK, Arnold R, Riecken EO, eds. Molecular and Cell biological Concepts of Gastroenteropancreatic Neuroendocrine Tumor Disease, Annals of the New York Academy of Sciences. Vol. 733. New York, 1994:416–424.

7. Jamar F, Fiasse R, Leners N, Pauwels S. Somatostatin receptor imaging with indium-111-pentetreotide in gastroenteropancreatic neuroendocrine tumors: safety, efficacy and impact on patient management. J Nucl Med 1995; 36:542–549.

8. Cardiot G, Lebtahi R, Sarda L, et al. Groupe d'etude du syndrome de Zollinger-Ellison. Gastroenterology 1996; 111:845–854.

9. Gibril F, Reynolds JC, Doppman JL, et al. Somatostatin receptor scintigraphy: its sensitivity compared with that of other imaging methods in detecting primary and metastatic gastrinomas. Ann Intern Med 1996; 125:26–34.

10. Goldsmith SJ, Macapinlac HA, O'Brien JP. Somatostatin-receptor imaging in lymphoma. Semin Nucl Med 1995; 25:262–271.

11. Slooter GD, Mearadji A, Breeman AP, et al. Somatostain receptor imaging, therapy and new strategies in patients with neuroendocrine tumors. Br J Surg 2001; 81:31–40.

12. Balon HR, Goldsmith SJ, Siegel BA, et al. Procedure guideline for somatostatin receptor scintigraphy with ^{111}In-pentetreotide. J Nucl Med 2001; 42:1134–1138.

13. Kwekkeboom DJ, Krenning EP. Somatostatin receptor imaging. Semin Nucl Med 2002; 32:84–91.

14. De Jong, Valkema R, Breeman WA, et al. Neuroendocrine neoplasia: scintigraphy and radionuclide therapy. In: Khalkhali I, Maublant JC, Goldsmith SJ, eds. Nuclear Oncology, Philadelphia: Lippincott, 2001.

15. Schillaci O, Spanu A, Scopinaro F, et al. Somatostatin receptor scintigraphy in liver metastasis detection from gastroenteropancreatic neuroendocrine tumors. J Nucl Med 2003; 44:358–368.

16. Briganti V, Matteini M, Ferri P, et al. Octreoscan SPET evaluation in the diagnosis of pancreas neuroendocrine tumors. Cancer Biother and Radiopharm 2001; 16:515–524.

17. Waldenstrom J. Clinical picture of carcinoidosis. Gastroenterology 1968; 54(suppl): 826–828.

18. Kvols LK. Metastatic carcinoid tumors and the malignant carcinoid syndrome. In: Wiedenmann B, Kvols LK, Arnold R, Riecken EO, eds. Molecular and Cell Biological Aspects of Gastroenteropancreatic Neuroendocrine Tumor Disease Annals of the New York Academy of Sciences vol. 733. New York 1994:464–470.

19. Forster GJ, Laumann C, Nickel O, et al. SPET/CT image co-registration in the abdomen with a simple cost-effective tool. Eur J Nucl Med Mol Imaging 2003; 30:32–39.

20. Amthauer H, Ruf J, Bohmig M, et al. Diagnosis of neuroendocrine tumors by retrospective image fusion: is there a benefit? Eur J Nucl Med Mol Imaging 2004; 31:342–348.

21. Patton JA, Delbecke D, Sandler MP. Image fusion using an integrated, dual-head coincidence camera with x-ray tube-based attenuation maps. J Nucl Med 2000; 41:1364–1368.

22. Even-Sapir E, Keidar Z, Sachs J, et al. The new technology of combined transmission and emission tomography in evaluation of endocrine neoplasms. J Nucl Med 2001; 42: 998–1004.

23. Pfannenberg AC, Fachmann SM, Horger M, et al. Benefit of anatomical–functional image fusion in the diagnostic work-up of neuroendocrine neoplasms. Eur J Nucl Med Mol Imaging 2003; 30:835–843.

24. Mirtcheva RM, Kostakoglu L, Goldsmith SJ. Hybrid Imaging using ^{111}In octreotide SPECT/CT in evaluation of somatostatin receptor positive tumors. J Nucl Med 2003; 44:73P.

25. Krausz Y, Keidar Z, Kogan I, et al. SPECT/CT hybrid imaging with ^{111}In-pentetreotide in assessment of neuroendocrine tumors. Clin Endocrinol 2003; 59:565–573.

6

Bone SPECT/CT in Oncology

Marius Stefan Horger
Department of Diagnostic Radiology, University Hospital Tuebingen, Tuebingen, Germany

Susanne Martina Eschmann and Roland Bruno Bares
Department of Nuclear Medicine, University Hospital Tuebingen, Tuebingen, Germany

INTRODUCTION

Despite the development of new bone imaging modalities, scintigraphy still plays a major role in the diagnosis of a variety of bone lesions. Since its introduction in 1971, bone scintigraphy with ^{99m}Tc-labeled phosphate compounds has been one of the most sensitive noninvasive methods for diagnosis of skeletal lesions. Skeletal involvement is a frequent and severe complication of various malignancies, ultimately affecting two-thirds of all cancer patients.

Technetium-99m (^{99m}Tc) methylene diphosphonate (MDP) is currently the most widely used radiopharmaceutical in bone scintigraphy. The phosphonate compounds bind to bone by chemiadsorption to the hydroxyapatite crystal. ^{99m}Tc decays to ^{99}Tc by releasing a 140 keV gamma ray, which is efficiently detected by a gamma camera. Bone scintigraphy reflects skeletal metabolic activity. Skeletal diphosphonate uptake is influenced by blood flow and osteoblastic activity related to new bone formation. Bone scintigraphy therefore has a high sensitivity for the detection of bone metastases, mainly osteoblastic lesions, and becomes positive before plain radiographs, due to the appearance of focal areas of increased bone turnover in response to tumor growth. Changes in bone formation of as little as 5% to 15% can be detected (1). Lesions showing only bone resorption may be detected as photopenic regions.

While highly sensitive, bone scintigraphy also has a much lower specificity as compared to radiography. Nevertheless, certain patterns of increased radiotracer uptake localized in the skeleton seen on scintigraphy can be more specific. Multiple asymmetric foci of abnormally increased activity, involving mainly the axial skeleton, and to a lesser extent the extremities, are highly suspicious of metastatic disease. The current state-of-the-art acquisition protocol for bone scintigraphy includes planar whole-body imaging using dual-detector gamma camera systems equipped with high or even ultra-high resolution collimators resulting in high image quality. Bone scintigraphy remains, at present, the method of choice in the evaluation of bone metastases because of its widespread availability, reasonable cost, and ability to show the entire skeletal system. In cases where focal abnormalities are detected, additional spot views or single-photon emission computed tomography (SPECT) may be performed. SPECT provides better contrast, edge definition, and separation of target from background compared to planar imaging. SPECT, which is often performed following equivocal planar scintigraphy findings, has superior resolution and better contrast, but is limited in its capability to further classify sites of increased tracer uptake on bone scintigraphy. For example, anatomical correlation is needed in the vertebral region due to the complexity of the structures and also because benign degenerative changes are very common there.

Image fusion using a hybrid SPECT/computed tomography (CT) system provides additional information with regard to lesion characterization and precise localization, which improves specificity of bone scintigraphy considerably. Bone lesions, initially classified as undetermined by SPECT, can be correctly diagnosed by SPECT/CT due to the demonstration of the underlying anatomy and applying known CT diagnostic criteria for their final definition.

Structural changes in the skeleton demonstrated by X-rays are often difficult to assess without corresponding functional information (2). On the other hand, in patients with known malignant disease in particular, interpretation of the significance of focally increased bone uptake detected on scintigraphy can be challenging. Further classification of abnormal foci as benign versus malignant is often based on the precise anatomic localization of a lesion. In the spine, for example, malignant lesions usually involve the vertebral body, possibly extending into the pedicles and rarely into the apophyseal joints or into the spinous process.

Combining functional and anatomical data can potentially improve the overall diagnostic accuracy of bone imaging. Attempts have been made to fuse anatomic and functional images, either by use of external or internal landmarks, or a combination of both (3). When the acquisition of functional and anatomical imaging data is performed on different devices, however, this may lead to errors in realignment due to respiratory motion, or patient movement. Furthermore, coregistration of separately performed studies is time-consuming and interferes with routine application (4–6). The currently available new hybrid imaging SPECT/CT modality represents a possible solution to this problem. Acquisition of transmission and emission data is performed during the same session without moving the patient. This allows accurate coregistration of both data sets.

The value of SPECT/CT as an add-on to bone scintigraphy should be compared to other imaging techniques used for the assessment of the skeleton. Plain radiographs are characterized by a relatively low sensitivity but relatively higher specificity for the detection of bone metastases. In the appendicular skeleton, the characterization of bone lesions into benign or malignant may be reliably performed using well-established diagnostic criteria. Plain radiographs have, however, several limitations in the axial skeleton, due to the high frequency of both malignant and degenerative bone changes, and also because the vertebral column is an anatomical region known as less accessible for plain X-ray diagnosis.

Computed tomography is a highly sensitive method for evaluating cortical and trabecular bone, being able to assess minimal bone changes in an early stadium of destruction, and also to classify lesions at risk for fracture or diagnose spine misalignment, because of the capacity of multiplanar reformats (7,8). Medullar metastases, however, may be missed by CT, except for those located in the medullar cavity of long-hole bones, such as the femoral or humeral metadiaphysis.

The CT pattern of osteolytic metastases may include different radiographic features such as moth-eaten or permeating bone infiltration, diffuse or otherwise ill-defined bone destruction, or large bubbly expansible solitary or multiple lesions. Osteoblastic metastases are characterized by ill-defined or mottled sclerosis. CT can identify bone lesions with 50% cortical destruction, which represents a fracture risk or osteolysis in an area with a diameter greater than 2.5 cm. Anatomic images also permit the assessment of the potential for dural compression when spine metastases are suspected.

Magnetic resonance imaging (MRI) is the method of choice for detecting intramedullary metastases, especially in bones with large marrow cavities such as vertebral bodies. This is important in early detection of small bone metastases, which are mostly located in the medullar cavity and are not accompanied by cortical involvement and thus do not cause sufficient bone remodeling to be detected by bone scans (9–13).

^{18}F-fluoride positron emission tomography (PET)/CT and ^{18}F-FDG PET are both sensitive and specific for detection of lytic and sclerotic malignant bone lesions (14–17).

CLINICAL INDICATIONS FOR BONE SCINTIGRAPHY, SPECT, AND SPECT/CT

Bone scintigraphy, performed as a whole-body study, is the standard procedure in cancer patients and is often complemented by SPECT studies. The main indications to perform bone scintigraphy in the oncological population includes the staging of disease—for detection of metastases and local invasion, assessment of prognosis, assessment of the etiology of pathologic fractures, evaluation of unexpected new abnormal chemical change or new musculoskeletal symptoms, and evaluation of response to therapy (18). When bone scintigraphy is negative, no further imaging is performed as a rule. Some findings, such as increased activity in the sternoclavicular, sternomanubrial, or costochondral joints in an elderly patient, are considered to represent benign foci secondary to degenerative changes. Adjacent lesions in two or more ribs, commonly suggesting benign fractures, do not require further evaluation, particularly if supported by the patient's history.

The lack of specificity is a particular problem in the evaluation of solitary bone lesions in patients with known malignant disease. There are many potential sources of error in the interpretation of bone scintigraphy; injection artefacts, urine contamination, prosthetic implants, restraint artefacts caused by soft-tissue compression, or pure lytic lesions may complicate image interpretation. Furthermore, extraskeletal uptake of bone-seeking radiopharmaceuticals due to chemisorption on the surface of calcium salts and hydroxyapatite crystals, although relatively rare, also needs to be considered in the differential diagnosis of abnormal foci on bone scintigraphy. These factors contribute to the need for correlation with anatomic images.

Combined SPECT/CT imaging offers a solution to these diagnostic challenges. The transmission images provided by the CT component of the SPECT/CT study enable better characterization of bone metastases compared with SPECT alone. CT images are superior to conventional plain X-rays despite the limited resolution of low-dose CT devices. Image fusion provides additional information with regard to lesion characterization and precise localization, which is of clinical significance in many cases. Very precise coregistration also improves guidance of bone biopsies, decreasing tissue sampling error and potentially reducing the number of invasive procedures.

BONE METASTASES: INCIDENCE, DISTRIBUTION, CLINICAL FEATURES, AND IMAGING CHARACTERISTICS

Metastatic involvement of the skeleton occurs in 30% to 70% of all cancer patients, with increasing frequency over recent decades due to prolonged overall patient survival. Four solid tumors account for 80% of all patients with bone metastases. Skeletal spread occurs with higher frequency in patients with breast carcinoma (73%), lung carcinoma (32%), prostate cancer in men (over 70%), and kidney cancer (24%) (19).

Skeletal metastatic involvement occurs primarily by hematogenous spread but occasionally by direct extension of a soft-tissue tumor to adjacent bone. The majority of metastatic lesions in adults are localized in the red bone marrow (20,21). The high frequency of metastases to the axial skeleton directs attention to the spine and pelvis. The appendicular skeleton is less frequently involved, accounting for 10% to 15% of metastatic bone lesions (22). In the long bones, metastases are located mainly in the metaphysis, the site of residual red marrow. Involvement of the long bones is often observed in patients with lung cancer and follicular thyroid carcinoma.

Initially, most bone metastases are asymptomatic. However, symptoms occur when metastases result in pathologic microfractures, vertebral collapse, or are invading neighboring soft tissues. As tumor cells proliferate, bone destruction occurs, leading to the appearance of osteolytic lesions. At the same time, new bone formation takes place, in order to balance destruction and repair the bone matrix defects. This leads to the appearance of osteoblastic bone metastases, and to a mixed pattern of lytic and sclerotic lesions. Some tumors, such as multiple myeloma, eosinophilic granuloma, or leukemia, are characterized by the presence of mainly lytic metastases.

BONE SCINTIGRAPHY AND SPECT/CT IN SPECIFIC MALIGNANT TUMORS

Breast cancer is characterized by a high incidence of bone metastases, which may be found in up to 70% of patients at autopsy (23). The diagnostic yield of bone scintigraphy is strongly influenced by the T- or N-stage, with no confirmed correlation with histological grading. Data about the sensitivity or specificity of skeletal scintigraphy vary substantially. Bone scintigraphy has shown very good results for diagnosis of metastatic breast cancer, with a sensitivity and specificity of 98% and 95%, respectively, positive and negative predictive values of 73% and 100%, and overall accuracy of 96% (24). Nevertheless, in the routine diagnostic setting, false positive and negative results may occur, leading to additional X-ray, CT, MRI, or PET evaluation. Coleman et al. reported a false-negative rate of 0.08% and 1.6% false-positives using plain radiographs for validation of bone scintigraphy (25).

There are several reasons for false-negative findings of bone scintigraphy. Many bone metastases, mostly in small bone lesions, are initially located at an intramedullar level, causing no cortical destruction or bone remodeling, and therefore may not be detected by bone scintigraphy. Small lytic lesions resulting in photopenic defects can easily be missed, particularly in the spine, because of poor contrast or the presence of superimposed activity. False-positive results may especially occur in elderly people due to nonmalignant tracer accumulation in processes such as osteoporotic fractures, osteochondrosis, spondylopathy, degenerative spondylarthrosis, or inflammatory conditions.

In an ongoing study in our facility, SPECT/CT was performed in 26 patients with known breast carcinoma and the results of SPECT, low-dose CT, and hybrid imaging were compared. Forty-four bone lesions were detected by bone scintigraphy in 26 patients. Planar scintigraphy and SPECT correctly classified 17/44 lesions, while SPECT/CT was true positive or true negative in 38/44 lesions. Most of the 44 lesions were caused by degenerative spine changes, which were identified on low-dose CT images (Fig. 1). The relationship between the intensity of uptake and the anatomical location of the foci played an important role in classifying these lesions. The initial poor specificity of SPECT in this patient cohort may have been due to the selection criteria and the fact that this group of patients had a high pretest probability for skeletal abnormalities due to osteoarthritic changes as well as bone metastases. However, following SPECT/CT, further diagnostic procedures were necessary in only 6/46 patients, therefore confirming the potential value of this one-step procedure in the diagnosis of metastatic bone disease.

Monitoring treatment response in breast cancer is complicated by persistent sclerotic changes on X-rays and the physiology of the healing response that results in increased radiotracer uptake. The presence of sites with increased uptake may indicate either tumor progression or treatment response (flare phenomenon). Reduction of tracer uptake after external beam radiation is highly indicative for a positive response (e.g., recalcification). In addition, tumor progression might be present in about 5% of the cases, despite improvement of scintigraphic findings in the skeleton (26). In this clinical setting, the value of SPECT/CT is limited.

In lung cancer, the patients benefit from accurate staging, which is essential for patient management and therapy. Although lung cancer is frequently associated with bone metastases, skeletal scintigraphy is performed mostly in symptomatic patients. The guidelines of the American Thoracic Society and the European Thoracic Society postulate that asymptomatic patients do not require skeletal diagnostic imaging even though the literature demonstrates that there is no correlation between symptoms and the presence of metastatic disease. Furthermore, chronic back pain is common (80%) in elderly patients and there is variability in the subjective perception of pain, making it difficult for the physician to differentiate between the different etiologies and severity of skeletal complaints (27,28). Although metastases occur in approximately 10% of surgically treated patients with stage I or II disease (29,30) who presumably were asymptomatic at the time of initial treatment, the role of preoperative screening for bony metastases in asymptomatic patients with lung cancer is still under discussion. The sensitivity of bone scintigraphy is greater than 95% for the detection of metastases, but this modality also has a high false-positive rate due to coincidental skeletal disorders (31). Schirrmeister et al. found an incidence of bone metastases of 14% to 22% among asymptomatic patients with stage I and II non small cell lung cancer who would have undergone unnecessary surgery or neoadjuvant chemotherapy if skeletal scintigraphy had not been performed (32).

Preliminary results of our group in an ongoing series in our department of patients with known lung cancer revealed significant improvement in lesion classification using SPECT/CT as compared to planar scintigraphy and SPECT. In five patients with known carcinoma, planar scintigraphy and SPECT correctly classified two lesions while 17/19 abnormal foci were defined as equivocal, making further diagnostic testing necessary. SPECT/CT correctly classified 17/19 lesions with focal tracer uptake, with only one false-positive and one equivocal lesion (Figs. 2 and 3). The correlation between physiologic and anatomic imaging information permitted correct characterization of bone changes. Benign bone lesions, mostly due to degenerative spine diseases, are easy to diagnose on CT, rendering the SPECT findings less significant. Also, recognition of extra-osseous tracer accumulations, as well

(A)

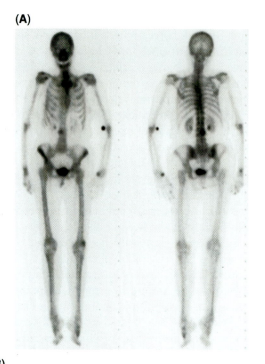

(B)

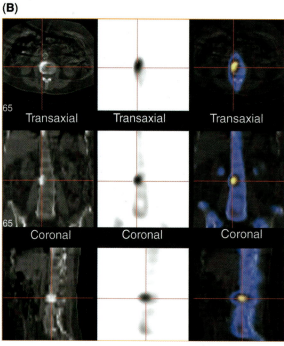

Figure 1 Seventy-five-year-old woman with breast cancer. (**A**) Planar whole-body bone scintigraphy shows focal tracer uptake in the lumbar spine. (**B**) SPECT/CT at the level of the lower thoracic and lumbar spine shows typical degenerative changes in the area of enhanced uptake in the L3 vertebral body. *Abbreviation*: SPECT/CT, single-photon emission computed tomography/computed tomography.

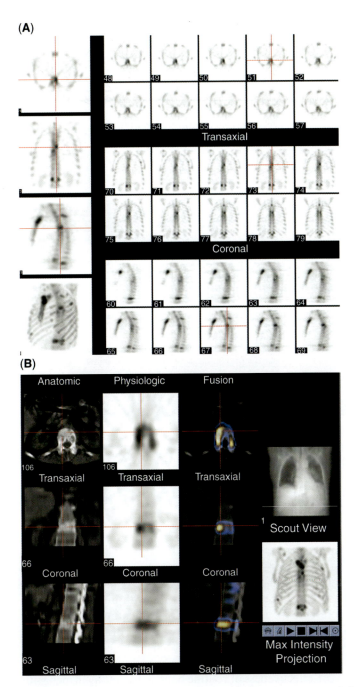

Figure 2 Fifty-eight-year-old patient with lung cancer revealed multiple foci of increased uptake in the axial skeleton on bone scintigraphy (not shown). (**A**): SPECT of the thoracic cage demonstrates the presence of sites of focally enhanced uptake in the thoracic and lumbar spine, sternum, and ribs. (**B**): SPECT/CT demonstrates the presence of a pathologic fracture in the L2 vertebra. Note the different density values in the second lumbar vertebral body, as seen on CT component. *Abbreviations*: SPECT, single-photon emission computed tomography; CT, computed tomography.

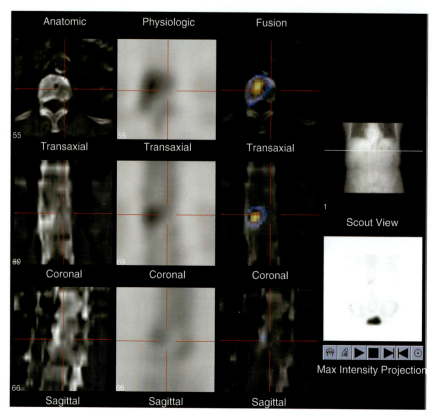

Figure 3 Seventy-four-year-old man with lung cancer showed a single focus of slightly increased tracer uptake on planar bone scintigraphy with no corresponding findings on plain radiography. SPECT/CT shows increased tracer uptake in the L1 vertebral body on SPECT with a pathological fracture in the body of the first lumbar vertebra with osteolysis, as demonstrated on low dose CT. *Abbreviations*: SPECT, single-photon emission computed tomography; CT, computed tomography.

as differentiation from paraneoplastic bone abnormalities such as hypertrophic osteoarthropathy, is facilitated by coregistration of anatomic–metabolic SPECT/CT data (33). These preliminary data indicate that SPECT/CT may substantially improve the diagnostic role of bone scintigraphy in patients with lung cancer.

Bone scintigraphy represents an important tool for evaluating patients with bone metastases of prostate cancer—the second leading cause of death from cancer among men. However, the role of bone scintigraphy has been lately redefined, in view of the event of new serum markers, including the prostate-specific antigen (PSA). An investigation regarding the cost-effectiveness of routine bone scintigraphy in men with a threshold PSA level of 10 ng/mL indicated a false-negative rate of 0.5% (34). It has been concluded, therefore, that staging including bone scan is unnecessary in patients with newly diagnosed and previously untreated prostate cancer, with a serum PSA level of < 10 ng/mL and no skeletal symptoms (35,36). However, the PSA level may differ significantly between patients with and without antiandrogeneic treatment. Bone scintigraphy has a high false-positive rate mainly due to degenerative disease but also to healing fractures, inflammatory processes, or Paget's disease. On the other hand, recognition of metastases on CT and conventional radiographs in patients with prostate cancer is facilitated by their osteoblastic pattern, showing focal or diffuse sclerosis of the bone. SPECT/CT image fusion is therefore expected to improve the diagnostic accuracy by increasing the specificity of positive findings on bone scintigraphy, by correctly classifying equivocal lesions, by identifying sclerotic bone changes due to metastatic prostate cancer or benign skeletal abnormalities (Fig. 4). In our series of 85 patients with prostate carcinoma, with 171 suspicious bone lesions detected with planar scintigraphy and SPECT, fused SPECT/CT correctly classified all bone abnormalities.

A variety of other malignancies may also benefit from the implementation of SPECT and SPECT/CT to bone scintigraphy (Fig. 5). Kim et al. found bone scintigraphy more sensitive than radiographs in the detection of metastatic renal cell carcinoma. There were 55 lesions showing increased tracer activity and seven lesions with decreased uptake (37). In patients with renal cell carcinoma, the CT component of SPECT/CT may allow for recognition of bubbly large bone metastases as well as for depiction of accompanying large soft-tissue tumors with bone scalloping.

Bone scintigraphy in combination with SPECT/CT can be also beneficial in assessing head and neck tumors (Fig. 6). However, assessment of this group of malignancies has become largely the domain of ^{18}F-FDG-PET.

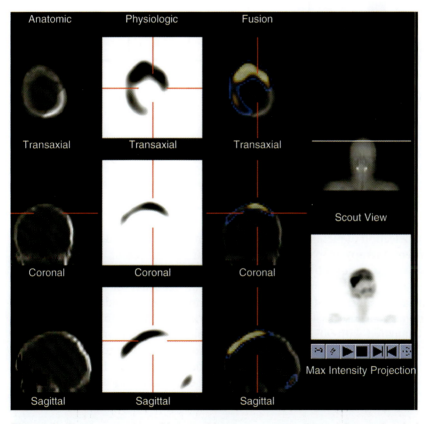

Figure 4 Forty-seven-year-old man with prostate cancer shows a focus of pathologic uptake in the skull on SPECT and decreased bone density accompanied by rarefaction of the trabecular structure on low dose CT, compatible with Paget's disease. Note the intact cortical bone, atypical for a large osteolysis. *Abbreviations*: SPECT, single-photon emission computed tomography; CT, computed tomography.

(A)

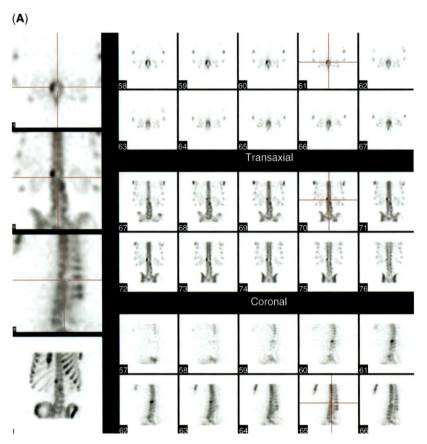

Figure 5 Seventy-four-year-old man with hepatocellular carcinoma. **(A)** SPECT shows a focus of increased tracer uptake in the T12 vertebra. **(B)** SPECT/CT localizes this focus to a large osteolytic lesion confirmed by **(C)** high-resolution CT of the corresponding vertebral body. *Abbreviations*: SPECT, single-photon emission computed tomography; CT, computed tomography.

Early detection of metastases is of value to ensure the efficient use of radiotherapy and chemotherapy in patients with testicular tumors. Bone scanning proved more sensitive than conventional radiography in the detection of bone metastases of seminomas or nonseminomas (38). Meijer et al. reported an incidence of 20% of bone metastases in patients with disseminated midgut carcinoids (39). The sensitivity for bone metastases is 44% for plain radiography and 70% for octreotide scintigraphy, compared with 90% for bone scintigraphy and 100% for MRI. The need for additional bone scintigraphy or MRI in patients with symptoms or suspected bone metastases on octreotide scintigraphy has been advocated. In one patient of our own series, there was a discrepancy between octreotide and bone scintigraphy with SPECT/CT precisely localizing bone metastases in the thoracic spine.

CONCLUSION

^{99m}Tc-MDP bone scintigraphy has been characterized as a procedure with a high sensitivity and low specificity. SPECT/CT appears to improve the overall diagnostic

(B)

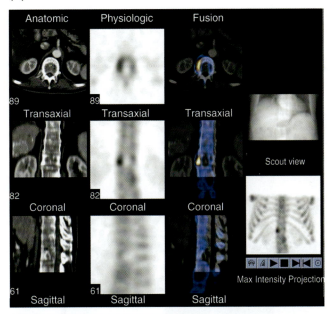

(C)

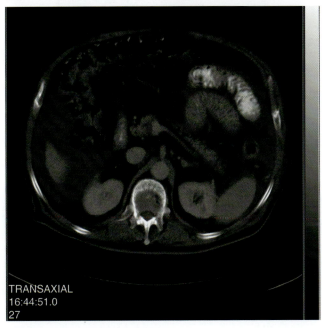

Figure 5 (*Caption on facing page*)

accuracy by excluding nonspecific causes of focal tracer uptake. The limitations of the technique are related to the nature of some osseous lesions, such as their small size, or the appearance of metabolic changes preceding anatomical findings. In the cases of very early skeletal tumor involvement without structural changes, MRI imaging is superior to CT.

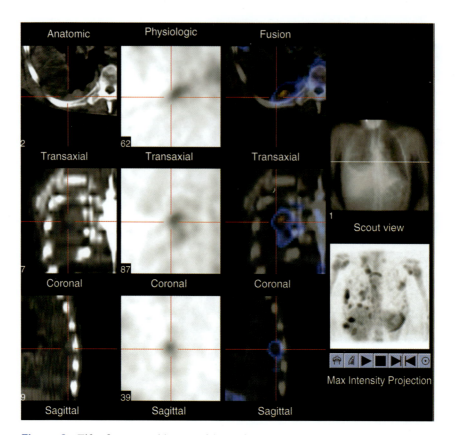

Figure 6 Fifty-five-year-old man with carcinoma of the head and neck. Bone scintigraphy showed multiple sites of abnormal tracer uptake in the thoracic cage, initially assumed to be localized in the ribs. SPECT/CT correctly identified them as extraosseous lesions in necrotic pleural metastases. No bone abnormalities were seen in the adjacent ribs. *Abbreviations*: SPECT, single-photon emission computed tomography; CT, computed tomography.

The development of new imaging devices combining SPECT and high-end CT may provide further improvement in the assessment of bone metastases. In addition, the role of MRI and PET should not be underestimated, since these imaging modalities have shown high potential for assessment of the skeleton. Skeletal scintigraphy is highly suitable for good quality image fusion. Bone SPECT/CT provides an improved procedure for the assessment of skeletal metastases in cancer patients.

REFERENCES

1. Murray IPC, Jell P, Hans van der Wall. Nuclear Medicine in Clinical Diagnosis and Treatment. 2nd ed. Churchill-Livingstone, 1998:1169–1170.
2. Edelstyn GA, Gillespie PJ, Grebell FS. The radiological demonstration of osseous metastasis: experimental observation. Clin Radiol 1967; 18:158–162.
3. Wahl RL, Quint LE, Cieslk RD, Aisem AM, Koeppe RA, Meyer CR. "Anatometabolic" tumor imaging. Fusion of FDG-PET with CT or MRI to localize foci of increased activity. J Nucl Med 1993; 34:1190–1197.

4. Shreve PD. Adding structure to function (invited commentary). J Nucl Med 2000; 41:1380–1382.

5. Eubank WB, Markoff DA, Schmiedl UT, et al. Imaging of oncologic patients: benefit of combined CT and FDG-PET in the diagnosis of malignancy. Am J Roentgenol 1998; 171:1103–1110.

6. Beyer T, Townsend DW, Brun T, et al. A combined PET/CT scanner for clinical oncology. J Nucl Med 2000; 41:1369–1379.

7. Malawer MM, Delaney TF. Treatment of cancer of metastatic cancer to bone. In: DeVita VT, Hellman S, Rosenberg SA, eds. Cancer Principles and Practice of Oncology. 4th ed. New york: JB Lippincott, 1993:2225–2245.

8. Durning P, Best JJ, Sellwood RA. Recognition of metastatic bone disease in cancer of the breast by computer tomography. Clin Oncol 1983; 9:343–346.

9. Taoka T, Mayr NA, Lee HJ, et al. Factors influencing visualization of vertebral metastases on MR imaging versus bone scintigraphy. Am J Roentgenol 2001; 176:1525–1530.

10. Colman LK, Porter BA, Redmond J, et al. Early diagnosis of spinal metastases by CT and MR studies. J Comput Assist Tomogr 1988; 12:423–426.

11. Porter BA, Shield AF, Olson DO. Magnetic resonance imaging of bone marrow disorders. Radiol Clin North Am 1986; 24:269–289.

12. Gold RI, Seeger LL, Bassett LW, et al. An integrated approach to the evaluation of metastatic bone disease. Radiol Clin North Am 1990; 28:471–483.

13. Aitchinson FA, Poon FW, Hadley MD, et al. Vertebral metastases and an equivocal bone scan: value of magnetic resonance imaging. Nucl Med Commun 1992; 13:429–431.

14. Even-Sapir E, Metser U, Flusser G, et al. Assessment of malignant skeletal disease: initial experience with ^{18}F-Fluoride PET/CT and comparison between ^{18}F-Fluoride PET and ^{18}F-Fluoride PET/CT. J Nucl Med 2004; 45:272–278.

15. Hawkins RA, Choi Y, Huang SC, et al. Evaluation of skeletal kinetics of fluorine ^{18}F-fluoride ion and PET. J Nucl Med 1992; 33:633–642.

16. Bar-Shalom R, Yefremov N, Guralnik L, et al. Clinical performance of PET/CT in evaluation of cancer: additional value for diagnostic imaging and patient management. J Nucl Med 2003; 44:1200–1209.

17. Sasaki M, Ichiya Y, Kuwabara Y, et al. Fluorine-18-Fluorodeoxyglucose positron emission tomography in technetium-99m-hydroxymethylenediphosphate negative bone tumors. J Nucl Med 1993; 34:288–290.

18. Hendler A, Hershkop M. When to use bone scintigraphy. It can reveal things other studies cannot. Bone Scintigraph 1998; 104(5):54–69.

19. Clain A. Secondary malignant disease of bone. Br J Cancer 1965; 19:15–29.

20. Krishnamurthy GT, Tubis M, Hiss J, et al. Distribution pattern of metastatic bone disease. A need for total body skeletal image. JAMA 1977; 237:2504–2506.

21. McNeil BJ. Value of bone scanning in neoplastic disease. Semin Nucl Med 1984; 14:277–286.

22. Abrams HL, Spiro R, Goldstein N. Metastases in carcinoma. Analysis of 1000 autopsied cases. Cancer 1950; 2:74–85.

23. Bares R. Skeletal scintigraphy in breast cancer management. Q Nucl Med 1998; 42:243–248.

24. Crippa F, Seregni E, Agresti R, et al. Bone scintigraphy in breast cancer: a ten-year follow-up study. J Nucl Biol Med 1993; 37:57–61.

25. Coleman RE, Rubens RD, Fogelman I. Reappraisal of the baseline bone scan in breast cancer. J Nucl Med 1988; 29:1045–1049.

26. Schneider JA, Divgi CR, Scott AM, et al. Flare on bone scintigraphy following Taxol chemotherapy for metastatic breast cancer. J Nucl Med 1994; 35:1748–1752.

27. Robertson JT. The rape of the spine. Surg Neurol 1993; 39:5–12.

28. Deyo RA, Tsui-Wu Y-J. Descriptive epidemiology of low-back pain and its related medical care in the United States. Spine 1987; 12:264–268.

29. Michel F, Soler M, Imhof E, et al. Initial staging of non-small-cell lung cancer: value of routine radioisotope bone scanning. Thorax 1991; 46:469–473.

30. Yano T, Yokoyama H, Inoue T, et al. The first site of recurrence after complete resection in non-small cell carcinoma of the lung. Comparison between pN0 disease and pN2 disease. J Thorac Cardiovasc Surg 1994; 108:680–683.
31. Citrin DL, Besssent RG, Greig WR. A comparison of the sensitivity and accuracy of the 99m-Tc-phosphate bone scan and skeletal radiograph in the diagnosis of metastases. Clin Radiol 1977; 28:107–117.
32. Schirrmeister H, Arslandemir C, Glatting G, et al. Omission of bone scanning according to staging guidelines leads to futile therapy in non-small cell lung cancer. Eur J Nucl Med Mol Imag 2004; 31:964–968.
33. Uchisako H, Suga K, Tanaka N, et al. Bone scintigraphy in growth hormone-secreting pulmonary cancer and hypertrophic osteoarthropathy. J Nucl Med 1995; 36:822–825.
34. Oesterling JE. Prostate specific antigen: a critical assessment of the most useful tumor marker for adenocarcinoma of the prostate. J Urol 1991; 145:907–923.
35. Wymenga LFA, Boomsma JHB, Groenier K, et al. Routine bone scans in patients with prostate cancer related to serum prostate-specific antigen and alkaline phosphatase. Br J Urol 2001; 88:226–230.
36. Chybowski FM, Larson Keller JJ, Bergstrahl EJ, et al. Predicting radionuclide bone scan findings in patients with newly diagnosed, untreated prostate cancer: prostate specific antigen is superior to all other parameters. J Urol 1991; 145:313–318.
37. Kim EE, Bledin AG, Guttierez C, et al. Comparison of radionuclide images and radiographs for skeletal metastases from renal cell carcinoma. Oncology 1983; 40:284.
38. Francisco J, Braga HN, Arbex MA, et al. Bone scintigraphy in testicular tumors. Clin Nucl Med 2001; 26:117–118.
39. Meijer WG, Van der Veer E, Jager PL, et al. Bone metastases in carcinoid tumors: clinical features, imaging characteristics, and markers of bone metabolism. J Nucl Med 2003; 44:184–191.

7

^{67}Gallium SPECT/CT in Imaging of Lymphoma

Rachel Bar-Shalom and Ora Israel
*Rambam Health Care Campus, B. Rapaport School of Medicine,
Technion—Israel Institute of Technology, Haifa, Israel*

INTRODUCTION

^{67}Ga scintigraphy (GS) plays an important role in the assessment of lymphoma (1,2), but its acceptance by the nuclear medicine community for routine clinical use has been slow, partly because ^{67}Ga is regarded as an old-fashioned radiotracer with unclear significance. Improved imaging devices, techniques, and protocols have resulted in a large amount of clinical data establishing ^{67}gallium as a sensitive agent to determine viability of lymphoma, providing functional information only recently matched by ^{18}F-Fluoro-deoxyglucose–positron-emission tomography (FDG-PET). The value of GS in monitoring response to therapy, detection of recurrent disease, and prediction of prognosis in Hodgkin's disease (HD) and non-Hodgkin's lymphoma (NHL) is now well established (3). In numerous studies comparing the performance of GS to conventional imaging such as regular X-ray, ultrasound (US), magnetic resonance imaging (MRI), and mainly computed tomography (CT), GS has proven superior in defining the response to treatment, in the characterization of posttherapeutic residual masses, and in early detection of relapse (4–7). GS has also been proven to be a good predictor of long-term prognosis in both HD and NHL. Performed as early as after the first cycle of chemotherapy, GS is predictive of long-term outcome (8–12).

Nevertheless, several shortcomings continue to limit the use of this modality. These limitations include (i) inherently low or non–gallium-avidity of several histological types of lymphomas, (ii) the spatial resolution capabilities of currently available gamma camera technology for detection of small lesions, (iii) the normal physiologic biodistribution of this radiotracer, and (iv) the nonspecific increased gallium uptake in benign or nonlymphomatous processes. Some of the latter drawbacks are enhanced by the inherent lack of clear anatomic detail of nuclear medicine images in general and GS in particular. Uncertainty in the anatomic localization of areas of increased tracer activity may hamper their characterization as normal or abnormal findings.

The complementary role of morphologic and functional imaging in the evaluation of lymphoma is well established. Use of an anatomic template for precise localization of functional imaging data has the potential of improving the accuracy of interpretation for both modalities (13). Accurate coregistration of different imaging data sets is of incremental value for the interpretation of both nuclear medicine and conventional imaging procedures, with further clinical impact on patient management. GS interpretation demands a thorough acquaintance with its normal and abnormal patterns (14,15). Anatomic localization of gallium uptake foci by fused imaging enhances the clinical significance of GS results.

TECHNICAL ASPECTS OF GALLIUM SPECT/CT IN LYMPHOMA

The quality of GS is dependent on several technical factors including the use of appropriate high doses of ^{67}gallium; the use of dual-head gamma cameras with appropriate collimation, always performing planar and single-photon emission computed tomography (SPECT) studies (at 48 or 72 hours after the injection); and the use of delayed images (up to seven days postinjection) to allow for the clearance of bowel tracer activity, thus enhancing image contrast as a result of decreased background activity (15–17).

SPECT improves image contrast with the enhancement of both tumor activity and nonspecific gallium uptake of nonlymphomatous origin (18). While SPECT can detect focal regions of gallium accumulation, SPECT/CT can improve specificity by excluding the presence of malignancy in anatomic sites of benign or physiologic tracer uptake and detection of lymphoma involvement in the close proximity of organs or anatomical regions of normal tracer biodistribution.

The concomitant acquisition of X-ray CT transmission with ^{67}gallium SPECT permits the production of CT-based attenuation-corrected GS images. Although this may enhance a nonlymphomatous uptake, the complementary use of attenuation-corrected SPECT fused with CT anatomical data assists in identifying nonspecific uptake (19).

Quantitation of gallium uptake in SPECT is difficult to perform and is of limited accuracy with routine reconstruction algorithms and due to the attenuation of tracer activity, particularly in the abdomen (20). With improved iterative processing protocols, combined with the attenuation correction provided by the hybrid SPECT/CT technology, gallium emission data used in conjunction with CT transmission data may be valuable in obtaining a reliable quantitative data for the assessment of gallium uptake in lymphoma lesions. The feasibility of quantitation has been described in a study of a phantom model that incorporated the structural information obtained during hybrid SPECT/CT into the reconstruction process of SPECT data, with limited improvement in quantitative measurements (21). Quantitation of GS is still challenging and has not been extensively explored. If successful, it can provide GS with additional information, in particular a more accurate differentiation between malignant and benign gallium-avid lesions, and better assessment of response to therapy and prediction of prognosis.

Interpreting GS may be difficult for the inexperienced reader. Relatively high interobserver interpretation variability has been reported in the GS assessment of lymphoma patients (22). SPECT/CT can expedite learning for inexperienced readers and increase interpretation confidence and imaging throughput in clinical routine.

Registration of GS with anatomic imaging modalities in lymphoma patients has been described using retrospective registration methods with both external (23) and internal (24) landmarks, as well as with nonimage-based registration techniques using integrated devices that combine SPECT and CT mechanically (25,26). The advantage

of retrospective coregistration lies in the ability to combine two studies of high diagnostic quality. The correlation of separately performed state-of-the-art GS with high-
resolution contrast-enhanced CT provides not only the precise anatomic localization
of GS findings but also their correlative morphologic characteristics for further data
confirmation. These methods, however, demand prospective logistics and patient preparation; they are also prone to misregistration due to differences in patient positioning and in organ location and volume between the two studies (27). Hybrid imaging
devices allow the sequential acquisition of GS SPECT and CT in a single session using
the same imaging table without changing the patient's position (27,28). This procedure reduces potential registration inaccuracies caused by patient motion or by organ
displacement. Hybrid SPECT/CT provides both high coregistration accuracy and
attenuation correction of emission data. The limitations of the first clinically available
hybrid devices are related to the use of low-dose CT scan acquired over a relatively
long time and covering a limited field of view in each acquisition session. The low-
resolution noncontrast-enhanced transmission images obtained are useful as an accurate anatomical map for localization of SPECT data but are usually not of diagnostic
quality. However, the precise coregistration obtained between SPECT and low-dose
CT can guide the performance and review of a diagnostic CT (28).

   Hybrid gallium imaging of lymphoma patients using the SPECT/CT device
equipped with a low-dose CT component is performed in two interchangeable steps.
The acquisition of the gallium emission part is performed using a variable dual-head
gamma camera with a field of view of 540×400 mm (27,28). Gallium-SPECT is performed with a 1.28 zoom in 360° rotation with a 6° angle step, for 55 seconds per
frame and a total acquisition time of about 30 minutes. A matrix of 128×128 is used
for reconstruction, using the iterative ordered subset expectation maximization
(OSEM) algorithm. The CT component consists of an X-ray tube with a working
power of 140 kV and 2.5 mA. A set of detectors is fixed on the opposite side of
the gamma camera gantry, which rotates around the patient together with the
gamma camera detectors. Low-dose CT is performed in a 220° half-scan acquisition
with a rotation speed of 2.8 r.p.m., at 16 seconds per slice. The axial scan of 40 slices,
1 cm wide, is completed in about 10 minutes. CT transmission images are reconstructed on a 256×256 matrix size with a pixel size of 4 mm. Transmission data
are integrated into the gallium emission database and used for attenuation correction
of SPECT images and for generation of fused images of matching pairs of X-ray and
SPECT images. Final SPECT and CT data can be reviewed separately and as fused
images in three planes.

   The typical effective radiation delivered by a routine tracer dose injected for
GS in adults is around 33 mSv. The additional radiation dose delivered to the
patients by a low-dose CT scan ranges from 0.13 cGy at the center to 0.5 cGy at
the body surface, which is about 20% of that of a diagnostic CT (13,28).

   There are few reports of the use of SPECT/CT in GS for assessment of lymphoma patients (23–26). The indication for hybrid imaging in this group of
patients is under evaluation. The types of lymphoma, the body regions, or clinical
problems for which hybrid imaging may be of most benefit remain to be defined.
It is currently suggested that for routine clinical use of GS in lymphoma, SPECT/
CT should replace SPECT only in case of suspicious findings in clinical examinations or in other imaging modalities, or when equivocal findings are detected on
planar whole-body GS.

   SPECT/CT has been shown to be of value for CT interpretation as well.
The precise localization of functional imaging data to an anatomic site can direct the

retrospective diagnosis of structural lesions, previously unrecognized within the load of anatomic details provided by high-resolution, contrast-enhanced CT (23–26).

CLINICAL ASPECTS OF GALLIUM SPECT/CT IN LYMPHOMA

SPECT/CT imaging can provide solutions to some of the drawbacks of GS in lymphoma patients. The advantages of SPECT/CT include (i) the differentiation between physiologic tracer uptake and foci of lymphoma involvement even when in close proximity; (ii) precise localization of abnormal gallium uptake to specific nodal stations or organs; (iii) delineation of viable residual lymphoma tissue within heterogeneous residual masses; and (iv) the retrospective identification of lesions that were initially ignored on GS and CT.

SPECT/CT Definition of Physiologic and Benign GS Patterns in Lymphoma Patients

In the region of the head and neck, lacrimal and salivary glands are the sites of physiologic gallium uptake. This may be a source of difficulty in interpretation and is a pitfall of GS. Uptake in salivary glands is usually symmetric and is of mild intensity. Inflammatory or obstructive processes, prior unilateral resection, or radiation therapy may all create focal, intense, and asymmetric uptake, which can be difficult to differentiate from the disease in neighboring regional lymph nodes. Comparison with Tc-pertechnetate uptake has been suggested to differentiate nodal from salivary gland uptake (15). It may be difficult to separate asymmetric benign lacrimal gland activity from retro-bulbar or orbital lymphoma. SPECT/CT fusion may pinpoint the precise anatomic location of uptake in salivary or lacrimal glands, further defining its benign nature.

In the chest, the mediastinum is a common site of disease involvement in both HD and NHL. Nevertheless, several nonlymphomatous patterns may appear on both initial and posttherapy GS studies. Bilateral hilar uptake of benign etiology has been described in 36% of 101 consecutive lymphoma patients, appearing mainly in elderly patients (29). Hilar uptake may be attributed to smoking, a history of previous granulomatous disease, and to effects of chemotherapy. Several methods have been suggested to resolve this diagnostic dilemma, including the use of [201]thallium scintigraphy to identify malignant hilar sites and semiquantitative methods for estimating the intensity of tracer uptake (30,31). A quantitative SPECT method has found benign hilar gallium uptake to be significantly lower than uptake in malignant hilar disease (29). SPECT/CT coregistration can facilitate the correlation of tracer uptake with CT findings (32). Exact definition of nonspecific gallium uptake in the hilar region was obtained with coregistration of GS SPECT and CT (33).

Increased tracer uptake in the thymus is another challenging problem for GS interpretation in the thorax, especially after treatment. Rebound thymic hyperplasia is common and has been described in children and young adults months after completing therapy (34). Benign thymic gallium uptake is usually characterized by its bilobar arrow-shaped pattern. The time sequence of the appearance of a gallium-avid thymus, in relation to prior therapy, is also usually suggestive of its benign nature. Nevertheless, defining the nature of abnormal gallium uptake in the anterior mediastinum may still be impossible, sometimes demanding tissue confirmation for final diagnosis. Fusion images may be helpful in this instance by providing precise localization of this type of uptake to the anatomic site of the thymus on CT, even when it appears in an unusual location. In such atypical locations, gallium SPECT/CT may point to a need for specific investigation with US or MRI before resorting to more invasive diagnostic procedures (35).

In the abdomen and pelvis, physiologic gallium uptake in several abdominal organs such as the liver and spleen, and its excretion via the urinary and more prominently the gastrointestinal tract, make this area difficult for interpretation. Bowel excretion is the main route of gallium excretion from the body after the first 24 hours following injection. Although usually of linear diffuse pattern and of mild intensity, bowel and gastric uptake of gallium may be focal and intense. Physiologic bowel activity may either obscure or mimic uptake in lymphoma sites. Cleansing enemas have been suggested as an aid, but this is an inconvenient procedure and may result in bowel irritation and inflammation leading to a further increase in tracer uptake. Delayed imaging of up to 7 to 14 days postinjection is currently the proposed solution for avoiding bowel uptake interference with image review (14,15). This protocol is noninvasive but it is inconvenient, costly, and not always effective. The precise coregistration of GS with anatomic CT data can define physiologic tracer activity in the bowel and at the same time separate it from uptake in adjacent lymphoma lesions. Fusion can distinguish physiologic bowel or gastric uptake from suspected adjacent regional mesenteric or peritoneal lymph node involvement. SPECT/CT imaging can therefore obviate the need for bowel preparation or for repeat delayed imaging. At baseline, defining both physiologic bowel uptake and abnormal activity in a known lymphoma lesion in its close vicinity can confirm the gallium-avidity of the lesions, which is important for future assessement of response to therapy. Even in the absence of overt structural pathology, SPECT/CT can direct further investigation to specific gastrointestinal sites in question of disease involvement.

Benign gallium uptake in the abdominal-pelvic region, in the presence of inguinal hernia, aortic aneurysm, abdominal hematoma, or vascular graft has been described as a source of misinterpretations of GS in lymphoma patients (15,36,37). Hybrid imaging can be valuable for the characterization of such uptake unrelated to lymphoma by demonstrating corresponding morphologic abnormalities of benign etiology on the CT. Renal uptake, being the main excretory route of gallium within the first 24 hours after injection, should not be seen on GS after this time. Precise SPECT/CT localization of focal renal uptake in hydronephrotic calyces or in an infected cyst can exclude the possibility of lymphoma in such sites.

In the evaluation of disease involvement in the musculo-skeletal system and in peripheral sites, one of the major advantages of GS is its whole-body scanning range, which provides information about peripheral regions not routinely screened by conventional morphologic modalities (15). ^{67}Gallium is a bone-seeking agent and is therefore taken up by various skeletal lesions, not necessarily of malignant origin. Uptake in bone fractures and in degenerative changes in the spine may cause false positive GS interpretations in lymphoma patients. Gallium uptake has been described also in regions of muscular strain (38). Precise SPECT/CT localization of gallium uptake to rib fractures, osteophytes, degenerative changes in vertebral articular facets, or the spinous process, within the joint space or along a muscle group, can exclude lymphomatous involvement of the bone or adjacent soft tissues (Fig. 1) (15).

Fused SPECT/CT can differentiate uptake in disease-involved peripheral nodal stations such as the axillary, inguinal, popliteal, and epitrochlear regions from benign uptake in the same regions. Axillary uptake may represent adenopathy as well as increased activity at skin folds (15). Uptake in the epitrochlear region may be the

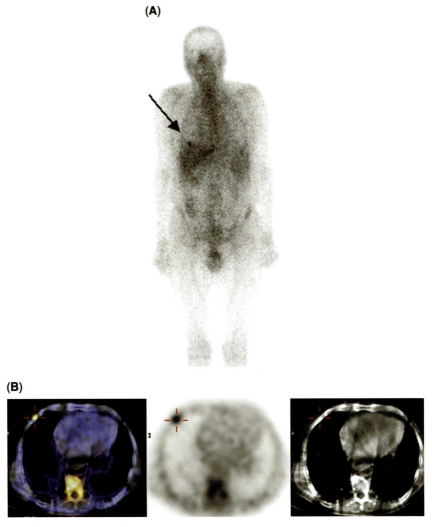

Figure 1 GS SPECT/CT for exclusion of disease. A patient with NHL following surgical removal of a tonsil considered as the single site of lymphoma involvement was referred for postsurgical restaging with GS. Planar whole-body GS in the anterior view (**A**) demonstrates a focus of abnormal uptake in the anterior lower chest (*arrow*). SPECT/CT (**B**, *left*) localizes the abnormal gallium focus seen on SPECT (**B**, *center*) to a fractured rib seen on CT (**B**, *right*). *Abbreviations*: CT, computed tomography; GS, ^{67}Ga scintigraphy; NHL, non-Hodgkin's lymphoma; SPECT, single-photon emission computed tomography.

artefactual result of faulty injection or tracer contamination. Nonmalignant uptake can also appear at sites of previous biopsy or surgery, and in regions where central lines, stents, or pacemakers are placed. The benign nature of all these suspicious foci can be ascertained by their precise anatomic localization on fused images. Other, previously unrecognized benign processes may be characterized by the use of hybrid imaging. Recently, a benign pattern of increased radiotracer uptake in metabolically active brown fat of lymphoma patients has been recognized by hybrid imaging using both gallium and FDG (39,40).

GS SPECT/CT in the Detection and Characterization of Lymphoma Lesions

At initial staging, lymphoma may involve multiple and variable nodal sites above and below the diaphragm, as well as extranodal regions. Accurate identification of disease sites enables accurate staging—a major determinant of therapy and prognosis in both HD and NHL (14). GS is considered inferior to CT at initial staging of lymphoma, showing relatively low site-based sensitivity, with the abdominal-pelvic region considered particularly challenging (14,41–43). The main function of GS before therapy is, therefore, to demonstrate the tracer avidity of a particular lymphoma in an individual patient. This is of utmost importance for enabling further assessment of response to therapy in sites of initial disease, for predicting prognosis, and for early detection of relapse during follow-up. Nevertheless, GS may detect additional disease sites not identified by conventional imaging, and may exclude disease in sites that appear suspicious on other imaging modalities. GS was reported to improve initial staging by upstaging or down-staging disease in 7% and 16% of lymphoma patients, respectively (41).

SPECT/CT can be useful in the precise identification and characterization of disease sites at various locations. In the region of the head and neck, gallium SPECT/CT can be of value in the diagnosis of lymphoma of the thyroid, a rare extranodal site of disease, and in its differentiation from involvement of adjacent cervical adenopathy (44). In the chest, SPECT/CT can define the precise relationship of gallium uptake to one of the various adjacent thoracic structures (45). Precise differentiation of hilar from mediastinal involvement may alter disease stage. SPECT/CT can differentiate between paramediastinal lung lesion (indicative of stage IV disease) and mediastinal adenopathy, as well as between peripheral lung involvement and a rib lesion.

Difficulties in GS interpretation in the abdominal-pelvic region are due to the fact that various organs can demonstrate both physiologic and pathologic gallium uptake. In the left hypochondrium, SPECT/CT can separate focal splenic, gastric, renal, or adrenal disease involvement from common adjacent lymphadenopathy. In the right hypochondrium, SPECT/CT may be of value in localizing abnormal gallium uptake to liver, renal, or neighboring nodal disease (15). SPECT/CT can localize tracer uptake in diseases involving mesenteric, retroperitoneal, or pelvic lymph nodes, and differentiate it from physiologic bowel activity (Fig. 2). Lymphomatous involvement of ovary, cervix, or prostate has been described anecdotally, suggesting a potential advantage of SPECT/CT for precise anatomic localization of gallium uptake in such rare sites of disease (46–48).

The availability of diagnostic contrast-enhanced CT for image coregistration may aid in the detection and differential diagnosis of abdominal-pelvic pathology. When this is achieved with retrospective image fusion of separately performed GS and CT, registration may be difficult because of displacement of the bowel loops and changes in bladder and bowel content between the two procedures. Non-enhanced low-resolution CT images provided by SPECT/CT systems cannot usually provide diagnostic morphologic data on gastrointestinal abnormalities. Nevertheless, the precise localization of uptake to the gastrointestinal tract may indicate the need for further endoscopic diagnostic investigation.

The lung–liver interface, a difficult region for the interpretation of GS, may also be challenging for coregistration with separately performed CT. This is due to the different diaphragmatic position on the diagnostic CT, performed during

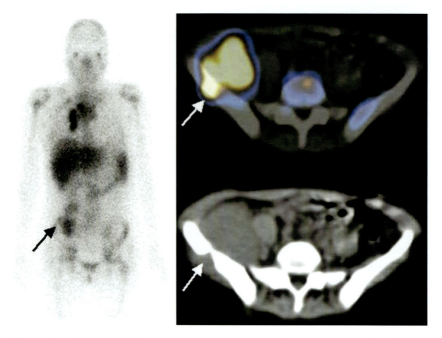

Figure 2 GS SPECT/CT for precise characterization of abdominal gallium uptake. A patient with HD was referred for staging with GS. Planar whole-body GS in the anterior view (*left*) performed seven days postinjection demonstrates several foci of abnormal uptake in the mediastinum, right hilum, and the right infraclavicular region. There is an additional region of increased tracer uptake in the right pelvis (*arrow*), which could represent either physiologic bowel activity or lymphomatous involvement of pelvic lymph nodes, bowel, or bone. Physiologic bowel activity is demonstrated, even on this delayed GS image, as linear mildly increased uptake along the upper abdomen. The SPECT/CT image (*right, top*) precisely characterizes gallium uptake as lymphomatous involvement in a right pelvic mass involving the soft tissues and adjacent iliac bone, as seen on CT (*right, bottom*). *Abbreviations*: CT, computed tomography; GS, ^{67}Ga scintigraphy; HD, Hodgkin's disease; SPECT, single-photon emission computed tomography.

breath hold, and on the GS acquired during a longer period of time. SPECT/CT systems may provide better registration in this region, given the similar shallow respiratory status during the acquisition of both imaging steps (27,28).

SPECT/CT can define lymphomatous bone involvement adjacent to or as a direct spread of disease in the soft tissues (Fig. 3). Hybrid imaging can exclude skeletal disease by differentiating vertebral from prevertebral retroperitoneal nodal involvement, both in the abdomen and in the chest where posterior mediastinal lymph-node involvement is frequent.

Bone marrow involvement is an important indicator of stage and prognosis in lymphoma. Its diagnosis can be impaired by inhomogeneous distribution, leading to biopsy sampling errors. Although GS is a sensitive tool for the detection of bone lymphoma, its role for detection of bone marrow involvement is questionable (49). SPECT/CT localization of gallium uptake within the medullary portion of the bone may suggest bone marrow disease (15).

With therapy, lymphoma is a curable malignancy, and complete remission is achieved in up to 80% of patients (14). Nevertheless, recurrent lymphoma is often unpredictable, involving old as well as previously uninvolved organs or nodal sites (7). CT

T stage. Melanoma patients are staged more accurately following SN assessment (4,33–35). A multicenter trial conducted by Morton et al. in 1784 melanoma patients reported a five-year survival of 90% to 95% if the SN was tumor-free compared with 65% if positive (24,25).

Assessment of lymph node involvement represents the best prognostic factor for breast cancer survival. Early and accurate identification of lymph node (LN) involvement enables the use of early systemic chemotherapy and local radiotherapy and may improve long-term survival. Incomplete SN resection results in inaccurate staging, and may lead to inadequate therapeutic decisions (36).

SENTINEL NODE IDENTIFICATION

The Unpredictability of Lymphatic Drainage

Scintigraphic lymph node mapping after radiocolloid injection has revealed unsuspected routes of lymphatic drainage. Lymphatic vessels may cross the midline to reach contralateral SNs in expected and in unexpected node fields (37–47). The density of the lymphatic system varies in different anatomic regions, and is highest in the skin of the head and neck region, followed by the shoulders and posterior trunk (48). Tumor drainage is complex due to the possible presence of in-transit or aberrant nodes. In-transit (interval) nodes lie along the course of lymphatic collecting vessels, between the primary tumor site and a draining node field. Aberrant nodes are located outside a standard drainage basin (49–51). Lymphatic drainage may be altered by infection, trauma, and previous surgery. SNs in unusual locations may be overlooked with surgical localization using blue-dye or probe techniques, which are limited to "predictable" drainage basins (2,32,51).

Identification of all true SN(s) of an individual tumor is a prerequisite for accurate nodal staging. Today, in most centers, SN identification is a multidisciplinary effort, combining the techniques of preoperative lymphoscintigraphy followed by the use of a hand-held gamma-probe to identify radioactive nodes during surgery, as well as injection of patent-blue-dye at surgery (1,4,11,37,52,53). Lymphoscintigraphy, the preoperative scintigraphic SN mapping, is the surgical "road-map" and can demonstrate the unpredictability of lymphatic drainage patterns in the individual patient (38). Scintigraphic SN mapping before surgery is of particular value in tumors located in regions with unpredictable lymph node drainage, such as the trunk, shoulder, and the head and neck (54,55).

Intraoperative Identification Techniques

A blue dye is injected intraoperatively to identify that the SN has also been used. This technique confirms the location of SN(s) by making the afferent lymphatic vessel coming from the injection site visible. The blue dye may, however, move also to more distally located nonsentinel (second-echelon) nodes (56).

The gamma-probe guides the dissection toward a SN before the surgeon can see it. Histopathological examination is performed on the lymph node with the highest counting rate and on additional lymph nodes showing uptake of the radiotracer. Some authors have suggested removal of nodes when the counting rate exceeds 20% of the hottest node (4). Complete removal of SNs is confirmed by reduction of the count rate in the surgical area to background levels.(4,13,57–60).

Lymphoscintigraphy

To facilitate topographic localization, the body contour may be outlined using various techniques (4,57,61). A hand-held point source has been used for many years. The method can achieve good, identifiable body contour. Its quality depends on the speed of movement of the source, the patient's habitus, and the type of source. The radiation dose delivered to the patient by a 2-MBq point source is negligible. A ^{57}Co flood source placed beneath the patient can also be used for transmission imaging providing a body silhouette, while simultaneously imaging the SN. The use of this outlining technique may occasionally lead to the loss of node identification. Some authors therefore suggest repeat imaging without the transmission source (37). The radiation dose to the patient for an SN study is approximately 200 to 400 μSv (62,63). The use of the ^{57}Co-flood source delivers up to 70 μSv (63), leading to a 25% increase in the total patient study dose. A ^{153}Gd line source is available in some new-generation cameras for attenuation correction of single-photon emission computed tomography (SPECT) images. The source can also be used to produce body-outline transmission images. Lymph node and transmission images are acquired simultaneously; images can be acquired in any desired angle. Alternatively, the use of the ^{153}Gd-line source is associated with a low additional radiation dose to the patient, approximately 3 μSv per view, which represents an increase of 1% in the total dose for the SN study (62).

SPECT imaging can potentially improve SN(s) identification and overcome some of the limitations of planar SN scintigraphic mapping because of contrast enhancement and improved resolution. However, reconstructed SPECT images following the radiocolloid injection are not informative because of the lack of anatomical landmarks, unless there is a corresponding computed tomography (CT) study.

In 29 melanoma patients, SPECT images have been fused with full-dose previously acquired diagnostic-CT images to assess the presence of hot pelvic lymph nodes (64). The use of fusion allowed for improved detection of SNs on SPECT, as well as better localization of pelvic lymph nodes based on the anatomical information provided by CT. Registration of separately performed SPECT and CT, however, may be a complicated task. In the pelvic region, registration of bony structures may facilitate the registration process. In the torso, however, errors in alignment of the two studies may evolve from different patient positioning, as well as movement of internal organs. Registration of separately performed SPECT and CT requires the use of external or internal landmarks, or a combination of both, as well as special registration software (65,66). The radiation dose associated with the addition of a diagnostic CT to presurgical SN identification may reach 20 mGy.

The introduction of a hybrid imaging device consisting of a low-dose CT and a gamma camera installed on a single gantry enables the transmission (low-dose CT) and emission (SPECT) to be performed in a single setting without changing the patient's position. Automatic fusion of images of both modalities is therefore obtained. The low-dose CT study results in an additional radiation dose ranging from 1.3 mGy at the center to 5 mGy at the surface of the body (67).

SPECT/CT acquisition has become an integral part of the lymphoscintigraphic procedure in our department, using ^{99m}Tc-rhenium colloid (TCK-17, CIS International) with a particle size of 50 to 100 nm. This tracer contains gelatin and has a relatively slow migration time from the injection site, a slow traveling time through the lymphatic vessels, prolonged accumulation in the SN(s), and tends less to move on to the more distally located nodes. Approximately 3.06% ± 0.10% of the injected dose is bound after one hour at the lymph node level, and this percentage reaches $3.83\% \pm 0.16\%$ at three hours after administration (13,68). Lymphoscintigraphy can be performed one day prior to surgery without concern about the operation room schedule. When nodes are not clearly depicted early during imaging, static acquisition can be repeated at 4 to 6 hours, or even 24 hours after injection, just before surgery.

SPECT/CT is performed following dynamic and static image acquisition using a hybrid system (Millennium VG^{TM} dual-head variable angle camera & Hawkeye^{TM}, GE Medical Systems, Milwaukee). SPECT acquisition parameters for SN detection include a matrix size of 128×128, an anterior L-mode rotation of $180°$ with a $3°$ angle step and a 25 second frame. For transmission data, a "half-scan" acquisition is performed over $220°$, with 16 seconds for each transaxial slice. The X-ray tube operates at 140 kV, 2.5 mA, with a copper filter to reduce the patient dose from soft X-rays. The full field of view consisting of 40 slices is completed in 10 minutes. Cross-sectional attenuation images in which each pixel represents the attenuation of the imaged tissue are produced. SPECT data are corrected and reconstructed using a filtered back projection with a Metz filter, or iteratively with the ordered subsets expectation maximization (OSEM) technique. Fused SPECT/CT images are available to localize the SN(s) (68).

INTERPRETATION OF LYMPHOSCINTIGRAPHY

Interpreting lymphoscintigraphy should allow the following:

  i. identify *all* draining basins at risk for metastatic disease as well as all hot SN(s). When several hot nodes are detected, a particular effort should be made to differentiate multiple nodes belonging to a single basin from multiple SN(s) belonging each to a separate basin;
  ii. precisely localize the SN (or nodes), so that limited incision biopsy may be performed in a well-defined surgical field; and
  iii. identify and localize in-transit, aberrant, and other uncommon nodal sites for possible harvesting and histopathological evaluation.

Correlation Between Intensity of Uptake and SN(s)

While not all hot lymph nodes are SN(s), the SN is also not necessarily the hottest node (37,69,70). A deeper node, for instance, may apparently be less active when compared with a superficial one on planar images. SN may also accumulate only a small amount of radiocolloid due to a massive metastatic infiltration (1,4,71). It is therefore important to record *all* detected lymph nodes, even if they show less intense uptake. It has been previously reported that if only the hottest nodes in each basin would have been removed, 13% of the basins with positive lymph nodes would have been missed (72).

Drainage to Multiple Nodal Basins

A primary tumor site may drain to more than one nodal basin (73). This is frequently the situation with melanoma located in the trunk or in head and neck, and in squamous cell carcinoma (SCC) of the head and neck region. A summary of data in 3059 melanoma patients emphasized the drainage pattern characterizing melanoma in different regions of the body. Lymphatic drainage to SNs in a single basin occurred in 64% of their patient population, in dual basins in 26%, in three basins in 7%, in four basins in 2%, and five node fields were found in 1% of the study population (74). The ability to differentiate multiple radioactive lymph nodes belonging to a single basin from multiple hot nodes belonging each to a separate basin is of clinical significance. The latter condition represents multiple true SN(s) that should be carefully identified, excised, and examined histologically. As previously reported in tumors with multiple draining basins, examination of one SN does not predict the tumor status of a different SN (75). The lymphatic channel draining the lesion is afferent to the SN. Dynamic acquisition may assist in the identification of multiple draining basins. This technique helps in distinguishing non-SN(s) from multiple SN(s) by visualizing multiple afferent channels in the latter (51,76). Different lymphatic channels may have different rates of lymph flow leading to a time lag in the detection of various SN(s). This is the reason for the definition of an SN by some authors as "any lymph node that receives lymphatic drainage directly from a primary tumor" without mentioning the word "first" (77). Occasionally, an SN may not be the node closest to the primary site since lymphatic vessels can bypass nodes before reaching the SN. Dynamic images may be of potential help in assessing this situation, by tracking the lymphatic channels that are afferent to the identified hot node (37,73).

Unexpected In-Transit and Aberrant Nodes Identification

In addition to drainage to usual nodal basins such as the axillary, inguinal, and cervical regions, some patients also have drainage to lymph nodes outside these regions. In-transit nodes that lie in a lymphatic channel between the primary tumor site and a usual lymph node basin have to be considered as representing SN(s) because they receive direct lymphatic drainage from the primary lesion. This is also the case with aberrant SN(s) lying outside of standard nodal basins and/or in unexpected locations, with an aberrant direct lymphatic channel draining the primary melanoma site (4,78).

In-transit and aberrant SN(s) have the same chance of being metastatic as SN(s) located in standard basins. Therefore, once they are identified, they should be harvested and further examined. If positive, lymphadenectomy of the aberrant nodal basin is indicated (49). The incidence of metastatic in-transit or aberrant nodes in melanoma has been reported to vary between 14% and 22% (49,51,78). In a multicenter study, in-transit nodes were positive for metastases at the same frequency as SN(s) in standard node fields in patients with melanoma. Of a total 13 subjects, in 11 melanoma patients (85%), the positive in-transit nodes were the only radioactive SN(s) (79). In another study, in-transit metastases were reported to occur in 7% of SN-negative and 23% of SN-positive melanoma patients (80). The overall incidence of in-transit SN(s) in melanoma is around 10%, with a varying incidence in different parts of the body (37,52). The highest incidence of in-transit SN(s) occurs in posterior trunk melanoma followed, in decreasing order, by tumors located in the anterior trunk, the region of the head and neck, and the upper and lower limbs. In-transit SN(s) may be found in more than 12% of patients with melanoma located in the posterior trunk compared with 0.5% to 2% of those with melanoma of the lower limbs (37,38).

It is assumed that aberrant SN(s) are probably encountered with similar frequency for extremity and trunk melanoma (49). Particular drainage patterns of in-transit and aberrant SN(s) have been described. The epitrochlear, supraclavicular, and popliteal regions are common sites for SN(s) in the case of limb melanoma.

Intramuscular nodes may be identified in trunk and in head and neck melanomas (1,49). For trunk lesions, in addition to both axillary and inguinal regions which are the standard SN(s) basins, draining may also take place to SN(s) located in the periscapular, costal, and umbilical regions. It is therefore recommended to image these regions at least on the delayed static images acquisition and, if possible, also in the early dynamic phase of lymphoscintigraphy (1,81). Failure to identify in-transit lymph nodes may result in nodal understaging and lead to a subsequent high recurrence rate (82,83). During surgery, a routine blue-dye injection is targeted for identification of SN(s) located in usual basins. When performed alone, without the guidance of a presurgical lymphoscintigraphy, the blue-dye injection technique, as well as the intraoperative gamma-probe technique, harbors the risk of high failure rate of in-transit SN(s) identification (51,76). These nodes need to be identified on lymphoscintigraphy and their location then marked on the patient's skin, in order to provide a guide for the surgeon to look for and assess histologically SN(s) in unexpected locations.

Potential Pitfalls in the Interpretation of Lymphoscintigraphy

The limited spatial resolution of the gamma-camera has been suggested as the main cause for discrepancies between the number of SN(s) identified at surgery by blue-dye injection and radioguided gamma probes, and those visualized with scintigraphy (71). An SN found at surgery can be missed or misinterpreted on scintigraphy when obscured by another node or retained uptake at the injection site, when it contains too little radioactivity, or when two adjacent lymph nodes are thought to represent a single node. The distance between two hot nodes needed in order to be identified as separate structures increases if one of the nodes is hotter than the other. In contrast, some of the nodes interpreted as SN(s) may in fact represent false-positive sites of uptake, such as uptake in a lymphangioma or uptake in skin folds. Also, a single elongated node can be depicted as two hot spots or two adjacent nodes may appear as a single lesion (37,71).

Nonvisualization of an SN due to proximity to the primary tumor, and therefore to the injection site, has been previously reported (84). In patients with breast cancer, this cause of false-negative results has been described mainly in association with peritumoral radiocolloid injection in tumors located in the outer breast quadrant, obscuring ipsilateral axillary hot nodes. It has been suggested therefore to perform a subareolar injection technique, rather than peritumoral. Hidden nodes are also frequent in tumors of the head and neck, including melanoma and mucosal tumors in that region, where draining nodes are commonly located in close vicinity to the primary site. Lead shielding of the injection site has been suggested in order to overcome the scattered radiation of the injection site that may hide an adjacent SN. However, the use of the lead shield itself carries the risk of masking draining lymph nodes (85).

LYMPHOSCINTIGRAPHY WITH SPECT/CT IN PATIENTS WITH MELANOMA, BREAST CANCER, AND HEAD AND NECK MALIGNANCY

The potential benefit of using SPECT/low-dose CT technology for SN identification is derived from the improved lesion detectability of SPECT, from the improved quality of the CT attenuation-corrected SPECT images, and from the improved anatomic localization of nodes, achieved by the 3-D data of SPECT as well as the anatomical data of CT. Some of the previously described pitfalls of planar lymphoscintigraphy interpretation may be avoided by adding SPECT/CT to the acquisition protocol of scintigraphic SN mapping. Two close SN(s) can be clearly separated on SPECT images. SN(s) showing only faint activity on planar images, for instance, if deeply located, may be better identified on SPECT. SPECT/CT may reduce the false-positive rate by differentiating between nodal and extranodal foci such as skin folds or radioactive contamination, which are occasionally misinterpreted as hot lymph nodes on planar images.

Failure to identify the SN due to scattered radiation from the injection site in the vicinity of the primary tumor has been previously described. 3-D SPECT/CT images may identify these hidden nodes due to a good separation between counts related to the tracer injection and those of a closely located hot lymph node (Fig. 1). The knowledge of a close proximity between the primary tumor and its

(A)

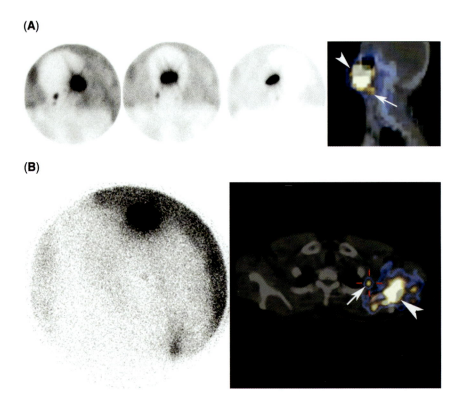

(B)

Figure 1 SNs obscured by the scattered radiation of the injection site. (**A**) Lymphoscintigraphy in a patient with upper lip melanoma including planar images in the right lateral, anterior, and left lateral views (*left*) and a sagittal slice of the corresponding SPECT/CT study (*far right*). The hot nodes identified on planar images are located in the right neck. An additional hot node located in the upper neck (*arrow*) immediately caudal to the tumor injection site (*arrowhead*) is identified on SPECT/CT. (**B**) Lymphoscintigraphy in a patient with melanoma in the left shoulder including an anterior planar view (*left*) showing an equivocal focus of faint uptake, which could potentially represent an axillary lymph node. However, the true SN (*arrow*) is located in proximity to the injection site (*arrowhead*) as detected on the transaxial fused SPECT/CT image (*right*). *Abbreviations*: SNs, sentinel nodes; SPECT/CT, single-photon emission computed tomography/computed tomography.

SN(s) is of value for the surgeon. This setting has been previously reported as a major cause for missed SNs not only on lymphoscintigraphy but also during surgery, resulting in failure of intraoperative SN harvesting and a higher rate of regional lymph node dissection (86).

Melanoma and SCC of the Head and Neck Region

Lymphatic drainage in the head and neck region is especially complex and unpredict-able. There are more than 350 lymph nodes and a rich lymphatic network in this region (87,88). The rich lymphatic network in this body part may account for the poor prognosis of melanoma originating in the head and neck (4). Lymphatic disse-mination, including micrometastases, was reported in approximately 30% of the patients without palpable lymphadenopathy (89). The lymphatic network of the head and neck region is three-dimensional. SN(s) may be located deep under mus-cles, in the parotid basin, or adjacent to vital nerves. Tumors located in the region of the head and neck frequently drain to multiple nodal basins with small lymph nodes. SN(s) are located in close vicinity to the primary tumor and are at times situ-ated immediately beneath the injection site (9,90,91). Bilateral drainage has been reported in up to 10% of patients (92). Draining lymph nodes may be missed during elective surgical neck dissection (Fig. 2). For instance, the skin of the face and ante-rior scalp may drain to postauricular nodes. Melanoma in the base of the neck may drain up to nodes in the occipital or upper cervical areas. Melanoma in the upper scalp may drain directly to nodes at the base of the neck or the supraclavicular region, bypassing the nodal stations in the upper and middle cervical, preauricular, occipital, and postauricular fields (37). Identification of SN(s) in patients with muco-sal head and neck malignancy and melanoma may be extremely difficult both on lymphoscintigraphy and at surgery. The rate of difficult localization of SN in this area is higher than in other parts of the body, reflecting the inherent difficult anat-omy of the head and neck, as well as the potential rapid washout of the dye via mul-tiple lymphatic drainage pathways that are present in this region (56). In a study on 30 patients with head and neck melanoma, Jansen et al. showed that only 53% of the

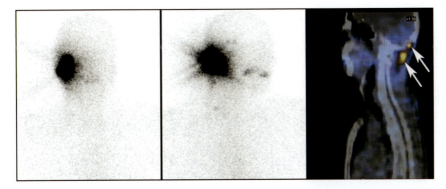

Figure 2 Nonvisualization of SN(s) on planar images. Lymphoscintigraphy in a patient with melanoma of the right auricle. N0 hot nodes are identified on planar anterior and right lateral images (*left*). Two occipital nodes, however, are identified on the sagittal slice of the corre-sponding SPECT/CT study (*right*). Note also the non-nodal activity in the oral region on the right lateral planar image. *Abbreviations*: SNs, sentinel nodes; SPECT/CT, single-photon emission computed tomography/computed tomography.

nodes were both blue-dyed and radioactive. In this study of melanoma in the head and neck, in 10% of patients the SN could not be identified while in 15% of patients SN(s) identified on lymphoscintigraphy could not be retrieved at surgery (28,46). Other authors have reported somewhat higher success rates for the identification of SNs, due to the use of a combined multidisciplinary assessment that included lymphoscintigraphy, blue-dye, and handheld gamma-probe techniques (93). In the clinical setting of close vicinity between the primary tumor and the SN, the planned surgical procedure may be modified and SNB may be performed following the removal of the primary tumor when counts from the injection site are no longer detected.

The anatomical mapping provided by the CT may assist in the precise localization of hot nodes detected by SPECT. In addition, it may allow for the precise definition of the specific basin related to the detected hot nodes. It should be emphasized, however, that SPECT/CT images cannot differentiate first- from second-echelon hot nodes and therefore should be complementary to dynamic and planar lymphoscintigraphy.

Trunk Melanoma

Presurgical lymphoscintigraphy identifies unpredictable drainage in approximately one-third of patients with trunk melanoma (46). The usual drainage regions of posterior trunk melanoma are the axilla and groins. However, melanoma in the posterior trunk may often drain to multiple basins, with some crossing the midline. Two unusual drainage pathways have been described in association with posterior trunk melanoma: one to the triangular intermuscular space lateral to the scapula, and the second through the posterior body wall to SNs in the para-aortic, paravertebral, and retroperitoneal region (Fig. 3) (37,74). These unusual pathways accompany, as a rule, the standard drainage pathways but may, on rare occasions, be the only lymphatic drainage route of the melanoma. Caprio et al. have assessed 241 patients with posterior trunk melanoma and have found triangular intermuscular space drainage in 12% of patients (104). Often the pathway then passes anteriorly, following the course of the circumflex scapular vessels into the posterior part of the axilla. If the true SN located in the intermuscular space is not accurately identified, the axillary lymph node, which is in fact a second-echelon node LN, may be mistakenly considered to represent an SN.

The second unusual drainage pattern is less common and has been observed in only 4% of patients with posterior trunk melanoma, mainly in the posterior loin skin area. About 24% of patients in this subgroup showing posterior loin melanoma may show drainage through the posterior wall of the torso to SNs in the para-aortic, paravertebral, and retroperitoneal region. If positive for tumor, it is important to recognize these as regional draining nodes and not systemic metastatic spread (74).

Lymphatic drainage from the skin of the anterior trunk is less variable and presents a lower incidence of in-transit and/or aberrant nodes as compared with tumors on the skin of the posterior trunk. However, an unusual drainage pathway has been described in association with melanoma located in the peri-umbilical area, including an initial SN at the costal margin and continuing further through the chest wall to internal ipsilateral mammary node (a second-echelon node) (74).

SPECT/CT for SN scintigraphic mapping in patients with melanoma and patients with SCC of the head and neck has been reported (68). SPECT/CT images were found to add clinically relevant data, identifying multiple draining basins in

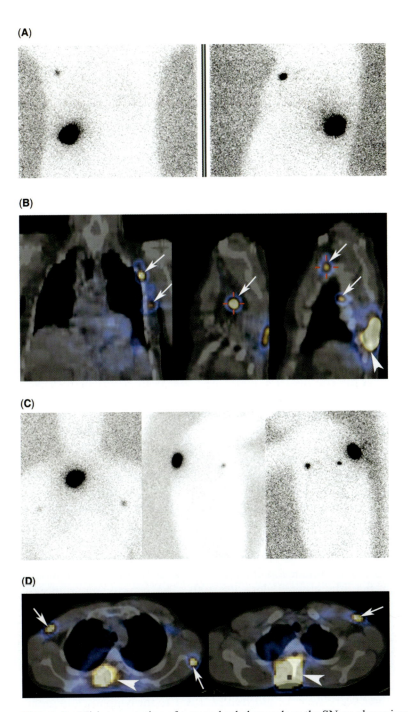

Figure 3 Misinterpretation of a second-echelon node as the SN on planar images. Patient with posterior trunk melanoma, (**A**) anterior (*left*) and left lateral (*right*) planar images show a hot node located high in the left axilla. (**B**) Fused transaxial SPECT/CT slices identify an additional axillary hot node located proximally in the central axillary basin. Additional patient with posterior trunk melanoma showing expected and unexpected lymphatic drainage. (**C**) Planar images in the posterior, right lateral, and left lateral views show bilateral axillary hot nodes. (**D**) Transaxial fused SPECT/CT slices identify an aberrant hot node in the triangular intermuscular space lateral to the scapula, in addition to the axillary hot nodes. *Abbreviations*: SNs, sentinel nodes; SPECT/CT, single-photon emission computed tomography/computed tomography.

50% of trunk melanomas and in 33% of head and neck malignancies. In 9 of 21 patients (43%) with a primary tumor located in the head and neck or trunk region, fused SPECT/CT images identified SN(s) that had been previously missed on planar images. In two patients, these SNs were involved with tumor. Missed lymph nodes on planar studies are further identified on SPECT/CT; images included nodes hidden by scattered radiation from the injection site as well as in-transit nodes. Some of the lymph nodes identified only on SPECT/CT were located in a different basin. The final results of the planar and SPECT/CT scintigraphic assessment guided the surgeon to the detection of more than a single SN. Unpredicted SN(s) that are defined by SPECT/CT included anterior nodes and SN(s) in the triangular intermuscular space lateral to the scapula, in the case of posterior trunk melanoma.

Breast Cancer

The main lymphatic drainage of the breast is through the axillary lymph nodes. Ipsilateral axillary lymph nodes of levels I and II located lateral to, and between the borders of, the pectoralis minor muscle represent the most common site of breast cancer metastases and are therefore usually harvested for SN staging (94). These nodes do not, however, represent the only drainage sites of the breast. Other draining nodes may be located at level III, medially to the pectoralis minor muscle, within the breast parenchyma (intramammary nodes), between the pectoralis muscles (interpectoral and Rotter's nodes), in the supraclavicular fossa, or in the internal mammary (IM) chain. Drainage to contralateral nodes has also been reported (68,95,96). Although the incidence of drainage to nodes in various locations may vary for tumors located in different breast quadrants, breast cancers may have an unpredictable drainage pattern, potentially to any of the above-mentioned basins (97). Lymphatic drainage to SNs outside levels I and II of the axilla has been previously described as being present in over 25% of the patients with breast cancer (69,70,90,95). An extra-axillary SN was the only tumor-positive lymph node in 5% of patients with lymph node metastases, leading to a change in the postoperative management in 17% of this patient group (90). There is large variability in the lymphoscintigraphic detection rate of IM nodes in various reports that range from 5% to 23%. Patient characteristics that affect the incidence of hot nodes in the IM basin include the patient's age, size of the breast, and the location of the primary tumor within the breast, as well as technical parameters such as the depth of the radiocolloid injection and the administered dose (98–100). The clinical importance of extra-axillary nodal drainage and its impact on patient management is yet to be determined (101–104). Harvesting and examining SN(s) outside the axilla may allow a more accurate nodal staging. It is not the routine, however, even when extra-axillary hot nodes are identified prior to surgery. Some surgeons suggest sampling extra-axillary hot nodes. In several centers, the IM chain is included in the radiation field if hot nodes are detected on scintigraphic SN mapping in this location (105,106).

For the last three years, we have added SPECT/CT to the lymphoscintigraphic acquisition protocol in patients with breast cancer referred for preoperative SN identification. When SPECT/CT findings were compared with those of planar lymphoscintigraphy in 157 consecutive patients with breast cancer prior to SNB, 46 of the total 361 hot nodes (13%) identified by lymphoscintigraphy were detected only on SPECT/CT, including both axillary and extra-axillary hot nodes, and had been previously overlooked on interpretation of planar images (107). Hot nodes detected only on SPECT/CT included lymph nodes hidden by the scattered radiation of

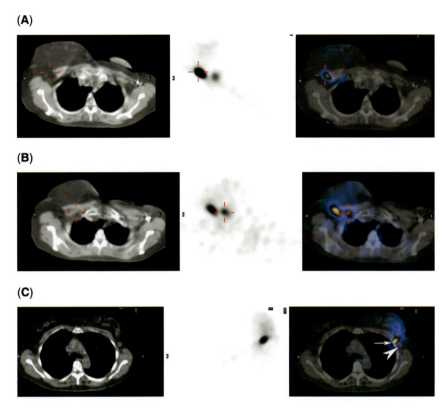

Figure 4 SPECT/CT in patients with breast cancer. Accurate localization of hot nodes to axillary level I (**A**) and III (**B**), due to the identification of the pectoralis minor muscle as an anatomical landmark on the CT component of the SPECT/CT study (**C**). Detection of a level I hot node (*arrow*). In addition, lateral to the focus of abnormal uptake in this SN, the CT component of the SPECT/CT study identifies an enlarged lymph node showing no uptake of the radiocolloid (*arrowhead*). This latter node was infiltrated with tumor as demonstrated by tissue sampling at surgery. *Abbreviations*: SNs, sentinel nodes; SPECT/CT, single-photon emission computed tomography/computed tomography.

the injection site and nodes located in the less common drainage as interpectoral and intramammary basins. Failure to identify hot LN(s) because of their proximity to the injection site is a well-known limitation when interpreting lymphoscintigraphy, mainly in patients with upper outer quadrant breast cancer, where counts originating at the injection site may obscure uptake in axillary lymph nodes. In order to overcome this "shine through" in tumors at this specific location, a change in the injection location, from peritumoral to the Sappey's subareolar plexus, has been suggested, thus increasing the distance between the site of the radioactive injection and the axilla (84). With the addition of SPECT/CT to routine lymphoscintigraphy, counts originating at the injection site are no longer a limitation resulting in failure to identify hot nodes. The usual peritumoral injection technique may be maintained regardless of the tumor location.

Although the anatomic data provided by the low-dose CT are inferior to a diagnostic CT, fused images were found to be satisfactory to delineate the pectoralis muscles, therefore providing accurate localization of the draining hot nodes to precise levels within the axilla. Standard axillary dissection removes only lymph nodes at levels I and II, since surgery affecting level III nodes is associated with an increased risk of lymphedema. Skip metastases to level III nodes without concomitant metastases to levels I and II nodes have been previously documented. Accurate localization of the hot nodes within the axilla is therefore clinically important (Fig. 4). The CT component of the SPECT/CT study identified occasionally enlarged lymph nodes suspected as being involved by tumor. This may represent an additional benefit of hybrid imaging since metastatic nodes may show no abnormal tracer uptake and should therefore be looked for on the CT component provided by the SPECT/CT study (Fig. 4) (108).

In conclusion, the addition of SPECT/CT to the acquisition protocol of scintigraphic SN mapping may be beneficial, providing clinically relevant data for further patient management. SPECT/CT detects hot nodes missed by planar lymphoscintigraphy, provides accurate localization of hot SNs to both common and uncommon drainage basins, and excludes non-nodal false-positive foci of increased radiotracer uptake.

REFERENCES

1. Alazraki N, Glass EC, Castronovo F, et al. Procedure guideline for lymphoscintigraphy and the use of intraoperative gamma probe for sentinel lymph node localization in melanoma of intermediate thickness. J Nucl Med 2002; 43:1414–1418.
2. Allan R. Sentinel node localization: do or dye alone? Br J Radiol 2001; 74:475–477.
3. Uren RF, Thompson JF, Howman-Giles RB. Lymphatic Drainage of the Skin and Breast. In: Locating the Sentinel Nodes. Amsterdam: Harwood Academic Publishers, 1999.
4. Mariani G, Gipponi M, Moresco L, et al. Radioguided sentinel lymph node biopsy in malignant cutaneous melanoma. J Nucl Med 2002; 43:811–827.
5. Nowecki ZI, Rutkowski P, Nasierowska-Guttmejer A, et al. Sentinel lymph node biopsy in melanoma patients with clinically negative regional lymph nodes—one institution's experience. Melanoma Res 2003; 13:35–43.
6. Morton DL. Sentinel node mapping and an International Sentinel Node Society: current issues and future directions. Ann Surg Oncol 2004; 11(3 suppl):137S–143S.
7. Goyal A, Mansel RE. Current status of sentinel lymph node biopsy in solid malignancies. World J Surg Oncol 2004; 24(2):9–11.
8. Nieweg OE, Tanis PJ, Rutgers EJ. Summary of the Second International Sentinel Node Conference. Eur J Nucl Med 2001; 28:646–649.
9. Acland KM, Healy C, Calonje E, et al. Comparison of positron emission tomography scanning and sentinel node biopsy in the detection of micrometastases of primary cutaneous melanoma. J Clin Oncol 2001; 19:2674–2678.
10. Johnson TM, Bradford CR, Gruber SB, et al. Staging Workup, Sentinel Node Biopsy, and Follow-up Tests for Melanoma. Update of Current Concepts. Arch Dermatol 2004; 140:107–113.
11. Cabanas RM. An approach for the treatment of penile carcinoma. Cancer 1977; 39:456–466.
12. Ell PJ. A revolution in surgical oncology: sentinel lymph node biopsy. Proceedings of the Continuing Education Courses of the SNM 2001, Toronto, 2001:72–79.
13. Schneebaum S, Even-Sapir E, Cohen M, et al. Clinical application of gamma-detection probes: radioguided surgery. Eur J Nucl Med 1999; 26:26S–35S.
14. Valdes-Olmos RA, Hoefnagel CA, Nieweg OE, et al. Lymphoscintigraphy in oncology: a rediscovered challenge. Eur J Nucl Med 1999; 24:2S–10S.

15. Zervos EE, Burak WE. Lymphatic mapping in solid neoplasms: state of the art. Cancer Control 2002; 9:189–202.
16. Morton DL, Wen DR, Wong JH, et al. Technical details of intraoperative lymphatic mapping for early malignant melanoma. Arch Surg 1992; 127:392–399.
17. Cochran AJ. The pathologist's role in sentinel lymph node evaluation. Semin Nucl Med 2000; 30:11–17.
18. Balch CM, Soong SJ, Gershenwald J, et al. Prognostic factors analysis of 17,600 melanoma patients: validation of the AJCC melanoma staging system. J Clin Oncol 2001; 19:3622–3634.
19. Balch CM, Soong SJ, Atkins, MB et al. An evidence-based staging system for cutaneous melanoma. CA Cancer J Clin 2004; 54:131–149.
20. Reintgen DS, Cruse CW, Glass F, et al. In support of sentinel node biopsy as a standard of care for patients with malignant melanoma. Dermatol Surg 2000; 26:1070–1072.
21. Messina JL, Glass LF. Pathologic examination of the sentinel lymph node. J Fla Med Assoc 1997; 84:153–156.
22. Cochran AJ, Balda BR, Starz H, et al. The Augsburg Consensus. Techniques of lymphatic mapping, sentinel lymphadenectomy, and completion lymphadenectomy in cutaneous malignancies. Cancer 2000; 89:236–241.
23. Palmieri G, Ascierto PA, Cossu A, et al. Detection of occult melanoma cells in paraffin-embedded histologically negative sentinel lymph nodes using a reverse transcriptase polymerase chain reaction assay. Melanoma Cooperative Group. J Clin Oncol 2001; 19:1437–1443.
24. Morton DL, Hoon DS, Cochran AJ, et al. Lymphatic mapping and sentinel lymphadenectomy for early-stage melanoma: therapeutic utility and implications of nodal microanatomy and molecular staging for improving the accuracy of detection of nodal micrometastases. Ann Surg 2003; 238:538–549 (discussion 549–550).
25. Morton DL, Chan AD. The concept of sentinel node localization: how it started. Semin Nucl Med 2000; 1:4–10.
26. Gershenwald JE, Colome MI, Lee JE, et al. Patterns of recurrence following a negative sentinel lymph node biopsy in 243 patients with stage I or II melanoma. J Clin Oncol 1998; 16:2253–2260.
27. Gershenwald JE, Tseng CH, Thompson W, et al. Improved sentinel lymph node localization in patients with primary melanoma with the use of radiolabeled colloid. Surgery. 1998; 124:203–210.
28. Jansen L, Nieweg OE, Peterse JL, et al. Reliability of sentinel lymph node biopsy for staging melanoma. Br J Surg 2000; 87:484–489.
29. Bieligk SC, Ghossein R, Bhattacharya S, et al. Detection of tyrosinase mRNA by reverse transcription-polymerase chain reaction in melanoma sentinel nodes. Ann Surg Oncol 1999; 6:232–240.
30. Shivers SC, Wang X, Li W, et al. Molecular staging of malignant melanoma. JAMA 1998; 280:1410–1415.
31. Uren RF, Howman-Giles R, Thompson JF. Patterns of lymphatic drainage from the skin in patients with melanoma. J Nucl Med 2003; 44:570–582.
32. Morton DL, Wen D-R, Foshag LJ, et al. Intraoperative lymphatic mapping and selective cervical lymphadenectomy for early stage melanomas of the head and neck. J Clin Oncol 1993; 11:1751–1756.
33. Krag DN, Meijer SJ, Weaver DL, et al. Minimal-access surgery for staging malignant melanoma. Arch Surg 1995; 130:654–658.
34. Bleicher RJ, Essner R, Foshag LJ, et al. Role of sentinel lymphadenectomy in thin invasive cutaneous melanomas. J Clin Oncol 2003; 21:1326–1331.
35. Mraz-Gernhard S, Sagebiel RW, Kashani-Sabet M, et al. Prediction of sentinel lymph node micrometastasis by histological features in primary cutaneous malignant melanoma. Arch Dermatol 1998, 134:983–987.

36. Porter GA, Ross MI, Berman RS, et al. How many lymph nodes are enough during sentinel lymphadenectomy for primary melanoma? Surgery 2000, 128:306–311.

37. Gershenwald JE, Thompson W, Mansfield PF, et al. Multi-institutional melanoma lymphatic mapping experience: the prognostic value of sentinel lymph node status in 612 stage I or II melanoma patients. J Clin Oncol 1999; 17:967–983.

38. Cascinelli N, Belli F, Santinami M, et al. Sentinel lymph node biopsy in cutaneous melanoma: the WHO Melanoma Program experience. Ann Surg Oncol 2000; 7: 469–474.

39. Cherpelis BS, Haddad F, Messina J, et al. Sentinel lymph node micrometastasis and other histologic factors that predict outcome in patients with thicker melanomas. J Am Acad Dermatol 2001; 44:762–766.

40. Balch CM, Soong S, Ross MI, et al. Long-term results of a multi-institutional randomized trial comparing prognostic factors and surgical results for intermediate thickness melanomas (1.0 to 4.0 mm). Intergroup Melanoma Surgical Trial. Ann Surg Oncol 2000; 7:87–97.

41. Balch CM, Soong SJ, Milton GW, et al. A comparison of prognostic factors and surgical results in 1,786 patients with localized (stage I) melanoma treated in Alabama, U.S.A., and New South Wales, Australia. Ann Surg 1982; 196:677–684.

42. Reintgen DS, Cox EB, McCarthy KSJ, et al. Efficacy of elective lymph node dissection in patients with intermediate thickness primary melanoma. Ann Surg 1983; 198: 379–385.

43. McMasters KM, Noyes RD, Reintgen DS, et al. Lessons learned from the Sunbelt Melanoma Trial. J Surg Oncol 2004; 86:212–223.

44. Balch CM, Wilkerson JA, Murad TM, et al. The prognostic significance of ulceration of cutaneous melanoma. Cancer 1980; 45:3012–3019.

45. Morton DL, Wanek L, Nizze JA, et al. Improved long-term survival after lymphadenectomy of melanoma metastatic to regional nodes: analysis of prognostic factors in 1134 patients from the John Wayne Cancer Clinic. Ann Surg 1991; 214:491–499.

46. Jansen L, Nieweg OE, Kapteijn AE, et al. Reliability of lymphoscintigraphy in indicating the number of sentinel nodes in melanoma patients. Ann Surg Oncol 2000; 7: 624–630.

47. Alex JC, Krag DN. Gamma-probe guided localization of lymph nodes. Surg Oncol 1993; 2:137–143.

48. Medina-Franco H, Beenken SW, Heslin MJ, et al. Sentinel node biopsy for cutaneous melanoma in the head and neck. Ann Surg Oncol 2001; 8:716–719.

49. Davison SP, Clifton MS, Kauffman L, et al. Sentinel node biopsy for the detection of head and neck melanoma: a review. Ann Plast Surg 2001; 47:206–211.

50. Stadelmann WK, Cobbins L, Lentsch EJ. Incidence of nonlocalization of sentinel lymph nodes using preoperative lymphoscintigraphy in 74 consecutive head and neck melanoma and Merkel cell carcinoma patients. Ann Plast Surg 2004; 52:546–549.

51. Mariani G, Erba P, Manca G. Radioguided sentinel lymph node biopsy in patients with malignant cutaneous melanoma: the nuclear medicine contribution. J Surg Oncol 2004; 85:141–151.

52. Shih WJ, Sloan DA, Hackett MT, et al. Lymphoscintigraphy of melanoma: lymphatic channel activity guides localization of sentinel lymph nodes, and gamma camera imaging/counting confirms presence of radiotracer in excised nodes. Ann Nucl Med 2001; 15:1–11.

53. Casara D, Rubello D, Rossi CR, et al. Sentinel node biopsy in cutaneous melanoma patients: technical and clinical aspects. Tumori 2000; 86:339–421.

54. Uren RF, Howman-Giles RB, Shaw HM, et al. Lymphoscintigraphy in high-risk melanoma of the trunk: predicting draining node groups, defining lymphatic channels and locating the sentinel node. J Nucl Med 1993; 34:1435–1440.

55. McMasters KM, Chao C, Wong SL, et al. Interval sentinel lymph nodes in melanoma. Arch Surg 2002; 137:543–549.

56. O'Toole GA, Hettiaratchy S, Allan R, et al. Aberrant sentinel nodes in malignant melanoma. Br J Plast Surg 2000; 53:415–417.

57. Uren RF, Howman-Giles RB, Thompson JF, et al. Variability of cutaneous lymphatic flow rates in melanoma patients. Melanoma Res 1998; 8:279–282.

58. Uren RF, Howman-Giles RB, Thompson JF. Demonstration of second-tier lymph nodes during preoperative lymphoscintigraphy for melanoma: incidence varies with primary tumor site. Ann Surg Oncol 1998; 5:517–521.

59. Haagansen CD, Feind CR, Herter FP, et al. Lymphatics of the trunk. In: Haagansen CD, ed. The Lymphatics in Cancer. Philadelphia, PA: WB Saunders, 1972:437–458.

60. Sugarbaker EV, McBride CM. Melanoma of the trunk: the results of surgical excision and anatomic guidelines for predicting nodal metastasis. Surgery 1976; 80:22–30.

61. Bergqvist L, Strand S, Hafstrom L, et al. Lymphoscintigraphy in patients with malignant melanoma: a quantitative and qualitative evaluation of its usefulness. Eur J Nucl Med 1984; 9:129–135.

62. Fee HJ, Robinson DS, Sample WF, et al. The determination of lymph shed by colloidal gold scanning in patients with malignant melanoma: a preliminary study. Surgery 1978; 84:626–632.

63. Meyer CM, Lecklitner ML, Logic JR, et al. Technetium-99m sulfur-colloid cutaneous lymphoscintigraphy in the management of truncal melanoma. Radiology 1979; 131:205–209.

64. Norman J, Cruse W, Espinosa C, et al. Redefinition of cutaneous lymphatic drainage with the use of lymphoscintigraphy for malignant melanoma. Am J Surg 1991; 162:432–437.

65. Sullivan DC, Croker BP Jr, Harris CC, et al. Lymphoscintigraphy in malignant melanoma: 99mTc antimony sulfur colloid. Am J Roentgenol 1981; 137:847–851.

66. Lubach D, Ludemann W, Berens D, et al. Recent findings on the angioarchitecture of the lymph vessel system of human skin. Br J Dermatol 1996; 135:733–737.

67. Lieber KA, Standiford SB, Kuvshinoff BW, et al. Surgical management of aberrant sentinel lymph node drainage in cutaneous melanoma. Surgery 1998; 124:757–761.

68. Bourgeois P, Fruhling J. Contralateral internal mammary node invasion in breast cancer: lymphoscintigraphic data. Breast 1999; 8:107–109.

69. Vidal-Sicart S, Pons F, Fuertes S. Is the identification of in-transit sentinel lymph nodes in malignant melanoma patients really necessary? Eur J Nucl Med Mol Imag 2004; 31:945–949.

70. Summer WE, Ross MI, Mansfield PF, et al. Implications of lymphatic drainage to unusual sentinel node sites in patients with primary cutaneous melanoma. Cancer 2002; 95:354–360.

71. Jansen L, Koops HS, Nieweg OE, et al. Sentinel node biopsy for melanoma in the head and neck region. Head Neck 2000; 22:27–33.

72. Rettenbacher L, Koller J, Kassmann H, et al. Reproducibility of lymphoscintigraphy in cutaneous melanoma: can we accurately detect the sentinel lymph node by expanding the tracer injection distance from the tumor site? J Nucl Med 2001; 42:424–429.

73. Nichols WS, Chisari FV. Structure and function of the lymphoid tissues. In: Williams WJ, Beutler E, Erslev AJ, et al., eds. Hematology. New York: McGraw-Hill, 1990:49–50.

74. Intenzo CM, Kim SM, Patel JI, et al. Lymphoscintigraphy in cutaneous melanoma: a total body atlas of sentinel node mapping. Radiographics 2002; 22:491–502.

75. Gennari R, Bartolomei M, Testori A, et al. Sentinel node localization in primary melanoma: preoperative dynamic lymphoscintigraphy, intraoperative gamma probe, and vital dye guidance. Surgery 2000; 127:19–25.

76. Uren RF, Howman-Giles RB, Thompson JF. Variation in cutaneous lymphatic flow rates. Ann Surg Oncol 1997; 4:279–280.

77. Strand SE, Persson BRR. Quantitative lymphoscintigraphy I: basic concept for optimal uptake of radiocolloids in the parasternal lymph nodes of rabbits. J Nucl Med 1979; 20:1038–1046.

78. Bergqvist L, Strand S-E, Persson BRR. Particle sizing and biokinetics of interstitial lymphoscintigraphic agents. Semin Nucl Med 1983; 8:9–19.
79. Even-Sapir E, Lerman H, Lievshitz G, et al. Lymphoscintigraphy for sentinel node mapping using a hybrid SPECT/CT system. J Nucl Med 2003; 44:1413–1420.
80. Clarke E, Notghi A, Harding K. Improved body-outline imaging technique for localization of sentinel lymph nodes in breast surgery. J Nucl Med 2002; 43:1181–1183.
81. Maza S, Valencia R, Geworski L. Influence of fast lymphatic drainage on metastatic spread in cutaneous malignant melanoma: a prospective feasibility study. Eur J Nucl Med Mol Imag 2003; 3:538–544.
82. Bailey DL, Robinson M, Meikle SR, et al. Simultaneous emission and transmission measurements as an adjunct to dynamic planar gamma camera studies. Eur J Nucl Med 1996; 23:326–331.
83. La Fontaine R, Graham LS, Behrendt D, et al. Personnel exposure from flood phantoms and point sources during quality assurance procedures. J Nucl Med 1983; 24:629–632.
84. Kretschmer L, Altenvoerde G, Meller J. Dynamic lymphoscintigraphy and image fusion of SPECT and pelvic CT-scans allow mapping of aberrant pelvic sentinel lymph nodes in malignant melanoma. Eur J Cancer 2003; 39:175–183.
85. Weber DA, Ivanovic M. Correlative image registration. Semin Nucl Med 1994; 24:311–323.
86. Perault C, Schvartz C, Wampach H, et al. Thoracic and abdominal SPECT-CT image fusion without external markers in endocrine carcinomas. J Nucl Med 1997; 38:1234–1242.
87. Keidar Z, Israel O, Krausz Y. SPECT/CT in tumor imaging: technical aspects and clinical applications. Semin Nucl Med 2003; 33:205–218.
88. McMasters KM, Reintgen DS, Ross MI, et al. Sentinel lymph node biopsy for melanoma: controversy despite widespread agreement. J Clin Oncol 2001; 19:2851–2855.
89. Jacobs IA, Chang CK, Salti GI. Significance of dual-basin drainage in patients with truncal melanoma undergoing sentinel lymph node biopsy. J Am Acad Dermatol 2003; 49:615–619.
90. Uren RF, Howman-Giles R, Thompson JF. Lymphoscintigraphy to identify sentinel lymph nodes in patients with melanoma. Melanoma Res 1994; 4:395–399.
91. Roozendaal GK, de Vries JD, van Poll D, et al. Sentinel nodes outside lymph node basins in patients with melanoma. Br J Surg 2001; 88:305–308.
92. Moloney DM, Overstall S, Allan R, et al. An aberrant lymph node containing metastatic melanoma detected by sentinel node biopsy. Br J Plast Surg 2001, 54:638–640.
93. Estourgie SH, Nieweg OE, Valdes Olmos RA, et al. Review and evaluation of sentinel node procedures in 250 melanoma patients with a median follow-up of 6 years. Ann Surg Oncol 2003; 10:681–688.
94. Dewar DJ, Powell BW. Sentinel node biopsy in patients with in-transit recurrence of malignant melanoma. Br J Plast Surg 2003; 56:415–417.
95. Thomas JM, Clark MA. Selective lymphadenectomy in sentinel node-positive patients may increase the risk of local/in-transit recurrence in malignant melanoma. Eur J Surg Oncol 2004; 30:686–691.
96. Maza S, Valencia R, Geworski L, et al. Temporary shielding of hot spots in the drainage areas of cutaneous melanoma improves accuracy of lymphoscintigraphic sentinel lymph node diagnostics. Eur J Nucl Med Mol Imaging 2002; 29:1399–1405.
97. Gipponi M, Di Somma C, Peressini A, et al. Sentinel lymph node biopsy in patients with Stage I/II melanoma: Clinical experience and literature review. J Surg Oncol 2004; 85:133–140.
98. Hyde N, Prvulovich E. Is there a role for lymphoscintigraphy and sentinel node biopsy in the management of the regional lymphatics in mucosal squamous cell carcinoma of the head and neck? Eur J Nucl Med 2002; 29:579–584.

99. Milton GW, Shaw HM, McCarthy WH, et al. Prophylactic lymph node dissection in clinical stage I cutaneous malignant melanoma: results of surgical treatment in 1,319 patients. Br J Surg 1982; 69:108–111.

100. O'Brien CJ, Uren RF, Thompson JF, et al. Prediction of potential metastatic sites in cutaneous head and neck melanoma using lymphoscintigraphy. Am J Surg 1995; 170:461–466.

101. Fincher TR, O'Brien JC, McCarty TM, et al. Patterns of drainage and recurrence following sentinel lymph node biopsy for cutaneous melanoma of the head and neck. Arch Otolaryngol Head Neck Surg 2004; 130:844–848.

102. de Wilt JH, Thompson JF, Uren RF, et al. Correlation between preoperative lymphoscintigraphy and metastatic nodal disease sites in 362 patients with cutaneous melanomas of the head and neck. Ann Surg 2004; 239:544–552.

103. Bostick P, Essner R, Sarantou T, et al. Intraoperative lymphatic mapping for early-stage melanoma of the head and neck. Am J Surg 1997; 174: 536–539.

104. Caprio MG, Carbone G, Bracigliano A. Sentinel lymph node detection by lymphoscintigraphy in malignant melanoma. Tumori 2002; 88:43–55.

105. Hare GB, Proulx GM, Lamonica DM, et al. Internal mammary lymph node (IMN) coverage by standard radiation tangent fields in patients showing IMN drainage on lymphoscintigraphy: therapeutic implications. Am J Clin Oncol 2004; 27:274–278.

106. Shahar KH, Hunt KK, Thames HD, et al. Factors predictive of having four or more positive axillary lymph nodes in patients with positive sentinel lymph nodes: implications for selection of radiation fields. Int J Radiat Oncol Biol Phys 2004; 59:1074–1079.

107. Lerman H, Metser U, Lievshitz G, et al. Lymphoscintigraphic sentinel node identification in patients with breast cancer: The role of SPECT-CT. Eur J Nucl Med Mol Imag 2005. In press.

108. Bernot-Rossi I, Houvenaeghel G, Jacquemier J, et al. Nonvisualization of axillary sentinel node during lymphoscintigraphy. J Nucl Med 2003; 44:1232–1237.

9

SPECT/CT Imaging for Prostate Cancer

Scott C. Bartley and Naomi P. Alazraki
*Emory University School of Medicine and the VA Medical Center, Atlanta,
Georgia, U.S.A.*

Stanley J. Goldsmith
*New York-Presbyterian Hospital/New York Weill Cornell Medical Center, New York,
New York, U.S.A.*

INTRODUCTION

In the United States, it is estimated that the lifetime risk of a man developing prostate cancer is one in six. It is currently the most common noncutaneous cancer
diagnosis in men and the second leading cause of death from cancer.

PROSTATE SPECIFIC ANTIGEN

The availability of a blood test for prostate specific antigen (PSA) is one of the
most important advances in prostate cancer screening in the past two decades
and has resulted in earlier detection. When initially introduced, there was an apparent increase in the incidence of the disease, followed a few years later by a decrease
in both the incidence and death rate. Between 1992 and 1996, deaths from prostate
cancer decreased by 2.5% per year. In addition to its role in early diagnosis, PSA
monitoring is a useful indicator of disease recurrence despite certain limitations.
The value of elevated serum PSA levels for detection of prostate cancer is limited
by similar results in patients with benign prostatic hypertrophy. In addition, when
patients are treated initially with less than a radical prostatectomy, normal prostate
tissue remains a source of PSA. Even in patients being evaluated for initial therapy
or during monitoring following therapy, serum PSA is not indicative of the tumor
aggressiveness, but does provide an indicator of the presence of tumor. Recently,
various algorithms have been developed to improve the utility of this serum marker
in the management of patients with prostate carcinoma, including calculating the
rate of increase of PSA levels. This determination is useful to raise the suspicion
of recurrent tumor in the patient with an enlarged prostate, as well as to detect
tumor recurrence in general, but it does not provide reliable information about
the site or extent of recurrent disease.

141

Given the propensity for prostate carcinoma to spread either by direct extension or metastasizing via lymphatic or blood vessels, and the poor prognosis associated with metastatic disease, accurate staging of the newly diagnosed patient and identification of the site(s) of recurrence following initial therapy when there is a significant rise in serum PSA are critically important.

CLINICAL FEATURES OF PROSTATE CANCER

The likelihood of metastases at various anatomic sites is complex. Prostate cancer spreads locally by direct extension or systemically by invasion into lymphatic or blood vessels. Locally, it invades the periprostatic fat or moves along the ejaculatory ducts into seminal vesicles. Spread to seminal vesicles is treated as stage III disease. The most frequent sites of metastases include lymph nodes, bones, lungs, and liver. Tumor may metastasize via periprostatic venous drainage to lower lumbar vertebral bodies (Batson's plexus) or spread via traditional hematogenous vena caval pathways to the lungs and other organs throughout the body. There is an inverse relationship between prevalence of spinal metastases and lung metastases suggesting that lung metastases and spine metastases occur independently; this support to the Batson pathway concept of metastases to the spine via local venous drainage versus systemic access. In addition, spinal metastases (hematogenous origin) usually precede lung and liver metastases, and there is a gradual decrease in frequency of spinal metastases from lumbar to thoracic to cervical consistent with the Batson pathway concept of dissemination of disease. Only 16% of the men with spinal metastases do not have pelvic or para-aortic lymph node metastases. In 84% of men with spinal metastases, therefore, there are coexisting para-aortic or pelvic lymph node metastases. Para-aortic lymph node involvement is more frequent than pelvic lymph node metastases; therefore, the coexistence of spinal metastases and para-aortic lymph node involvement is more frequent than spinal and pelvic lymph node metastases. The greater frequency of para-aortic lymph node metastases compared with pelvic lymph node metastases in association with spinal metastases has led to the notion that para-aortic lymph node involvement may evolve from tumor in spinal sites.

Accurate staging of patients and appropriate selection of treatment remains challenging in spite of the availability of modern anatomic imaging techniques such as computed tomography (CT) and magnetic resonance imaging (MRI). In patients with prostate cancer, prognosis and treatment planning are based on the Gleason score, the PSA level, and the stage (determined by imaging studies). The Gleason score is based on a summation of the histological grade (scale 1 to 5) of the two most prevalent prostate cancer cell types seen on the biopsy samples. Thus, Gleason scores are in the range of 2 to 10; the most common Gleason score is 7. There is good correlation between higher Gleason scores (>7) and increased likelihood of recurrence of disease, presumably based on the fact that the more aggressive histopathology indicates that there is a greater likelihood that the disease has already metastasized at the time of initial presentation. Urologists tend to use a composite of Gleason score and PSA to identify those patients who are at high risk for lymph node metastases. Thus, newly diagnosed patients with PSA > 40 ng/mL and Gleason > 7 or Gleason > 8, or clinical T3 disease and Gleason > 6, would qualify as high risk for lymph node disease (1).

IMAGING OF PROSTATE CANCER

Various imaging modalities have been used in an attempt to assess both primary and metastatic diseases. CT has been found to be of limited value. Hricak et al. found that CT imaging had an accuracy of only 65% in the preoperative staging of patients with prostatic malignancies (2). MRI was able to detect only 35% of occult lymph node involvement in prostate cancer (3). The limitations of conventional imaging techniques, such as CT or MRI, are mainly due to their dependence on lymph node enlargement (>1 cm) as evidence of disease, resulting often in false-negative studies.

Nuclear medicine techniques rely on assessment of the physiology rather than on the presence of anatomic changes. Various scintigraphic techniques have been employed for the assessment of the presence of metastatic prostate cancer. Bone scintigraphy has been the main tool for diagnosis of bone metastases for over four decades. Although highly sensitive, bone scintigraphy is not specific and cannot detect soft tissue metastases.

PET imaging using different radiotracers may provide in the future a more comprehensive evaluation of patients with cancer of the prostate. To date, fluoro-deoxy glucose positron emission tomography (FDG-PET) has not been effective in imaging prostate cancer for several reasons. Renal excretion of the ^{18}F-FDG results in normal high levels of activity in the bladder, which masks and interferes with visualization of the prostate, prostate bed, and pelvic lymph nodes. This can be addressed by bladder catheterization with or without irrigation but this is difficult to do in the PET imaging environment. Furthermore, prostate cancer is a relatively well-differentiated cancer, which typically does not express increased anaerobic glucose metabolism early in the clinical course as this is more characteristic of tumors with a lesser degree of differentiation. Aggressive forms of prostate cancer show a higher degree of FDG avidity and are therefore more effectively imaged by FDG-PET.

Other PET agents have been investigated as tools to image primary and metastatic prostate cancer, including ^{11}C- and ^{18}F-fluorocholine. Following their intravenous injection, malignant prostate cancer cells rapidly take up these tracers. Because very little ^{11}C-choline is excreted into the urine, this appears to be a potentially excellent agent for imaging locoregional prostate cancer involvement. The short physical half-life of ^{11}C of 20 minutes is a major drawback to the routine use of this agent in the clinical setting and requires on-site cyclotron and rapid labeling techniques. ^{18}F-fluorocholine is another version of this tracer with potentially better clinical applicability because of the longer physical half-life of ^{18}F of 118 minutes. ^{18}F-fluorocholine is, however, excreted into the urine. If imaging is performed very early (i.e., at about one minute following intravenous injection, before urinary excretion is evident in the bladder), it may be possible to utilize ^{18}F-choline effectively due to the very rapid uptake into prostate cancer cells. Another agent, ^{11}C-acetate, which is not excreted into the urine, is currently under investigation for clinical applications. The ^{11}C-acetate label has the same disadvantage of a short, 20-minute physical half-life, as described earlier for ^{11}C-choline.

IMAGING OF PROSTATE CANCER USING RADIOLABELED MONOCLONAL ANTIBODY

^{111}In capromab pendetide (ProstaScint®, Cytogen) is a murine (mouse origin) monoclonal antibody labeled to ^{111}indium, and is therefore imaged using routine gamma-camera single-photon emission computed tomography (SPECT) techniques. The labeling process involves attaching the [111]indium to a chemically stable linker, a chelator called GYK DTPA, to which the monoclonal antibody molecule is attached. The labeled monoclonal antibody binds to the 100 kDa glycoprotein [prostate specific membrane antigen (PSMA)], which is found in normal, benign, and malignant epithelium of the prostate, but with overexpression in primary and metastatic prostate cancer cells. The antibody, capromab pendetide, binds to an intracellular epitope—an antigen that is located on the intracellular side of the cell membrane. Thus the labeled monoclonal antibody needs to penetrate the cell membrane of a prostate cancer cell to bind to its targeted antigen. Capromab pendetide has been shown to react with >95% of prostate adenocarcinoma samples. The human immune system reacts to mouse protein by producing human anti mouse antibodies (HAMA). HAMA causes the antibodies to be rapidly bound and eliminated from the host and can cause kidney damage and even, rarely, anaphylaxis. There is a reported 4% incidence of adverse reactions, none serious, to administration of ^{111}In capromab pendetide and an incidence of 8% frequency of HAMA identified after one infusion of ^{111}In capromab pendetide. After repeat infusions, the frequency of HAMA increases to 19% (4). HAMA can be measured in the laboratory from blood samples. It is not routinely done in practice because experience suggests that HAMA is not a sufficiently significant clinical issue to warrant preinjection measurements. For patients who are having repeat ProstaScint® injections, precautions might include obtaining a serum HAMA measurement and/or premedicating the patient with an antihistamine, such as diphenylhydramine (Benadryl®).

Overexpression of PSMA protein in the primary tumor determined by immunohistochemical staining is an independent predictor of outcome (5) and increased intracellular PSMA independently predicts biochemical recurrence. PSMA expression is increased in metastatic foci of high-grade disease (6). It is also upregulated after hormone therapy.

[111]In capromab pendetide is the best imaging procedure presently available to detect lymph node metastases from prostate cancer. CT and MRI have reported sensitivities for detecting pelvic and abdominal lymph node metastases ranging between 4% and 51%, when compared with 62% to 92% for ^{111}In capromab pendetide scintigraphy (7). It is estimated that 40% to 60% of patients with clinically localized prostate cancer actually have lymphatic or extraprostatic extension of the disease at the time of presentation. Clinical staging has a suboptimal performance. [111]In capromab pendetide can differentiate between patients with localized disease versus those with metastatic disease. Compared to pelvic lymph node dissection as a gold standard, ^{111}In capromab pendetide SPECT imaging has a sensitivity of 62% and a specificity of 72% for detecting lymph node metastases (7). The low sensitivity likely reflects the inability of ^{111}In capromab pendetide SPECT to detect small lymph nodes (<1.0–1.5 cm) with early metastatic tumor. Hinkle et al. reported slightly better results in a multicenter evaluation of ^{111}In capromab pendetide in patients with newly diagnosed disease (8). The study included 51 patients with high risk for metastatic prostate cancer. After performing the scintigraphic studies, the patients underwent bilateral open pelvic lymph node dissection. Additionally, if imaging modalities demonstrated any extrapelvic nodal lesions, these sites were explored and biopsied. Histopathology results were compared to the imaging data of CT, MRI, ultrasound, and ^{111}In capromab pendetide. With regard to extrapelvic sites, the sensitivity and specificity for ^{111}In capromab pendetide were 75% and 86%, respectively. The overall accuracy and negative predictive value were 81% and 79%,

respectively. [111]In capromab pendetide significantly outperformed CT, MRI, and ultrasound (Table 1) (8).

Ellis et al. reported that ^{111}In capromab pendetide images fused to CT scans enhanced identification of foci of adenocarcinoma within the prostate, correlating well with biopsy results (overall accuracy 80%, sensitivity 79%, specificity 80%, positive predictive value 68%, and negative predictive value 88%). [111]In capromab pendetide has been mostly used to detect nodal or metastatic disease in those patients who have had prior definitive therapy, who present with rising PSA levels (9).

The potential of ^{111}In capromab pendetide imaging to identify pelvic and extrapelvic metastases may have an impact on staging and restaging in newly diagnosed and recurrent prostate cancer. [111]In capromab pendetide scintigraphy is recommended as a complementary diagnostic and prognostic tool in addition to PSA and Gleason score, which are each independent predictors of prostate cancer recurrence and metastases. In patients who have previously undergone prostatectomy with positive surgical margins, and in whom adjunctive radiotherapy is therefore a consideration, ^{111}In capromab pendetide imaging can be used as a guide to radiotherapy planning. Patients following surgery or radiotherapy in whom the PSA level does not decrease satisfactorily may benefit from ^{111}In capromab pendetide imaging to detect sites of unsuspected metastases.

Patients under consideration for radical prostatectomy and those suspected of recurrent disease due to rising PSA levels following prior therapy such as prostatectomy or radiotherapy are the clinical groups where decisions of further clinical management are most likely to benefit from ^{111}In capromab pendetide imaging. While a combination of manual rectal examination and PSA blood level measurements lead to diagnosis of prostate cancer at an earlier stage than in the past, these procedures do not identify metastatic disease. SPECT imaging of ^{111}In capromab pendetide, enhanced by fusion with CT, improves staging of the disease. Patients who present with disease localized to the prostate (T_{1-2}, N_0, M_0) have a much better prognosis and are treated differently than patients whose disease has spread to para-aortic and/or pelvic lymph nodes, or distant sites (N_1 to N_4 and M_1). Thus, patients who are considered to be high risk for metastases because of a particularly high PSA or high Gleason might be candidates for ^{111}In capromab pendetide staging. Patients who had definitive surgery or radiotherapy and who present with rising PSA are managed differently if the recurrent disease is localized to the prostate as compared

Table 1 Multicenter Trial of [111]In Capromab Pendetide

	Prostate carcinoma		Extraprostatic carcinoma	
	[111]In capromab pendetide (%)	CT, MRI, and ultrasound (%)	[111]In capromab pendetide (%)	CT, MRI, and ultrasound (%)
Sensitivity	92	51	75	20
Specificity	0	0	86	68
Accuracy	92	51	81	48
Positive predictive value	100	100	79	31

Abbreviations: CT, computed tomography; MRI, magnetic resonance imaging.
Source: From Ref. 8.

to disease involving lymph nodes or distant sites. [111]In capromab pendetide is useful in identifying and differentiating these groups of patients.

Autopsy data have demonstrated that para-aortic lymph node involvement may occur even in the absence of pelvic lymph node involvement. Detection of multiple central abdominal para-aortic and mesenteric lymphadenopathy showing abnormal ^{111}In capromab pendetide uptake on scintigraphy (Fig. 1) is associated with poor prognosis. In a study of the outcomes of patients showing abnormal central abdominal multifocal uptake, 341 patients with the finding had an 85.5% survival at 14.6 years. A matched group of patients showing no central abdominal tracer uptake had a statistically significantly higher survival of 94.8% at 4.1 years ($P = 0.0067$) (10).

SCINTIGRAPHIC IMAGING PROTOCOLS

[111]In capromab pendetide imaging is not a simple task for either the patient or the nuclear medicine physician. An ideal prostate cancer imaging agent would be simple to image and very specific. In the FDA approval process, Cytogen specified a particular technique for imaging ^{111}In capromab pendetide. Various other methods that improve the ^{111}In capromab pendetide images, and are therefore more diagnostically accurate, have been introduced and validated. The combination of SPECT and CT is the latest variation, which provides a powerful tool for increasing accuracy in diagnosis and providing additional resources for therapy planning.

The capromab pendetide antibody is an intact antibody and therefore takes 48 to 72 hours before sufficient background clearance provides adequate lesion to background contrast. This limits the choice of gamma emitters that can be used. [99m]Tc is not a practical choice because of the short physical half-life of six hours. The longer half-life of ^{111}In of 67.1 hours allows for both lesion identification based on uptake of the radiolabeled antibody and background clearance, as well as the ability of performing repeated delayed imaging when needed. The acquisition protocol calls for the use of a large field of view gamma camera, and because of the two medium energy gamma emissions of the radioisotope, a parallel hole, medium-energy collimator is required. The acquisition parameters should include 15% to 20% energy windows, at 172 and 247 keV.

One of the accepted clinical protocols involves imaging on two separate visits. The first imaging session is performed at 72 hr/3 days after administration of the tracer and the second session at 120 hr/5 days postinjection. The initial session should include planar whole-body or spot images of the pelvis, abdomen, and thorax. For the whole-body study, imaging is performed from the skull through the midfemur, with a 128×512 or 256×1024 matrix, for a minimum scanning time of 35 minutes. Spot images are acquired in the anterior and posterior views, with seven and one-half minutes per view, using a 128×128 or 256×256 matrix.

Imaging of the region of the liver is challenging. Because of the nonspecific liver uptake of ^{111}In capromab pendetide, scintigraphy must contain a sufficient number of counts to allow for detection of adjacent abdominal and pelvic lesions. This may result, however, in pixel overflow of the hepatic counts, which may further lead to image degradation in this area.

The use of bladder catheterization and cathartics is recommended. Enema can be administered one hour (or as close in time as possible) prior to imaging. Administering enemas in an outpatient setting is not convenient in most nuclear

(A)

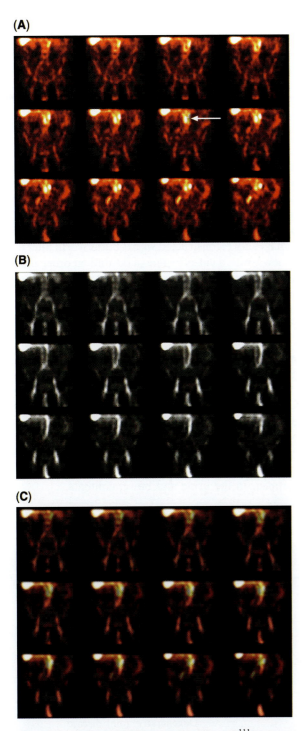

(B)

(C)

Figure 1 SPECT/SPECT coregistration of ^{111}In capromab pendetide and ^{99m}Tc-labeled RBC. Coronal SPECT slices of (**A**) [111]In capromab pendetide, (**B**) ^{99m}Tc-labeled RBCs, and (**C**) fused In/Tc images showing ^{111}In capromab pendetide tumor uptake in para-aortic nodes (*arrow*), not visualized on the labeled red cell images. *Abbreviations*: SPECT, single-photon emission computed tomography; RBC, red blood cell.

medicine facilities; administration of cathartics should be performed on the evening before each imaging session. The bladder needs to be catheterized and irrigated prior to imaging. As an alternative, the patient may void immediately prior to imaging, which then starts with the region of the pelvis and progresses in a cranial direction.

SPECT should be performed in all patients. The first SPECT study of the pelvis is performed approximately 30-minutes after injection of the tracer in order to delineate blood pool activity. These early 30-minute SPECT images can be eliminated if blood pool identification is obtained at the time of the delayed, three to five days imaging, using additional ^{99m}Tc-labeled red blood cell (RBC) scintigraphy (see discussion below). The second set of SPECT images is performed at 72 to 120 hours after injection, and include, as a rule, both the pelvis and abdomen, extending from the lower liver margin through the pelvis. SPECT acquisition parameters should include a matrix of 64×64 or 128×128, 60 to 120 steps over a total $360°$ rotation. For early SPECT studies, an imaging time of 25 seconds per step is recommended, and this changes to 50 seconds per step for the delayed acquisition. SPECT images are reconstructed with a Butterworth filter in the transverse, coronal, and sagittal views, with a cut-off value of 0.5 and an order of 5. Reconstructed SPECT images have a slice thickness between 6 and 12 mm. Additional delayed imaging with full patient preparation can be performed if tracer activity in the blood pool, urinary bladder, or bowel persists.

Interpretation of repeat planar and SPECT ^{111}In capromab pendetide studies can be difficult. Alignment of the images done over multiple days is very challenging, as the patients are not likely to be in identical positions. Automatic algorithms cannot resolve this limitation although an experienced reader may be able to overcome this problem to some extent. In addition, uptake of ^{111}In capromab pendetide is not exclusively specific to PSMA. The physiologic biodistribution of the radiopharmaceutical includes the liver, bone marrow, spleen, genitourinary system, and bowel. Increased tracer uptake has also been reported in the presence of inflammatory processes related to the presence of colostomy sites, abdominal aneurysms, postoperative bowel adhesions, degenerative joint disease, and local inflammatory lesions (Table 2) (11). Labeled fragments appear in the urine and bowel contents.

Table 2 Potential Causes for False-Positive Findings in ^{111}In Capromab Pendetide Scintigraphy

Physiologic tracer uptake	Liver
	Bone marrow
	Colon
	Kidneys
	Spleen
	Reproductive organs
Inflammatory processes	Colostomy
	Postoperative colon adhesions
	Abdominal aneurysms
	Degenerative joint disease
Pathology other than prostate cancer	Meningioma
	Myelolipoma
	Non-Hodgkin's lymphoma
	Neurofibroma
	Renal cell carcinoma

There have been numerous reports with respect to ^{111}In capromab pendetide uptake in normal tissue and other neoplasms. The kidney has a gene that expresses PSMA, which may lead to reactivity with the antibody (12). Increased tracer uptake has been reported in a meningioma (13), in a benign myelolipoma (14), in non-Hodg-kin's lymphoma (15), in neurofibromas (16), as well as in renal cell carcinoma (17). Endothelial cells of tumor-associated neovasculature synthesize and express PSMA protein, and therefore any malignancy with sufficient angiogenesis expression may potentially take up ^{111}In capromab pendetide (18).

Blood pool activity can be confounding, hence the recommended initial SPECT at 30 minutes can be used for comparison with images acquired later. If the delayed SPECT images are acquired with the patient in a slightly different position, the ability to differentiate between a node and vessels by comparison of early and late images may be compromised, and as stated, "apparent localization to sites of tortuous blood vessels" may occur (4). These causes for false-positive results will reduce the accuracy and specificity of the test, although simultaneous acquisition of ^{99m}Tc-labeled red blood cells can correct for those potential false positives (see below).

Gastrointestinal activity can also cause erroneous interpretation. Administration of cathartics and enema may reduce some of this increased physiologic uptake. However, even these procedures may not completely clear the gastrointestinal tract of tracer uptake and the patient may have to come back the next day after repeating the cathartic regimen.

Since the radiolabeled antibody localizes physiologically to some degree in the bone marrow, the sensitivity for detecting metastases in the bone is reduced and is inferior to that of a bone scan.

It is the goal of alternative imaging protocols to reduce the rate of false-positive results. One of proposed methods relies on the use of dual-isotope imaging proto-cols. Sychra et al. (19) described dual-isotope scintigraphy using the addition of ^{99m}Tc-tagged RBCs to help delineate the vascular structures. This was further opti-mized by eliminating the initial blood pool SPECT acquisition (20). The final mod-ified protocol starts with the intravenous injection of ^{111}In capromab pendetide on day zero and acquisition of a dual-radioisotope SPECT on day five, with ^{99m}Tc-labeled RBC images performed in a 140 keV window, and ^{111}In capromab pendetide images in the 247 keV window. If three energy windows are available, one is assigned to the ^{99m}Tc window and the others to the two ^{111}In energy peaks. On day four after injection, prior to imaging, the patient undergoes cathartics preparation. If the bowel activity is still high, imaging is discontinued and the patient is rescheduled following additional preparation including cathartics and/or enema.

Prior to day five SPECT imaging, blood is withdrawn from the patient and labeled with ^{99m}Tc. While labeling the blood, whole-body and planar spot images of the chest, abdomen, and pelvis are acquired. After the whole-body images have been acquired, 5 mCi ^{99m}Tc-labeled RBCs are reinjected. SPECT imaging of the pel-vis and abdomen are performed. SPECT images are acquired using a 128×128 mat-rix, 60 seconds per step, for 120 projections obtained over a 360° rotation. The set-up includes two 15% windows around the 172 and 247 keV energy peaks for the ^{111}In and one 10% window at 140 keV for ^{99m}Tc. First-order Chang attenuation correction is applied to the filtered SPECT data sets. The ^{99m}Tc RBCs SPECT data set can be subtracted from the ^{111}In images using the manufacturers software (Fig. 1).

Even with the dual-isotope ^{99m}Tc RBC/^{111}In capromab pendetide subtraction method, bone marrow and bowel activity are not removed from the SPECT images of the pelvis (20). Sanford et al. (21) reported a good correlation between ^{99m}Tc sulfur colloid bone marrow scintigraphy and ^{111}In CYT-356 capromab pendetide studies. The addition of intravenous ^{99m}Tc sulfur colloid along with the ^{99m}Tc RBCs may be useful in providing information for subtraction algorithms.

The dual-isotope acquisition method addresses and attempts to solve some of the ^{111}In capromab pendetide imaging challenges. An additional immediate benefit is the reduced total imaging time to the patient, since the dual tracer data is collected during a single imaging session. As the ^{111}In and ^{99m}Tc SPECT images are acquired simultaneously, misalignment does not represent a major technical problem. Blood pool images provide a road map to help interpret monoclonal antibody images using either side-by-side comparison or the image subtraction technique. If ^{99m}Tc sulfur colloid is also used, liver, spleen, and bone marrow images will also be mapped to identical locations on the indium-labeled antibody images.

However, even following the simultaneous administration and imaging of one or two ^{99m}Tc-labeled agents, there are still difficulties that need to be overcome. Gastrointestinal activity persists and may be difficult to distinguish from metastatic tumor. Additionally, cross-reactivity with bowel adhesion or inflammatory processes cannot be identified as such. Neither the ^{99m}Tc-tagged RBCs nor the ^{99m}Tc sulfur colloid can differentiate uptake by other neoplasms, which may cross-react with the antibody. Threshold values for the ^{99m}Tc-tagged RBCs have been selected for improved identification of blood vessels (19,22). Because the volume, shape, and counts in the urinary bladder change over the time during image acquisition, care must be taken when subtracting images if they are not acquired near-simultaneously in order to not remove abnormal ^{111}In tumor activity (22). Nevertheless, using the dual-isotope method, Sychra et al. found detection rate of disease increased by 31% as compared to the original FDA-approved protocol (19). Quintana et al. also reported improved sensitivity for those patients who had the dual-isotope protocol compared with those who did not (20).

SPECT/CT IMAGING

One of the earliest methods for prostate cancer image fusion was the registration of the ^{111}In capromab pendetide data onto CT images. Hamilton et al. initially reported this in 1999, using the combined protocol of ^{111}In-labeled antibody and ^{99m}Tc-tagged RBCs (23). They suggested that this modality might provide some anatomical reference points to help with the alignment for registration of the images. The CT images were performed as part of routine radiation therapy planning. Both oral and intravenous contrast administration was used.

The vessels visualized on both CT and SPECT images were segmented with all images transformed to a 256×256 matrix. A tool for automatically segmenting a structure by connecting all the pixels that had the same values with line segments was used for this purpose. Since arteries and veins cannot be distinguished on ^{99m}Tc-RBC SPECT imaging, these were combined for the purpose of segmentation. The main vessels were delineated and the actual registration was a software operator-dependent procedure [Advanced Visual Systems, Inc. (AVS)]. The [99m]Tc- SPECT images were then aligned with the CT images using all six degrees of freedom to achieve the desired alignment. The final alignment was used to create the transformation matrix providing a full 3-D rigid-body map from SPECT to CT. The CT images were resliced and the alignment was again verified. Since the ^{111}In SPECT and the ^{99m}Tc SPECT were in identical alignment, the transformed ^{111}In images could be

(A)

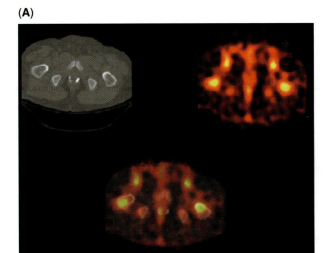

(B)

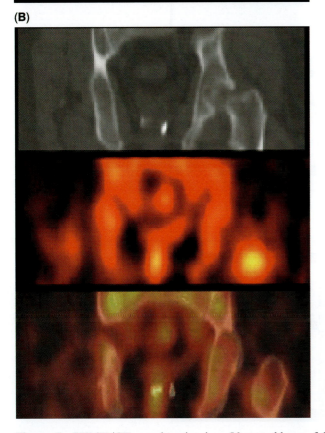

Figure 2 SPECT/CT coregistration in a 56-year-old man following prostatectomy five years earlier, who presented with elevated serum PSA levels of 0.95 ng/mL. (**A**) Transaxial images of CT (*upper left*), ^{111}In capromab pendetide SPECT (*upper right*), and manually coregistered images (*lower panel*) show increased ^{111}In capromab pendetide activity in the area of the right surgical clip, indicating the presence and localization of active malignancy. (**B**) Coronal images of CT (*top*), ^{111}In capromab pendetide SPECT (*center*), and manually co-registered images (*lower panel*) confirm the localization of the abnormal uptake to the region of the right-sided surgical clip. Note the operator-related misalignmen of images. *Abbreviations*: CT, computed tomography; PSA, prostate specific antibody; SPECT, single-photon emission computed tomography.

(A)

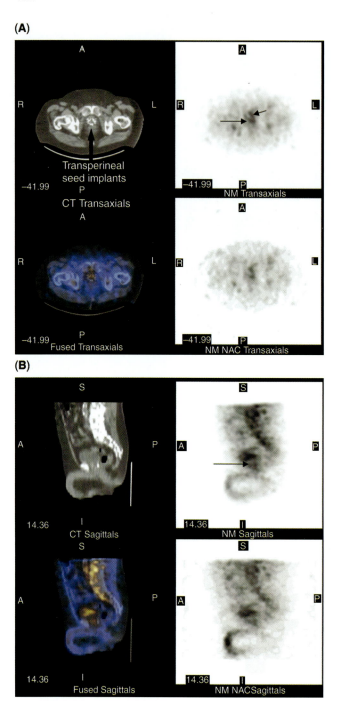

Figure 3 (*Caption on facing page*)

overlaid on the resliced CT images for registration (Fig. 2). This is an extremely operator-dependent procedure.

Following the development and clinical availability of combined SPECT/CT devices, the registration process became an automated procedure. Carroll et al.

described the use of an integrated SPECT/CT device (Millennium Hawkeye™, General Electric) for successful hybrid imaging at 48 hours after the tracer administration, with good success (24). Wong et al. published the first complete protocol using a dual-isotope SPECT/CT method at four days after injection (25). The dual-isotope SPECT protocol was used to provide the interpreter with anatomic reference points for the scintigraphic SPECT component to confirm the quality of the registration. This appears to be necessary since even with hybrid imaging cameras, which acquire the CT and SPECT sequentially and not simultaneously, patient movement while lying on the table may still be a source of misalignment.

The combination of SPECT and CT (or another anatomic imaging technique) may help to reduce or eliminate the problems of lack of anatomic correlation with prior imaging methods (Fig. 3). The Hamilton protocol for manual image registration resulted in changes in interpretation in two of five patients (23). Specifically, in each of these two patients, a focus identified on the SPECT as suspicious of representing metastatic lymphadenopathy was demonstrated to represent bowel activity after the registration.

Schettino et al. (26) assessed the value of anatomic registration of ^{111}In capromab pendetide SPECT studies with CT or MRI images and concluded that registration provides a significant improvement in diagnostic accuracy. There were 161 suspicious sites in the 58 patients studied. In 46% of these 161 sites (74 foci), the suspicion of malignancy was excluded and the findings were considered as being unrelated to prostate cancer, following interpretation of image fusion. The majority of these sites represented physiologic bowel, vascular, or bone marrow uptake of the tracer. In 25 of the 58 patients initially suspected of having nodal disease, the final diagnosis following the interpretation of the SPECT/CT fusion images was of only local, limited disease. By changing the definition of these 46% suspicious sites and by reducing the number of patients with suspected nodal disease by 43%, the addition of the anatomic images appears to be a powerful enhancement in the evaluation of prostate cancer with capromab pendetide.

[111]IN CAPROMAB PENDETIDE SPECT/CT IN THERAPY PLANNING

The use of SPECT/CT with ^{111}In capromab pendetide has been described in radiation therapy planning. DeWyngaert et al. reported on the value of ^{111}In capromab pendetide with either CT or MRI fusion for radiation therapy planning (22). Using manual coregistration of SPECT and structural data, these authors found that

---

Figure 3 (*Facing page*) Hybrid ^{111}In capromab pendetide SPECT/CT imaging in a 72-year-old man with a history of prostate cancer, Gleason score of six, status transperineal seed implantation, and external beam radiation, who presented with rising PSA levels up to 42.8. (**A**) Transaxial and (**B**) sagittal images generated following sequential imaging using a hybrid SPECT/CT device (Millenium Hawkeye, General Electric) and processing on the dedicated workstation (Xeleris, General Electric) show abnormal ^{111}In capromab pendetide uptake in the region of the prostate consistent with local recurrent malignancy (*arrows*). *Note*: Attenuation-corrected (*upper right*) and noncorrected (*lower right*) images show the same findings, although brachytherapy metal seed implants in the prostatic bed are visualized. This indicates that the abnormal uptake represents recurrence and is not a metal-related artifact introduced by the CT. *Abbreviations*: CT, computed tomography; PSA, prostate specific antibody; SPECT, single-photon emission computed tomography.

fusion of anatomic images with ^{111}In capromab pendetide SPECT decreased the radiation fields and subsequent dose to normal tissues. [111]In capromab pendetide SPECT/CT images led to alterations in the volume of the target zone while also minimizing non-target tissue damage (27). In fact, Jani et al. reported that radioimmunoscintigraphy with ^{111}In capromab pendetide resulted in significant modification in postprostatectomy clinical target volume (CTV) definition (28). These modifications did not result in increased radiation effects to the rectum or bladder, but did affect the volume of the bladder-receiving dose. In this report, radioimmunoscintigraphy with ^{111}In capromab pendetide was instrumental mainly in patients postprostatectomy in improving the ability to identify disease unlikely to benefit from radiotherapy. Ellis et al. also reported outcomes of prostate cancer patients treated with external beam therapy for which radioimmunoscintigraphy guidance using ^{111}In capromab pendetide SPECT/CT fusion imaging was performed (29).

Brachytherapy in which radioactive seeds are implanted into prostate tissue is another means of irradiating tumor as either first-line or salvage therapy if there is disease recurrence. Ellis et al. (29) have documented a four-year biochemical outcome after ^{111}In capromab pendetide image-guided transperineal brachytherapy for patients with adenocarcinoma of the prostate. Results showed that biochemical disease-free survivals for patients treated following ^{111}In capromab pendetide image guidance matched or exceeded outcome reported by other studies that did not use radioimmunoscintigraphy guidance (30,31). Longer term follow-up comparative data for these patient groups are yet to be reported.

Cryotherapy, a procedure that destroys tumor by freezing, is another technique used for patients with stage I and II diseases. [111]In capromab pendetide image guidance may be beneficial in directing cryotherapy either as an initial therapy, in the treatment of recurrent disease in the prostate bed, or as an alternative to salvage prostatectomy in patients who have failed radiation therapy (32). More studies need to validate this application for a therapeutic approach.

CONCLUSION

[111]In capromab pendetide is a radiolabeled murine monoclonal antibody. [111]In capromab pendetide scintigraphy has been shown to be superior to CT and MRI in identifying metastatic prostate carcinoma. The technique is complimentary to the Gleason score, PSA, and clinical staging in the assessment of prognosis. SPECT/CT appears to be the optimal modality to detect occult metastases in high-risk patients with presumed localized disease. Results of ^{111}In capromab pendetide SPECT/CT imaging may influence the choice and planning of the appropriate therapy approach. Furthermore, ^{111}In capromab pendetide SPECT/CT imaging has also been helpful in defining treatment fields in patients with recurrent disease referred for cryosurgery and brachytherapy.

REFERENCES

1. Bubendorf L, Schopfer A, Wagner U, et al. Metastatic patterns of prostate cancer: an autopsy study of 1,589 patients. Hum Pathol 2000; 31:578–583.
2. Hricak H, Dooms GC, Avallone A, et al. Prostatic carcinoma: staging by clinical assessment, CT, and MR imaging. Radiology 1987; 162:331–336.

3. Harisinghani MG, Barentsz J, Hahn PF, et al. Noninvasive detection of clinically occult lymph node metastases in prostate cancer. N Engl J Med 2003; 348:2491–2499.

4. ProstaScint Package Insert, Cytogen, 1997.

5. Ross JS, Sheehan CE, Fisher HAG, et al. Correlation of primary tumor prostate-specific membrane antigen expression with disease recurrence in prostate cancer. Clin Cancer Res 2003; 9:6357–6362.

6. Chang SS, Gaudin PB, Reuter VE, et al. Prostate-specific antigen: present and future applications. Urology 2000; 55:622–629.

7. Blend MJ, Sodee DB. ProstaScint: an update. In Freeman LM, ed. Nucl Med Annual 2001, Philadelphia: Lippincott Williams and Wilkins, 2001.

8. Hinkle GH, Burgers JK, Neal CE, et al. Multicenter radioimmunoscintigraphic evaluation of patients with prostate carcinoma using indium-111 capromab pendetide. Cancer 1998; 83:739–747.

9. Ellis RJ, Kim EY, Conant R, et al. Radioimmunoguided imaging of prostate cancer foci with histopathological correlation. Int J Radiat Oncol Biol Phys 2001; 49: 1281–1286.

10. Haseman M, Rosenthal SA, Polascik TJ, et al. Capromab pendetide imaging of prostate cancer. Cancer Biother Radiopharm 2000; 15:131–140.

11. Freeman LM, Krynyckyi Y, Li G, et al. The role of ^{111}In capromab pendetide (ProstaScint®) Immunoscintigraphy in the management of prostate cancer. Q J Nucl Med 2002; 46:131–137.

12. O'Keefe DS, Bacich DJ, Heston WD. Comparative analysis of prostate-specific membrane antigen (PSMA) versus a prostate-specific membrane antigen-like gene. Prostate 2004; 58:200–210.

13. Zucker RJ, Bradley YC. Indium-111 capromab pendetide (ProstaScint) uptake in a meningioma. Clin Nuc Med 2001; 26:568–569.

14. Scott DL, Halkar RK, Fischer A, et al. False-positive 111 indium capromab pendetide scan due to benign myelolipoma. J Urol 2001; 65:910–911.

15. Zanzi I, Stark R. Detection of non-Hodgkin's lymphoma by capromab pendetide scintigraphy (ProstaScint) in a patient with prostate carcinoma. Urology 2002; 60:514–518.

16. Khan A, Caride VJ. Indium-111 capromab pendetide (ProstaScint) uptake in neurofibromatosis. Urology 2000; 56:154–158.

17. Michaels EK, Blend M, Quintana JC. [111]Indium capromab pendetide unexpectedly localizes to renal cell carcinoma. J Urol 1999; 161:597–598.

18. Chang SS, O'Keefe DS, Bacich DJ, et al. Prostate-specific membrane antigen is produced in tumor-associated neovasculature. Clin Cancer Res 1999; 5:2674–2681.

19. Sychra JJ, Lin KQ, Blend MJ. On-line publication in the Electronic Journal RSNA 1997, currently at http://www.uic.edu/com/uhrd/nucmed/rsnaej/prost_3.htm, 2004.

20. Quintana JC, Blend MJ. The dual-isotope ProstaScint imaging procedure: clinical experience and staging results in 145 patients. Clin Nucl Med 2000; 25:33–40.

21. Sanford E, Grzonka R, Heal A, et al. Prostate cancer imaging with a new monoclonal antibody: a preliminary report. Ann Surg Oncol 1994; 1:400–404.

22. DeWyngaert JK, Noz ME, Ellerin B, et al. Procedure for unmasking localization information from ProstaScint scans for prostate radiation therapy treatment planning. Int J Radiat Oncol Biol Phys 2004; 60:654–662.

23. Hamilton RJ, Blend MJ, Pelizzari AC, et al. Using vascular structures for CT–SPECT registration in the pelvis. J Nucl Med 1999; 40:347–351.

24. Carroll MJ, El-Megadmi H, Elnaas S, et al. Prostate cancer: combined ProstaScint SPET/CT/blood pool imaging. Nucl Med Commun 2003; 24:473–477.

25. Wong TZ, Turkington TG, Polascik TJ, et al. ProstaScint (Capromab Pendetide) imaging using hybrid gamma camera-CT technology. Am J Roentgenor 2005; 184:676–680.

26. Schettino CJ, Kramer EL, Noz ME, et al. Impact of fusion of indium-111capromab pendetide volume data sets with those from MRI or CT in patients with recurrent prostate cancer. Am J Roentgenor 2004; 183:519–524.

27. Hamilton IU, Blend MJ, Calvin D, et al. Use of ProstaScint-CT image registration in the post-operative radiation therapy of prostate cancer. Int J Radiat Oncol Biol Phys 1995; 33:979–983.

28. Jani AB, Spilbring D, Hamilton R, et al. Impact of radioimmunoscintigraphy on definition of clinical target volume for radiotherapy after prostatectomy. J Nucl Med 2004; 45:238–246.

29. Ellis RJ, Vetocnik A, Kim E, et al. Four-year biochemical outcome after radioimmuno-guided transperineal brachytherapy for patients with prostate adenocarcinoma. Int J Radiat Oncol Biol Phys 2003; 572:362–370.

30. DiBiase SJ, Hosseinzadeh K, Gullapalli RP, et al. Magnetic resonance spectroscopic imaging-guided brachytherapy for localized prostate cancer. Int J Radiat Oncol Biol Physics 2002; 1:95–101.

31. Zelefsky MJ, Cohen G, Zakian K, et al. Intraoperative conformal optimization for pre-ansperineal prostate implantation using magnetic resonance spectroscopic imaging [abstr.]. Cancer J 2000; 6:249–255.

32. Touma NJ, Izawa JI, Chin JL. Current status of local salvage therapies following radiation failure for prostate cancer. J Urol 2005; 173:373–379.

10

SPECT/CT for Dosimetry Calculations

Eric C. Frey
*Division of Medical Imaging Physics, Russell H. Morgan Department of Radiology and
Radiological Science, Johns Hopkins University, Baltimore, Maryland, U.S.A.*

INTRODUCTION

Estimating radiation dose to organs or tumors in the body is important in determining injected activity limits for radiopharmaceuticals and in targeted radiotherapy treatment (TRT) planning. This chapter will concentrate on the latter application and the potential improvements in the accuracy of dose estimation that single-photon emission computed tomography (SPECT)/computed tomography (CT) provides. This improved accuracy results from the existence of a CT image registered to planar or SPECT images that can both provide the improved image quantitation and allow the use of improved dose estimation methods.

As background to the discussion of the potential improvements, this chapter first reviews the concepts of radionuclide internal dosimetry, including a discussion of the well-known Medical Internal Radiation Dose (MIRD) schema and newer voxel-based dosimetry methods. The description of the dosimetry methods includes a discussion of the parameters that must be obtained from the imaging procedure as inputs to the dosimetry calculations. Next, planar imaging and processing methods currently used for dosimetry applications as well as their limitations are discussed. This is followed by a discussion of the ways SPECT/CT imaging can provide improved dose estimation accuracy. One major improvement facilitated by SPECT/CT systems is that the CT image enables the use of quantitative planar processing methods that model the 3-D nature of the activity distribution as well as quantitative SPECT (QSPECT) methods that include compensation for physical image degrading factors. The improvement in quantitative accuracy from these methods is demonstrated by phantom and simulation experiments. Results from a clinical trial using a combination of quantitative SPECT and planar processing methods are presented, demonstrating that there are large differences in estimated organ doses obtained with these two methods. Based on the results of the phantom and simulation experiments where QSPECT provided very good accuracy, these differences are likely due to errors in planar processing.

REVIEW OF DOSIMETRY CONCEPTS AND DOSE ESTIMATION METHODS

The ultimate goal of dosimetry is to provide an estimate of whether exposure to a given amount of radiation will result in some biological effect. In TRT planning, it is desirable to estimate the activity of a therapeutic agent necessary to kill a tumor and the maximum activity that will not cause unacceptable harm or toxicity to normal tissues. Since the effects of radiation on a biological system are complex, the assessment of the appropriate activity is usually calculated as the product of a physical parameter, the absorbed dose, and a parameter that takes into account the biological properties, i.e., the relative biological effectiveness or quality factor of a radiation source. This chapter will deal only with estimating the absorbed dose and with the type of information the imaging systems and processing methods need to provide to perform this estimation. Detailed reviews of dosimetry concepts that provide more data, especially with respect to biological effects, have recently been published (1,2).

The absorbed dose is defined as the energy that is deposited per unit mass of tissue by a radiation source. The international units are the gray (Gy), defined as 1 J/kg. The traditional unit is the rad (R), where 1 R = 0.01 Gy. The task of internal dosimetry is to estimate in vivo the absorbed dose deposited in some tissues due to an internal radionuclide source. Because direct measurement of the absorbed dose in patients is impossible, it is computed based on measurements of the temporal and spatial activity distribution in the patient and the physical properties of the radionuclide using a dose estimation method.

While the physical properties of the radioisotope are well known and independent of radiopharmaceutical, the physical properties of the patient (i.e., body and organ sizes, shapes, and positions) are quite variable. This variability greatly complicates the calculation of the absorbed dose. As a result, the most commonly used dose estimation methods make approximations about the patient's anatomy that will be discussed in more detail later.

There are two classes of dose estimation methods, i.e., methods for calculating the absorbed dose based on the activity distribution information obtained from an imaging procedure. The first class of methods is organ-based such as the MIRD schema. In these methods, every organ is treated as a radiation source when estimating the total absorbed dose to a specific organ. The second class of methods is voxel-based, where the patient's body is voxelized and the absorbed dose in each voxel is estimated. The total absorbed dose and parameters describing the dose distribution in the organ, such as the dose–volume histogram, are then computed from the voxelized dose distribution.

In all dose estimation methods, the concept of the cumulated activity, $\tilde{A}$, in some region (organ or voxel) is important. The cumulated activity is defined as the time integral of the activity and is given by:

$$\tilde{A} = \int_0^\infty A(t)\, dt, \qquad (1)$$

where $A(t)$ is the activity in the region as a function of time. The cumulated activity can be estimated from a time series of nuclear medicine scans.

Given the cumulated activity in the various regions in the body, estimating the dose for a particular target can be performed using the following general equation:

$$\bar{D}_j = \sum_i \tilde{A}_i \Delta\Phi_{j\leftarrow i}. \qquad (2)$$

In Eq. (2), $\bar{D}$ is the mean absorbed dose in the region j, $\tilde{A}_i$ is the cumulated activity in region i, Δ is the mean energy per disintegration for the radionuclide, and $\Phi_{j\leftarrow i}$, the specific absorbed fraction, is the amount of energy emitted in region i that is absorbed in region j divided by the mass of region j. Note that the specific absorbed fraction depends on the radionuclide used and the relationship of the regions with respect to one another. The various dose estimation methods discussed below differ based on which regions are used (i.e., organs or voxels) and on the method used to estimate the specific absorbed fraction [i.e., tabulations based on assumed composition and geometries or Monte Carlo (MC) simulation].

Two other quantities closely related to the cumulated activity and absorbed dose are the residence time (defined as the cumulated activity divided by the administered activity) and the specific absorbed dose (equal to the absorbed dose divided by the administered activity).

The most widely used dose estimation method is the MIRD schema. In the MIRD schema, the organs serve as the regions for dose estimation. To simplify the estimation of the specific absorbed fractions, a set of idealized standard phantoms are used. Phantoms are available for adult males, adult females, and children of various ages. These phantoms use organs with relatively simple shapes (combinations of planes and quadratic surfaces such as ellipsoids, and spheres). For a given phantom, the organs have fixed sizes, compositions, and spatial relationships to the other organs. The specific absorbed fractions are calculated for a given isotope using MC simulation. Because the specific absorbed fractions are isotope-specific, they are usually combined with the factor Δ and are referred to as S-factors. Tables of S-factors for various isotopes and the sets of standard phantoms are available in the literature and, for convenience, have been included in dose estimation software such as MIRDOSE and Olinda (2–5). A major limitation of the MIRD schema is the use of standard phantoms (6). In particular, this means that organs are assumed to have standard masses. However, in patients, there is a great deal of variability in organ size and mass. For radionuclides where the dominant particles are charged particles (instead of photons), this can be corrected using the measured organ volume. Assuming a standard density, the volume can be used to correct the S-factor by multiplying the MIRD dose by the ratio of the standard organ volume to measured organ volume. Anatomical imaging modalities such as magnetic resonance imaging (MRI) and X-ray CT can be used to obtain organ volume estimates.

Even with volume correction, the MIRD method is impaired by the use of standard sizes and assumption of standard organ shapes and spatial relationships. For charged particles, where the majority of the dose for a given organ is due to decay occurring within that organ, this is not a major limitation. However, for radionuclides where photons are an important mechanism of dose deposition, these approximations may significantly affect the accuracy of dose estimates. Other factors, such as the assumption that the activity in the organ is uniform and the fact that only the total organ dose is estimated, may represent significant limitations, especially when applied to treatment planning for TRT agents. In addition, organs containing tumors may show variable amounts of tracer uptake, and tumors themselves can show a highly inhomogeneous dose distribution. These factors can lead to very nonuniform dose distributions and can complicate the prediction of the biological response of the organ or tumor.

More recently, largely in response to demands for improved dose estimates for TRT planning, voxel-based dose estimation methods have been developed. There are two basic subclasses of these voxel-based methods: methods using convolutions with dose point-kernels (or, when integrated over a voxel, voxel *S*-factors) (7) and those using MC simulations.

In dose point-kernel methods (e.g., the 3-D-ID software package) (8–10), point-kernels represent the absorbed dose fraction as a function of distance from a point source in a uniform medium (usually water) with infinite extent. These kernels depend on the radionuclide and represent the specific dose fraction due to radiation at the center of the kernel to the positions around the center. Equation (2) can be implemented by convolving a 3-D map of the cumulated activity in the patient (estimated from imaging) with the dose kernel to provide voxel-by-voxel estimates of the absorbed dose. These estimates can be summed over the voxels in an organ to give a total organ dose estimate or used to compute dose–volume histograms. The dose point-kernel method accommodates patient-specific organ sizes and spatial relationships as well as nonuniform activities inside organs. The limitation of dose point-kernel methods is the assumption of uniform media with infinite extent. For photon emitters where a significant fraction of the energy is deposited far from the point of emission, this approximation can lead to significant inaccuracies in the dose estimates. For radionuclides emitting predominantly charged particles, the approximation is in agreement with dose measurements using other methods. However, even in this case, it can lead to poor accuracy in organs such as the lungs and at boundaries of organs. In order to implement this dose kernel–based dosimetry, the imaging procedures must provide a 3-D map of the cumulated activity along with 3-D volumes of interest (VOIs) registered to this cumulated activity map.

In order to overcome the problems with dose kernel methods, a variety of researchers have implemented MC systems for dose estimation (11–16). In MC dose estimation, charged particles and photons are tracked through a voxelized version of the body. MC methods are used to simulate the interaction of these particles with the body and a record is kept of how much dose is deposited in each voxel. The composition of each voxel is usually estimated from an anatomical image. MC dose estimation has the advantage that the MC simulations can be very accurate—the only approximations are the voxelization and estimation of the voxel composition from anatomic images. The major limitation is that the MC simulations are very computationally intensive and currently not practical for routine clinical use.

Table 1 provides a summary of the dose estimation methods discussed above. It lists the major approximations, what is required from the imaging procedures, and the dose information that is provided. It is clear that all the methods benefit from improved accuracy in the determination of cumulated activities; the following will discuss in more detail how SPECT/CT can provide these improved estimates. In addition, SPECT/CT can be used to provide high-resolution and well-registered organ VOIs. These can provide organ volume estimates for volume correction and accurate delineation of organs over which average absorbed doses and dose volume histograms can be computed. Finally, the registered X-ray CT image can be used to provide the requisite voxel compositions for MC simulation dose estimation.

CONVENTIONAL METHODS FOR ESTIMATING CUMULATED ACTIVITY FROM PLANAR IMAGES

As discussed above, dose estimation requires accurate estimation of the cumulated activity in organs (for organ-based methods) or on a voxel grid (for voxel-based methods). The conventional approach to estimating the cumulated organ activities

Table 1 Summary of Dose Estimation Methods

Method	Regions	Approximations	Required from imaging	Information provided
MIRD	Organs	Standard anatomy, uniform organ uptake	Cumulated activities per organ	Organ doses
MIRD w/volume correction	Organs	Standard anatomy w/modified organ sizes, uniform organ activity distribution	Cumulated activities per organ, organ volumes	Organ doses
Dose Kernel (voxel S-factors)	Voxels	Fixed organ composition	Cumulated activities per voxel (organ VOIs)	Voxel and organ doses, organ dose–volume histograms
MC simulation	Voxels	Voxelization and composition estimation	Cumulated activities per voxel Voxel-densities (organ VOIs)	Voxel and organ doses, organ dose–volume histograms

Abbreviations: MIRD, Medical Internal Radiation Dose; VOIs, volumes of interest; MC, Monte Carlo.

is the use of a time series of conjugate view whole-body (planar) scans. Typically, a series of planar scans are performed at five or more time points over the expected time activity curves (TACs). Regions of interest (ROIs) are defined over the organ projections and quantitative planar processing methods are applied to the series of scans to extract estimates of activity in each organ of interest at each time point. The resulting TAC is then integrated, usually by fitting it with a curve such as an exponential. Each of these steps and their inherent limitations are discussed below.

The first step in quantitative analysis of the conjugate view planar images is scatter compensation. This is usually done by subtracting an estimate of the scatter in the projections obtained from images measured in nonphotopeak energy windows (17–24). One simple and commonly used method is the triple energy window (TEW) method (17,19). In this method, images in narrow satellite energy windows surrounding each photopeak window are acquired. The estimate of the scatter contribution image in pixel i, S_i, is given by:

$$S_i = \frac{w_p}{2}\left[\frac{U_i}{w_u} + \frac{L_i}{w_l}\right], \qquad (3)$$

where w_p, w_l, and w_u are the widths of the primary, lower satellite, and upper satellite energy windows, respectively, and L_i and U_i are the counts in the ith pixel of the lower and upper satellite energy windows, respectively.

The next step in activity quantitation is attenuation compensation, which compensates for the fact that some emitted photons interact with the body. This is typically done using the geometric mean method (25). In this method, the attenuation-compensated counts in the pixel, G_i are given by:

$$G_i = e^{\mu T_i/2}\sqrt{A_i^s P_i^s} \qquad (4)$$

where μ is the attenuation coefficient of water for the isotope used in the imaging, T_i is the water-equivalent thickness of the patient in the region under pixel i, and A_i^s and P_i^s are the counts in pixel i of scatter-compensated anterior and flipped posterior images, respectively. The exponential term is often referred to as a thickness correction. Also, note that the water equivalent thickness, T_i, is the integral of the attenuation coefficient through the object divided by the attenuation coefficient of water. It can be measured using a transmission source or estimated from a registered X-ray CT image. A major limitation of the geometric mean method is that it is valid only for a point source of activity. More complicated formulas are available for use with extended sources of activity. However, these are seldom used in practice.

Because obtaining the water equivalent thickness under each pixel in the planar image is difficult, an alternate strategy is to compute a single geometric mean count-to-activity conversion factor (26). This conversion factor is computed as the ratio of the administered activity decay corrected to the time of the first planar image divided by the sum of the geometric mean of the counts in the anterior and posterior images in the first planar scan. Note that this assumes that all the administered activity is still inside the patient during the first planar scan. It also implicitly assumes that, when applied to subsequent conjugate view scans, the activity distribution does not change. Finally, note that this method implicitly includes the camera sensitivity factor that will be discussed in more detail later.

The next step is to compute the total counts in each organ by defining ROIs over the organs in the planar images and summing the counts in these ROIs. It is this step that is perhaps the most serious limitation in estimating the cumulated activity from planar images. Figure 1 shows planar images of a patient injected with ^{111}In ibritumo-mab tiuxetan at 1 and 24 hours postinjection. Note the significant overlap between organs such as the heart and lungs, liver and lungs, and liver and right kidney. Thus, defining ROIs on the planar images requires a great deal of experience and involves significant subjectivity. To reduce the effect of overlapping activity, background regions are often defined and the counts in them are subtracted from the counts in the organ ROI. Again, this process involves significant subjectivity.

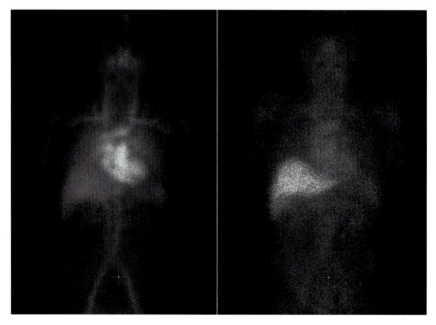

Figure 1   Anterior planar images obtained at 1 and 24 hours after injection of ^{111}In ibritumo-mab tiuxetan into a patient with non-Hodgkin's lymphoma.

The steps described earlier will provide estimates of the attenuation and scatter-compensated total counts in each organ. Converting to activity requires dividing the total number of counts in each organ by the product of the sensitivity of the collimator–detector system and the acquisition time. The sensitivity is usually measured in close temporal proximity to the patient scan using a source with known activity. For geometrically collimated photons (i.e., ones passing through the collimator holes), the sensitivity is independent of distance from the collimator face. However, this is no longer true when using radionuclides emitting medium- or high-energy photons such as ^{111}In or ^{131}I, isotopes that are commonly used in TRT applications. In this case, a significant fraction of the photons detected are from photons that penetrate or scatter in the collimator septa.

While at first it may seem that the other factor needed to estimate activity, the acquisition time, is trivial to obtain, this is not always the case. This is especially true when the whole-body scan is obtained with continuous motion. In that case, the product of the acquisition time and sensitivity can be obtained directly by imaging a standard source with known activity. It is calculated as the ratio of the counts in the image in a region over the source divided by the activity of the source at the scan time.

A time series of activity estimates in the organs of interest is obtained by applying the previously described methods to a time series produced by the conjugate view whole-body scan. In order to obtain the estimate of the cumulated activity, the TAC is further integrated after fitting the time series data to a sum of exponentials or a kinetic model. A schematic depiction of the steps involved in estimating the cumulated organ activities is shown in Figure 2.

As previously mentioned, conventional planar processing methods have significant shortcomings that limit the accuracy of cumulated activity estimates.

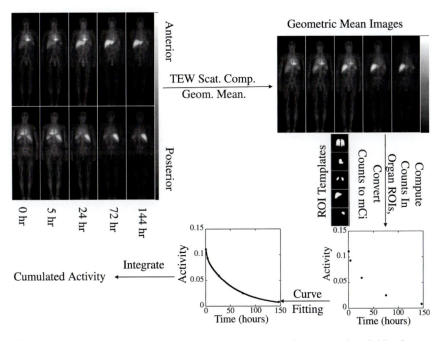

Figure 2 Schematic showing steps used in estimation of cumulated activities for organs from the anterior/posterior planar whole-body scans. *Abbreviations*: TEW, triple energy window; scat. comp., scatter compensation; geom. mean., geometric mean.

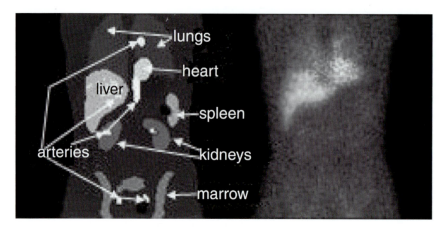

Figure 3 Image showing a coronal slice through the NURBS-based cardiac-torso phantom with a simulated ibritumomab tiuxetan activity distribution (*left*) and MC-simulated anterior planar view. *Abbreviation*: MC, Monte Carlo.

The limitations in quantitative accuracy can be demonstrated using data from a simulation experiment. In this experiment, uptake of ^{111}In ibritumomab tiuextan was simulated using the NURBS-based cardiac-torso (NCAT) phantom—a mathematical model that simulates human anatomy (27,28). Anterior and posterior projection data were MC-simulated (29,30) assuming acquisition parameters appropriate for a GE Millenium VH SPECT/CT system with a 2.54 cm thick crystal, including the use of a medium-energy general-purpose (MEGP) collimator, 15% energy windows centered at 171 and 245 keV, and 10% energy resolution. The individual organs were simulated separately. The projections of the organs were scaled assuming exponential decay of organ activities and summed to simulate acquisition of projections at 1, 5, 24, 72, and 144 hours postinjection. Figure 3 shows a coronal image through the phantom and a sample anterior projection image.

The simulated conjugate view time series were processed using the previously described methods for obtaining cumulated activities. The resulting cumulated activities were divided by the injected activity to obtain the residence time for each organ. It should be mentioned that the ROIs represented the simple projections of the true 3-D organ volumes on the planar projections. Therefore, they represented a worst case example in terms of overlap, but a best case example in terms of including the entire organ volume. Table 2 shows a comparison of the true and estimated residence times from this experiment, with negative values indicating underestimated values. A wide range of errors was found, largest for organs with low activity concentrations (such as the lungs and kidneys) and small organs (such as the marrow and blood).

Table 2 Errors in Residence Time Estimates Using Conventional and CAMI Planar Processing Methods

Organ	Heart	Lung	Liver	Kidney	Spleen	Marrow	Blood
Conventional % error	36	345	−7.5	119	121	157	442
CAMI % error	10.4	12.2	9.9	27	1	37	−4.6

Abbreviation: CAMI, computer-assisted matrix inversion.

POTENTIAL IMPROVEMENTS IN DOSE ESTIMATION USING SPECT/CT SYSTEMS

It appears that conventional planar processing methods are confounded by several weaknesses—mainly compensating for organ overlap and the effects of attenuation. However, even with planar images, SPECT/CT systems have the potential to provide improved quantitative accuracy. The availability of CT images registered with the planar scintigraphy would allow the use of newer, more accurate quantitative planar processing methods that may handle overlap and attenuation in a better way. A SPECT/CT system does have the potential to allow for convenient acquisition of registered CT and whole-body nuclear medicine studies.

While these improved planar processing methods allow only for the estimation of activity at the organ level, voxelized dose estimation methods can also model nonuniform activity distributions and potentially provide better information about the likely biological effects of radiation. Voxelized estimation methods require, however, the 3-D cumulated activity distribution as an input. Conventional SPECT can provide 3-D activity estimates with a poor quantitative accuracy when using standard reconstruction techniques such as filtered backprojection.

Improving the quantitative accuracy of SPECT requires compensating for image-degrading effects such as scatter and attenuation. Attenuation compensation, especially in regions with nonuniform attenuation distributions like the thorax, requires a registered attenuation map; this can be obtained easily using a SPECT/CT system. Scatter compensation using energy-window–based methods such as TEW has limited accuracy and tends to increase image noise. Reconstruction-based scatter compensation, where photon scatter is modeled inside an iterative reconstruction algorithm, has been shown to provide improved quantitative accuracy and image quality compared with energy-window–based methods. These reconstruction-based methods also require an attenuation map, so the use of a SPECT/CT system also facilitates their use. Even with compensation for the blurring in SPECT images due to the collimator–detector response function (CDRF), the resolution of SPECT images remains relatively poor and partial volume effects (PVEs), especially for small objects, are significant. The availability of a high-resolution CT image registered with the SPECT image enables high-resolution definition of organs and tumors and thus the use of region- or voxel-based partial volume compensation (PVC) methods to further improve quantitative accuracy.

The availability of a registered CT image combined with a 3-D cumulated activity distribution enables the use of MC simulation dose-estimation methods. This provides the ultimate accuracy in dose estimation, allowing patient-specific modeling of nonuniform activity and density distributions, organ size, shape, and spatial relationships.

The following sections provide more details of the improved quantitative imaging methods enabled by the use of SPECT/CT systems, data from simulation experiments demonstrating the efficacy of these methods, and results from a clinical trial comparing planar and QSPECT/planar imaging studies.

QUANTITATIVE PLANAR METHODS

Several approaches have been proposed as alternatives to the geometric mean method (31–33). One such method is computer-assisted matrix inversion (CAMI) (31), a minor modification of which is described below.

In this method, CT images registered with the planar images are used to provide both the attenuation along the path connecting pixels in the anterior and posterior views as well as definitions of the 3-D organ VOIs. From these 3-D VOIs, the attenuation map, and assumption of uniform activity concentration in the organs, it is possible to compute the attenuated projections of the organ VOIs with activity concentrations of unity. The attenuated projection for each organ VOI is performed both in the anterior and posterior directions, producing an anterior and posterior attenuated projection image of the organ VOI. Repeating this for each organ of interest (including the background region that contains voxels inside the body that do not belong to an organ of interest) gives $2*N$ projection images, each with M pixels, where N is the number of organ VOIs and M is the total number of pixels in each whole-body image. This may be represented as a $2*M$ by N matrix, T. In this matrix, the number of rows is equal to the number of pixels in each whole-body image times two (because there are both anterior and posterior images) and the number of columns equals the number of organs. We now represent the activity concentration in each organ by a constant, a_j, where j goes from 1 to N. This resulting vector, **a**, is the unknown that must be determined and, when multiplied by the organ volume, gives the total activity in the organ. Now define the scatter corrected counts in the anterior and posterior projection images as p_i, where i goes from 1 to $2*M$. Note that this vector includes the counts from pixels in the anterior and posterior images. The measured projections thus form a vector, **p**. The net result is that we obtain a matrix equation relating the measured whole-body images to the unknown organ activities:

$$Ta = p. \tag{5}$$

Solving this equation by inverting the matrix **T** and multiplying this inverse by the vector p will give an estimate of the organ activities. This method has the advantage that it simultaneously compensates for attenuation in the organs as well as organ overlap. The major limitation is the assumption that the activity concentration in each organ (as well as the background) is uniform and the fact that the matrix in Eq. (5) may not be easily invertible.

To illustrate the potential improvement in accuracy that can be obtained using planar processing methods such as CAMI, we performed a simulation experiment similar to the one discussed earlier for conventional planar processing. In this experiment, the planar images were scatter-compensated using TEW. The 3-D VOIs used in the method were based on the true organ volumes. As a result, this data represents something of a best-case result for the CAMI method. The errors, as shown in Table 2, are substantially reduced compared to conventional planar processing, largely due to the fact that the method implicitly incorporates background and overlap compensation, but also due to the improved attenuation compensation.

QUANTITATIVE SPECT METHODS

As discussed above, obtaining QSPECT images, where the voxel values are in units of activity per unit of volume, requires compensation for image degrading factors including attenuation, scatter, and the CDRF. Much research has been devoted to this area, and software packages implementing these methods are beginning to emerge for commercial SPECT systems.

Quantitative SPECT methods are usually based on iterative reconstruction algorithms such as ordered subsets-expectation maximization (OSEM) (34). Iterative reconstruction involves the refinement of an initial estimate of the 3-D activity distribution in the patient by comparing projections computed from the estimated activity distribution with measured projections, as illustrated in Figure 4. This projection process is able to model the previously mentioned physical image-degrading effects. The computed and measured projections are then compared and, since the measured projections are corrupted by statistical noise, the comparison is based on a statistical criterion such as maximizing the statistical likelihood that the measured data came from the estimated object. The current estimate is then updated, usually using a backprojection algorithm that also models the physics of the imaging system. The key to compensating for effects such as scatter and attenuation is to model them in the projection and backprojection steps.

A variety of iterative reconstruction methods have been proposed. The methods differ in the soundness of their theoretical basis, noise properties, ease of implementation, and the number of iterations required to produce useful images. While its theoretical soundness is somewhat limited and its noise properties are not optimal, the OSEM algorithm provides ease of implementation and produces useful images in a relatively small number of iterations. It is thus the most widely available iterative reconstruction algorithm on commercial SPECT systems.

The most important factor degrading image-quantitative accuracy is attenuation, which results from the interaction of photons with the patient's body. Attenuation results in a depth-dependent reduction in the number of photon counts compared to the same source in air. For a point $\bar{r}$ in the detection plane and a source at point $\bar{x}$ in the 3-D activity distribution, the reduction factor of the photon counts equals:

$$\exp\left[-\int_{L(\bar{x},\bar{d})} \mu(\bar{r})\,d\bar{r}\right] \tag{6}$$

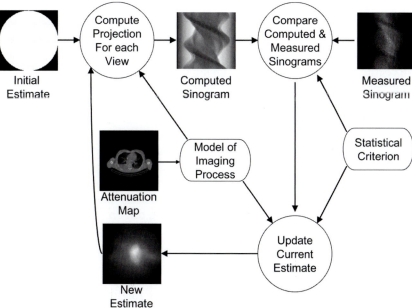

Figure 4 Schematic illustrating iterative reconstruction process.

where $\mu(\bar{r})$ is the 3-D attenuation distribution in the body and $L(\bar{x}, \bar{d})$ indicates that the integral is over the line in 3-D space from the point of emission, $\bar{x}$, to the point of detection, $\bar{d}$. Note that if the object is uniform and has a constant attenuation coefficient, then the attenuation factor is

$$e^{-\mu D} \tag{7}$$

where D is the depth of the source and μ is the attenuation coefficient of the material. It is clear from Eqs. (6) and (7) that the attenuation factor increases with source depth. Thus the effect of attenuation on SPECT images is greatest near the center of the object.

It is also clear from Eq. (6) that modeling the attenuation process requires knowledge of the attenuation distribution, $\mu(\bar{r})$, inside the patient. This is usually obtained using a radioisotope-based transmission scan or an X-ray CT scan. As the attenuation coefficient of a material depends on the photon energy, there are some difficulties in translating the CT images that result from use of a continuous-spectrum X-ray tube to an attenuation map appropriate for attenuation compensation. This problem, however, has largely been solved and registered CT images provide a fast and convenient way to obtain attenuation maps.

A second image-degrading factor related to interaction of photons with the patient is scatter. A significant fraction of these scattered photons will pass through the collimator and be detected. Owing to the fact that these scattered photons have lower energy than unscattered ones, the use of energy windows to reject lower energy photons can significantly reduce the number of scattered photons counted in the projection image. However, the finite energy resolution of gamma cameras limits the effectiveness of energy discrimination in reducing scatter in projection images, and typically 20% to 60% of detected photons are scattered at least once. The fraction of scattered photons detected increases with source depth. Thus, the quantitative effects of scatter are greatest in the center of the body. Note that without attenuation compensation, scatter tends to reduce the effects of attenuation.

There has been a great deal of work to develop scatter compensation methods for SPECT. Perhaps the simplest method is the use of altered attenuation coefficients to overcome the quantitative effects. The attenuation coefficient that describes the reduction in beam due to interaction with the object is known as the narrow beam attenuation coefficient. This coefficient is well understood and tabulated for different materials and energies. As alluded to above, scatter partially offsets the quantitative effect of attenuation. Therefore, one can approximately compensate for the effects of scatter by reducing the attenuation coefficient used in attenuation compensation by 20% to 30%. This coefficient, which depends on the object and detector geometries, is known as a broad beam attenuation coefficient. The use of broad beam attenuation coefficients is approximate, however, and does not take into account the very different spatial distributions of scattered and unscattered photons. In addition, the broad beam attenuation coefficient is patient-dependent and only partially effective in reducing quantitative inaccuracies in the SPECT images due to scatter.

Another simple and effective method is to use the ability of gamma cameras to simultaneously acquire images in multiple energy windows. These images from non-photopeak energy windows can be used to estimate the scatter in the photopeak energy window. An example of this method is the TEW method already discussed in the context of planar imaging. The simplest way to use this energy-window–based scatter estimate is to subtract it from the measured projections and then perform the reconstruction, treating the projections as if they were scatter-free. However, the subtraction process can lead to negative values in the scatter-compensated projections. These negative values cause problems for iterative reconstruction algorithms such as OSEM and, even if handled, result in suboptimal noise-properties. A better way to use the scatter estimate is to add it to the computed sinogram before comparing it to the measured one in the iterative loop shown in Figure 4.

As discussed in the context of planar image scatter compensation, there are limitations in accuracy when using energy-window–based scatter estimates. These result from the fact that the distribution of scattered photons in the projection image depends on energy, and is thus different for photopeak and nonphotopeak energy windows. In addition, the scatter estimate from an energy-based method is noisy, resulting in increased noise in the reconstructed image.

The last class of methods involves modeling the scatter during the reconstruction process. This modeling is done in the projection and backprojection steps (though sometimes omitted from the backprojection for efficiency reasons) (35) of an iterative reconstruction algorithm and involves calculating the probability that a photon emitted in a particular voxel of the reconstructed image would be scattered and detected in a given projection bin. This probability can be estimated using MC methods, direct numerical integration (36–38), or using a scatter model (39–45). MC methods are intermediate in both complexity to implement and computation time, but have the best accuracy. Direct numerical integration of the equations governing photon scatter, similar to model-based scatter compensation used in 3-D PET, avoids the noise generated in MC methods. However, the direct methods require very large computation times, especially if multiple scattering is considered. Scatter models are the most computationally efficient, but are not as accurate as MC methods. The model-based methods, even when using scatter models, have advantages in terms of accuracy and noise properties compared to energy-window–based methods. Model-based methods require an attenuation map such as the one provided by a SPECT/CT system. Thus SPECT/CT systems enable the use of model-based scatter compensation.

The resolution in a SPECT image is determined largely by the spatially varying CDRF. The CDRF is simply the image of a point source in air. For this paper we confine discussion to parallel-hole collimators. There are several important components of the CDRF that are distinguished by how photons interact (or do not interact) with the collimator septa: the intrinsic, geometric, septal scatter, and septal penetration response components. The geometric response is the portion of the CDRF due to photons that pass through the collimator holes and largely determines the full-width at half-maximum (FWHM) of the CDRF. The width of the geometric response is essentially constant in the planes parallel to the collimator face, but increases with distance. The sensitivity of the collimator is independent of distance from the collimator face. The intrinsic component of the CDRF is due to the imprecision in estimating the position of photons incident on the detector crystal. For typical collimator designs and imaging distances, the intrinsic response has little effect on the resolution of the SPECT images. The final two components of the CDRF are usually important only for isotopes emitting medium- (e.g.,[111]In) or high- (e.g.,[131]I) energy photons. For these isotopes, a significant fraction of photons penetrate through or scatter in the collimator septa. These components give rise to very long, low intensity tails in the CDRF. They also result in a sensitivity that varies as a function of distance in a somewhat complicated way that depends on the details of the collimator design.

Compensation for the CDRF can result in improved image resolution (44) and thus provide improved quantitation for small organs and objects of interest such as tumors; it can be performed by modeling the CDRF in the projection and backprojection steps of an iterative reconstruction algorithm. Most commonly, only the geometric and intrinsic components, which can be predicted using analytic formulas, are modeled. However, modeling the full CDRF, including penetration and septal scatter, can be performed using tables of measured or MC-simulated CDRFs. To the extent that the sensitivity is independent of distance from the collimator (e.g., after full CDRF compensation), the effects of sensitivity on image quantitation can be modeled using a system sensitivity measured at a single distance.

Even with CDRF compensation, the resolution of SPECT images is limited and in the range of 1 to 2 cm. Thus, quantitation of small objects or objects with low activities in close proximity to objects with high activities is degraded by these resolution effects. Collectively, these effects are often referred to as PVEs. PVEs result in both spill-out of activity from an object and spill-in of activity from neighboring objects. The magnitude of these effects depends on the (spatially varying) system resolution and the sizes of the various objects. Several methods have been developed to compensate for the effects of PVEs on quantitation (45,46). State-of-the-art methods are based on accurate definitions of VOIs in the SPECT images. These accurate VOI definitions can be obtained from registered high-resolution anatomical images such as those obtained from MRI or X-ray CT. Thus, SPECT/CT systems, which provide the requisite registered CT image, enable the use of PVC methods to provide further improvements in organ activity estimates.

Another limitation of SPECT is the noise inherent in the images. For large VOIs such as organs, as well as typical SPECT acquisition times, the imprecision in the activity measurements due to statistical noise will be relatively small due to averaging. However, if the SPECT images are used to estimate cumulated activities in each voxel, the distribution of cumulated activities will be affected by the noise and there will likely be errors in dose–volume histograms.

As described earlier, SPECT/CT systems allow the reconstruction of quantitative SPECT images with compensation for attenuation, scatter, and the CDRF. In addition, the registered CT images can be used to define organ or tumor VOIs much more accurately than would be possible using only the SPECT images. These registered VOIs are useful for improving the accuracy of the activity estimates using PVC methods. Figure 5 illustrates the use of a SPECT/CT image for improved VOI definition. This figure shows a single slice from a patient ^{111}In ibritumomab tiuxetan image obtained with a GE Millenium VG SPECT/CT system. The image shows the SPECT, CT, and fused SPECT/CT images. It is much easier to accurately define the lung boundaries using the CT image

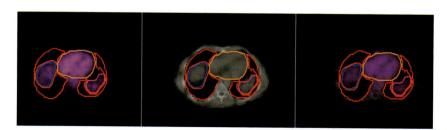

Figure 5 (*Left to right*): SPECT, CT, and fused SPECT/CT images from patient injected with ^{111}In ibritumomab tiuxetan imaged with GE Millenium VG SPECT/CT system. The red, orange, blue, and violet lines indicate lines drawn to define the portions of the lung, heart, liver, and spleen VOIs, respectively, in this slice. *Abbreviations*: SPECT, single-photon emission computed tomography; CT, computed tomography; VOIs, volumes of interest.

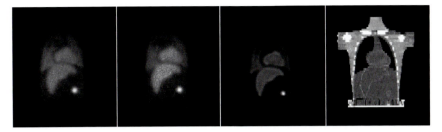

Figure 6 Coronal images of Radiology Support Devices (RSD) Heart/Thorax Phantom from different SPECT reconstructions. The images are (*left to right*) SPECT with no compensation, SPECT with attenuation compensation, SPECT with attenuation, collimator–detector response, scatter compensation, and X-ray CT. *Abbreviations*: SPECT, single-photon emission computed tomography; CT, computed tomography.

than the SPECT image alone. This figure also illustrates how SPECT inherently eliminates the difficulties with organ overlap found in planar images.

To illustrate the quantitative accuracy that can be obtained using QSPECT methods, we performed a phantom study using the RSD Heart/Thorax Phantom (Radiology Support Devices, Inc., Long Beach, California, U.S.A.). The heart, lung, liver, and background compartments were filled with a solution containing ^{111}In chloride with activities in the ratios of 19:5:20:1, respectively. A 3.5 cm sphere was also placed in the phantom and had an activity concentration of 156-times the background. The projection data were reconstructed using OSEM with no compensation, with attenuation compensation only, and using model-based compensation for attenuation, scatter, and the CDRF. The X-ray CT image was used to obtain an attenuation map. Sample coronal slices through SPECT images reconstructed with various compensations and the X-ray CT image are shown in Figure 6. These images demonstrate the significant improvement in image resolution and image contrast obtained using QSPECT reconstruction.

To demonstrate the quantitative accuracy of QSPECT, we used the SPECT images reconstructed with the various compensations and manually defined VOIs obtained using the X-ray CT image to obtain activity estimates. Table 3 shows the errors between the estimates and the true activities in the phantom's compartments. Note that, for all the organs, there is a very large underestimation when no compensation was used. Attenuation compensation reduced the magnitude of the errors, but resulted in significant overestimates of the activity due to the presence of scattered photons. Addition of compensation for scatter and the CDRF resulted in errors in activity estimates smaller than or equal to 5%.

Table 3 Percent Errors[a] in Activity Estimates for RSD Phantom

Compensations	Heart	Lung	Liver	Sphere
None	−75.5	−62.1	−70.8	−74.8
Attenuation	31.7	38.5	42.8	−1.1
Attenuation, CDRF, and scatter	−5.2	2.5	4.1	0.7

[a]Negative errors indicate underestimates.
Abbreviations: CDRF, collimator–detector response function; RSD, radiology support devices.

QUANTITATIVE SPECT METHODS FOR CUMULATED ACTIVITY ESTIMATION

There are two potential ways to use QSPECT to obtain the voxel or organ cumulated activity estimates for use in dosimetry calculations. The first is to obtain the TAC from a series of planar scans and combine this with a single QSPECT study and the second is to obtain a time series of QSPECT scans.

For the hybrid planar/QSPECT method, the TAC obtained from the planar time series is then rescaled so that the curve passes through the data point from the SPECT study:

$$\tilde{A} = \frac{A_{\mathrm{SPECT}}(t_{\mathrm{SPECT}})}{A(t_{\mathrm{SPECT}})} \int\limits_{0}^{\infty} A(t)\,\mathrm{d}t \tag{8}$$

In Eq. (8), t_{SPECT} is the time of the SPECT acquisition, $A_{\mathrm{SPECT}}(t_{\mathrm{SPECT}})$ is the activity in the voxel or organ obtained from the QSPECT image, and $A(t_{\mathrm{SPECT}})$ is the activity in the organ as determined by the planar study at the time of the SPECT acquisition. The SPECT/CT acquisition used to obtain the QSPECT image can be used to obtain the 3-D VOIs required for quantitative planar methods such as CAMI. A limitation of this hybrid method when used as the basis for voxel-based dose estimation is that only a single residence time will be estimated for each organ; spatial variation of the shape of the TAC is not modeled.

Using a QSPECT time series has the potential to allow estimation of the cumulated activity in each voxel. One difficulty is that the SPECT/CT images from the different time points must be registered so that the corresponding pixel in each image represents the same portion of the body. This clearly requires a combination of careful patient-positioning and registration of the images after acquisition. However, this registration process will likely be easier for SPECT/CT images than it would be for SPECT images alone. The cumulated activity images obtained from a QSPECT time series can be combined with voxel-based dose estimation methods to provide information about the spatial variation of the dose or the time and spatial variation of dose rate in an organ or tumor. It should be noted that the dose distributions will have limited accuracy for volumes less than 1 to $4\,\mathrm{cm}^3$ due to the image resolution and noise even after CDRF compensation.

CLINICAL APPLICATION OF SPECT/CT FOR DOSIMETRY

As shown earlier, QSPECT methods that are facilitated by SPECT/CT systems can provide significant improvement in the accuracy of organ quantitation. To demonstrate the differences this can have on the dose estimates to critical organs, we present data from an ongoing clinical trial of high dose ibritumomab tiuxetan therapy for non-Hodgkin's lymphoma. The data presented are from the first two cohorts of a dose escalation phase. Prior to therapy, stem cells were harvested and saved for later autologous stem cell transplantation. A 5 mCi dose of ^{111}In ibritumomab tiuxetan was administered for treatment-planning purposes. A series of conjugate view whole-body planar scans were acquired at approximately 1, 5, 24, 72, and 144 hours after injection with scan speeds of 10, 10, 7, 5, and 5 cm per minute, respectively. A standard source with activity of ~500 μCi at injection time was placed lateral to the patient's feet and imaged in all scans. At 24 hours, thoracic and abdominal

SPECT/CT scans were obtained. The acquisition of each SPECT scan included a total of 120 projections over 360°, with one minute per view. A one- to five-minute static image of the standard source with the source-to-collimator distance approximately equal to the average radius of rotation for the SPECT study was obtained and used to compute the system sensitivity. All acquisitions were performed using a GE Millenium VG SPECT/CT system with a 1.59 cm (5/8") thick crystal. A GE VP-5 MEGP collimator and a pixel size of 0.442 cm were used and 15% wide energy windows were centered over the 171 and 245 keV ^{111}In photopeaks. As a result of limitations on the number of allowed energy peaks, only two additional scatter images were acquired using 147 to 159 keV and 183 to 227 keV energy windows. These were used to perform TEW scatter compensation for the whole-body scans assuming that there were no counts in an energy window above the 245 keV photopeak window. The attenuation maps obtained from the CT scans were rescaled so that the values were appropriate for the average energy of the ^{111}In photopeaks.

The whole-body scan images were processed by performing TEW scatter compensation on both the anterior and posterior planar images, then computing the geometric mean of these two images. The images were also scaled by sensitivity scale factors obtained from the standard source in each whole-body scan. Thickness correction was not performed, but a conversion factor from geometric mean counts to activity was obtained from the image acquired one hour postinjection and the assumption that the total injected activity was still in the patient.

The SPECT images were reconstructed using custom software that modeled attenuation and scatter using the CT-derived attenuation map (41). Conversion to activity was performed using the sensitivity obtained from the static acquisition of the standard source.

VOIs were defined manually slice-by-slice using overlaid SPECT and CT images. The CT-defined organ boundaries were used where possible. However, for some organs, and due to the relatively coarse axial slice thickness (1 cm) of the CT image, the SPECT image was also used as a guide for the region definition. The VOIs for each organ were then converted into a mask image (voxels inside the VOI set to 1 and those outside to 0) and simple projections of this mask image to the anterior and posterior projection views were then computed. The projections of these 3-D organ ROIs were used as the basis for the ROIs for the planar image, but were adjusted manually to avoid overlap of the ROIs. No explicit background or overlap correction was performed. The VOIs and ROIs were applied to the SPECT and processed planar images, respectively, to obtain the organ activity estimates. Planar geometric mean images and the organ ROIs from one patient at each of the five time points are shown in Figure 7. Sample (noncontiguous) slices of the 24-hour SPECT image are shown in Figure 8 along with the portion of the VOIs for the organs lying in each slice.

The planar activity values for each organ were fitted with mono- or bi-exponential curves and integrated to obtain the cumulated activity. For the QSPECT/planar hybrid method, the cumulated activity estimate was rescaled by the ratio of the QSPECT activity values to the planar ones at the time of the SPECT scan. Figure 9 shows the relative difference in the cumulated activity estimated with the QSPECT/planar hybrid method as compared to the planar method. The average relative difference and standard deviations of the differences divided by the mean organ activity over six patients are shown in Table 4. Since the actual cumulated activity is not known, it is not possible to say definitively which activity estimate is better. However, in the simulation and phantom experiments, QSPECT provided very accurate

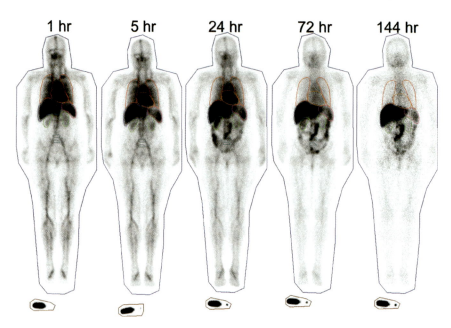

Figure 7 Sample time series of planar whole-body scan images for a patient injected with ^{111}In ibritumomab tiuxetan. The ROIs drawn over the lungs, liver, heart, spleen, kidneys, bone marrow, whole body, and standard source are illustrated. *Abbreviation*: ROIs, regions of interest.

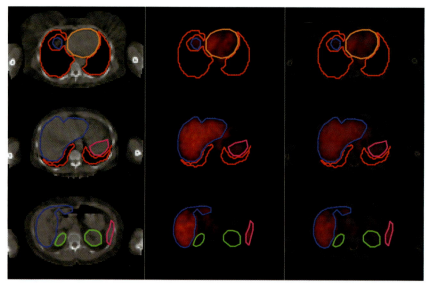

Figure 8 Sample CT (*left* column), SPECT (*center* column), and fused SPECT/CT (*right* column) images from same patient shown in Figure 7. The images were acquired 24 hours postinjection. The slices are noncontiguous and show portions of the 3-D volumes of interest for the lungs (*red*), liver (*blue*), heart (*orange*), spleen (*fuchsia*), and kidneys (*light green*). *Abbreviations*: SPECT, single-photon emission computed tomography; CT, computed tomography.

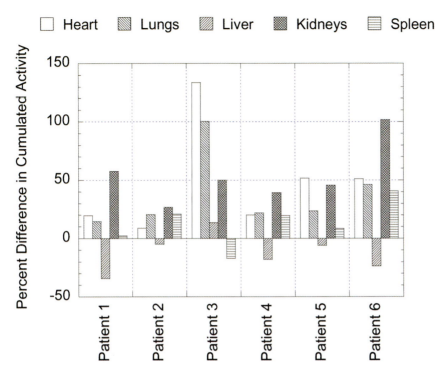

Figure 9 Relative difference between cumulated activities estimated using conventional planar and QSPECT/planar hybrid processing. Positive percent differences indicate that the estimate using the planar method was larger than that obtained when using the QSPECT/planar method. *Abbreviation*: QSPECT, quantitative single-photon emission computed tomography.

estimates of organ activity, superior to those obtained with planar processing. Thus it is reasonable to conclude that the largest part of the difference between the two methods was due to errors in the estimates made with conventional planar methods.

Making the assumption that the differences mostly represent errors in the planar cumulated activity, we can make the following observation: for the liver, the planar method significantly underestimated the cumulated activity by an average of 17%. Since the liver was the dose-limiting organ for these patients, giving a therapeutic dose based on the planar estimate of the liver activity would have resulted in the patients receiving, on average, a 17% higher liver dose than the target dose.

Table 4 Relative Difference in Cumulated Activity Estimated Using QSPECT/Planar and Conventional Planar Methods

Organ	% Difference	% Standard deviation
Heart	36	20
Lungs	33	19
Liver	−17	22
Kidneys	63	70
Spleen	−7	34

Abbreviation: QSPECT, quantitative single-photon emission computed tomography.

However, almost as significant as this average error, there is significant variation in the dose estimated with planar imaging. This is reflected in the percent standard deviations shown in Table 4. For the liver, the use of a therapeutic activity based on the planar cumulative activity estimate for this population of patients would have resulted in actual liver doses ranging from 34% larger to 14% smaller than the target dose. For other organs, with the exception of the spleen, the planar methods resulted in overestimation of the cumulated activity (and would thus have underestimated the dose to the organ) with the differences even larger than for the liver. There was also large variation in the errors for the other organs.

In addition to the large differences in organ activities, the QSPECT/planar method would allow the use of voxel-based dose estimation, though only allowing a single TAC shape (scaled to a different height based on the QSPECT image) for each voxel in an organ. The addition of voxel-based dose estimation has the potential to provide more accurate estimates of organ dose by explicitly accounting for differences in organ mass and spatial relationship compared to standard phantom-based approaches.

SUMMARY

This chapter describes the potential improvement in dose estimation provided by the use of SPECT/CT systems. Dose estimation methods may be organ-based methods and voxel-based methods. Organ methods are based on the use of standard phantoms, such as the MIRD phantoms, with an estimate of the cumulated activity in each organ of interest. These methods do not take into account variability in patient organ size and spatial relationship or variations in the spatial or temporal distribution of activity within each organ. For radionuclides emitting charged particles, the accuracy of dose estimates can be improved using volume-compensation based on volumes measured in CT images.

Voxel-based methods require the estimation of the cumulated activity in each voxel in the patient. They explicitly model each patient's anatomy and the variation in activity inside organs. They can thus provide improved estimates of organ dose as well as information about the spatial distribution of dose and dose rate inside the organ. This extra intraorgan information has the potential for improving predictions of the maximum permissible doses (MPDs) and the ability of the TRT to deliver sufficient radiation to the target.

The limitation in accuracy of dose estimates obtained using conventional planar activity estimation and standard phantom dose estimation are largely due to the limited information about the 3-D organ shapes, activity distributions, and compositions provided by planar nuclear medicine imaging alone. As a result, there are a number of potential improvements in accuracy of dose estimates that are enabled by the use of SPECT/CT systems. Conventional planar imaging and processing methods can be used to estimate the organ-cumulated activities needed for organ-based dose estimation. To accurately estimate organ activities, a prerequisite for estimating cumulated activity, it is essential to compensate for organ overlap and the background in the planar projections. Compensating for physical factors such as attenuation, object scatter, and penetration and scatter in the collimator–detector system is also important. However, with conventional planar methods, only approximate compensation methods are available. The relatively poor quantitative accuracy of conventional planar imaging and processing methods was demonstrated using a realistic phantom simulation study for the case of imaging ibritumomab tiuxetan.

On the basis of the background in dose estimation methods and conventional planar processing methods, there are potential improvements in dose estimates available with the use of SPECT/CT systems. The use of a registered CT image allows one to model the 3-D nature of the activity distribution, provides accurate delineation of organs and objects (such as tumors) of interest, and also provides information with respect to the radiation-attenuating properties within the patient. The CT images thus enable the use of quantitative planar and SPECT processing methods to produce more accurate organ activity estimates. QSPECT methods require an attenuation map (which can be readily obtained from the registered CT image) and benefit from the accurate, high-resolution delineation of organ boundaries available from the CT. These QSPECT methods can yield accurate estimates of the 3-D activity distributions used to estimate the 3-D cumulated activity distribution and thus enable the use of voxel-based dose estimation methods.

Improved planar processing methods are available when a CT image is registered with the whole-body scintigraphy. A simulation study was used above to demonstrate the large improvement in accuracy that such methods can provide.

QSPECT methods are used to compensate for attenuation, scatter, the collimator–detector response, and PVEs. Of these, attenuation, model-based scatter, and PVC can only be applied when a registered CT image is available. While it is possible to obtain the CT image by registration of images from separate SPECT and CT systems, this is a difficult process and the accuracy of registration is limited. A phantom study was used to show that QSPECT methods provide excellent accuracy in organ activity estimation, providing error margins of 5% or less. In dosimetry applications, QSPECT images are used for the estimation of cumulated activity estimates. The QSPECT methods can be used at each time point or at a single time point in conjunction with a time series of whole-body scans to estimate the cumulated activity in organs. Using QSPECT at each time point allows voxel-based dosimetry with a different TAC in each voxel, providing full generality. However, this requires careful registration of the QSPECT scans obtained at the different time points. This registration is facilitated by the use of SPECT/CT systems, which can provide registered SPECT and CT images at each time point.

Results from a set of six patient dosimetry studies using ^{111}In ibritumomab tiuxetan have been used to illustrate the potential impact of the QSPECT methods on TRT planning. Cumulated activities in the spleen, liver, lungs, heart, kidneys, and bone marrow were estimated using planar and QSPECT/planar methods. There were large differences between the two estimates and the size and sign of the differences varied among the patients. Assuming that the QSPECT cumulated activity estimates are accurate, observed differences in the planar and QSPECT estimates can be attributed to errors in the planar quantitation process. The errors and variations in the errors among patients of the planar cumulated activity estimates could result in errors in estimating the MPD for TRT applications.

In conclusion, the registered CT image provided by SPECT/CT systems allows significant improvements in dose estimates both by providing more accurate inputs and by enabling the use of more advanced dose estimation methods that are likely to provide better predictions of biological effects. While clinical studies demonstrating the impact of improved dosimetry on patient outcomes are not yet available, the increasing availability of SPECT/CT systems, as well as voxel-based dose estimation methods and quantitative SPECT and planar processing methods enabled by these systems should lead to such studies in the near future.

ACKNOWLEDGMENTS

The author gratefully acknowledges Bin He for help in preparing the data and figures for this chapter and Dr. Yong Du and Martin Stumpf for their assistance in proofreading the manuscript.

REFERENCES

1. Zanzonico PB. Internal radionuclide radiation dosimetry: a review of basic concepts and recent developments. J Nucl Med 2000; 41:297–308.
2. Sgouros G. Dosimetry of internal emitters. J Nucl Med 2005; 46(suppl 1):18S–27S.
3. Stabin MG. MIRDOSE: personal computer software for internal dose assessment in nuclear medicine. J Nucl Med 1996; 37:538–546.
4. Stabin MG, Siegel JA. Physical models and dose factors for use in internal dose assessment. Health Physics 2003; 85:294–310.
5. Stabin MG, Sparks RB. OLINDA—PC-based software for biokinetic analysis and internal dose calculations in nuclear medicine. J Nucl Med 2003; 44:103P–103P.
6. Cristy M, Eckerman K. Specific absorbed fractions of energy at various ages from internal photon sources (I-VII). Oak Ridge National Laboratory Report ORNL/TM8381/V1-7. 1987, National Technical Information Service, United States Department of Commerce Springfield, VA.
7. Bolch WE, Bouchet LG, Robertson JS, et al. MIRD pamphlet No. 17: the dosimetry of nonuniform activity distributions—Radionuclide S values at the voxel level. J Nucl Med 1999; 40:11S–36S.
8. Kolbert KS, Sgouros G, Scott AM, et al. Implementation and evaluation of patient-specific three-dimensional internal dosimetry. J Nucl Med 1997; 38:301–308.
9. Sgouros G, Barest G, Thekkumtahala J, et al. Treatment planning for internal radionuclide therapy—3-dimensional dosimetry for nonuniformly distributed radionuclides. J Nucl Med 1990; 31:1884–1891.
10. Sgouros G, Chiu S, Pentlow KS, et al. 3-Dimensional dosimetry for radioimmunotherapy treatment planning. J Nucl Med 1993; 34:1595–1601.
11. Clairand I, Ricard M, Gourion J, et al. DOSE3D: EGS4 Monte Carlo code-based software for internal radionuclide dosimetry. J Nucl Med 1999; 40:1517–1523.
12. Furhang EE, Chui CS, Sgouros G, et al. Implementation of a Monte Carlo dosimetry method for patient-specific internal emitter therapy. Med Phys 1997; 24:1163–1172.
13. Furhang EE, Chui CS, Sgouros G. A Monte Carlo approach to patient-specific dosimetry. Med Phy 1996; 23(9):1523–1529.
14. Liu A, Williams LE, Wong JYC, et al. Monte Carlo-assisted voxel source kernel method (MAVSK) for internal beta dosimetry. Nucl Med Biol 1998; 25(4):423–433.
15. Yoriyaz H, dos Santos A, Stabin MG, et al. Absorbed fractions in a voxel-based phantom calculated with the MCNP-4B code. Med Phys 2000; 27(7):1555–1562.
16. Yoriyaz H, Stabin MG, dos Santos A. Monte Carlo MCNP-4B-based absorbed dose distribution estimates for patient-specific dosimetry. J Nucl Med 2001; 42(4):662–669.
17. Ichihara T, Ogawa K, Motomura N, et al. Compton scatter compensation using the triple-energy window method for single-isotope and dual-isotope SPECT. J Nucl Med 1993; 34(12):2216–2221.

18. King MA, Hademenos GJ, Glick SJ. A dual-photopeak window method for scatter correction. J Nucl Med 1992; 33(4):605–612.
19. Ogawa K, Harata Y, Ichihara T, et al. A practical method for position-dependent compton-scatter correction in single photon-emission CT. IEEE Trans Med Imag 1991; 10(3):408–412.
20. Mas J, Hannequin P, Benyounes R, et al. Scatter correction in planar imaging and SPECT by constrained factor-analysis of dynamic structures (Fads). Phys Med Biol 1990; 35(11):1451–1465.
21. Gagnon D, Toddpokropek A, Arsenault A, et al. Introduction to holospectral imaging in nuclear-medicine for scatter subtraction. IEEE Trans Med Imag 1989; 8(3):245–250.
22. Koral KF. SPECT compton-scattering correction by analysis of energy spectra. J Nucl Med 1988; 29:195–202.
23. Haynor DR, Kaplan MS, Miyaoka RS, et al. Multiwindow scatter correction techniques in single-photon imaging. Med Phys 1995; 22(12):2015–2024.
24. Jaszczak RJ, Greer KL, Floyd CE, et al. Improved SPECT quantification using compensation for scattered photons. J Nucl Med 1984; 25(8):893–900.
25. Sorenson JA. Quantitative measurement of radioactivity in vivo by whole-body counting. In: Hine GJ, Sorenson JA, eds. Instrumentation in Nuclear Medicine. New York: Academic Press, 1974:311–348.
26. van Reenen PC, Lotter MG, Heyns AD, et al. Quantification of the distribution of ^{111}In-labelled platelets in organs. Eur J Nucl Med 1982; 7(2):80–84.
27. Segars WP, Lalush DS, Tsui BMW. A realistic spline-based dynamic heart phantom. IEEE Trans Nucl Sci 1999; 46(3):503–506.
28. Segars WP, Lalush DS, Tsui BMW. Modeling respiratory mechanics in the MCAT and spline-based MCAT phantoms. IEEE Trans Nucl Sci 2001; 48(1):89–97.
29. Du Y, Frey EC, Wang WT, et al. Combination of MCNP and SimSET for Monte Carlo simulation of SPECT with medium- and high-energy photons. IEEE Trans Nucl Sci 2002; 49(3):668–674.
30. Song X, Segars WP, Du Y, et al. Fast modeling of the collimator-detector response in Monte Carlo simulation of SPECT imaging using the angular response function. Phys Med Biol 2005. In Press.
31. Liu A, Williams LE, Raubitschek AA. A CT assisted method for absolute quantitation of internal radioactivity. Med Phys 1996; 23(11):1919–1928.
32. Miller C, Filipow L, Jackson S, et al. Planar imaging quantification using 3D attenuation correction data and Monte Carlo simulated buildup factors. Phys Med Biol 1996; 41(8):1401–1423.
33. Sjogreen K, Ljungberg M, Strand SE. An activity quantification method based on registration of CT and whole-body scintillation camera images, with application to 131I. J Nucl Med 2002; 43(7):972–982.
34. Hudson HM, Larkin RS. Accelerated image-reconstruction using ordered subsets of projection data. IEEE Trans Med Imag 1994; 13(4):601–609.
35. Kamphuis C, Beekman FJ, van Rijk PP, et al. Dual matrix ordered subsets reconstruction for accelerated 3D scatter compensation in single-photon emission tomography. Eur J Nucl Med 1998; 25(1):8–18.
36. Welch A, Gullberg GT, Christian PE, et al. A transmission-map based scatter correction technique for SPECT in inhomogeneous-media. Med Phys 1995; 22(10):1627–1635.
37. Riauka TA, Gortel ZW. Photon propagation and detection in single-photon emission computed-tomography—an analytical approach. Med Phys 1994; 21(8):1311–1321.
38. Wells RG, Celler A, Harropo R. Analytical calculation of photon distributions in SPECT projections. IEEE Trans Nucl Sci 1998; 45(6):3202–3214.
39. Beekman FJ, Viergever MA. Fast SPECT simulation including object shape dependent scatter. IEEE Trans Med Imag 1995; 14(2):271–282.

40. Frey EC, Tsui BMW, Kadrmas DJ. A new method for modeling the spatially-variant, object-dependent scatter response function in SPECT. IEEE Nuclear Science Symposium, 1996, Anaheim.

41. Kadrmas DJ, Frey EC, Karimi SS, et al. Fast implementations of reconstruction-based scatter compensation in fully 3D SPECT image reconstruction. Phys Med Biol 1998; 43(4):857–873.

42. Frey EC, Tsui BMW. A Practical method for incorporating scatter in a projector-backprojector for accurate scatter compensation in SPECT. IEEE Trans Nucl Sci 1993; 40(4):1107–1116.

43. Frey EC, Ju ZW, Tsui BMW. A fast projector-backprojector pair modeling the asymmetric, spatially varying scatter response function for scatter compensation in SPECT Imaging. IEEE Trans Nucl Sci 1993; 40(4):1192–1197.

44. Tsui BMW, Frey EC, Zhao X, et al. The importance and implementation of accurate 3D compensation methods for quantitative spect. Phys Med Biol 1994; 39(3):509–530.

45. Rousset OG, Ma YL, Evans AC. Correction for partial volume effects in PET: principle and validation. J Nucl Med 1998; 39(5):904–911.

46. Meltzer CC, Zubieta JK, Links JM, et al. MR-Based correction of brain PET measurements for heterogeneous gray matter radioactivity distribution. J Cereb Blood Flow Metab 1996; 16(4):650–658.

11

SPECT/CT Attenuation Compensation in Nuclear Cardiology

Michael K. O'Connor
Division of Medical Physics, Department of Radiology, Mayo Clinic, Rochester, Minnesota, U.S.A.

INTRODUCTION

Myocardial perfusion imaging, first using [201]TI and later ^{99m}Tc-based radiopharmaceuticals, has been in clinical use since the early 1980s. While myocardial imaging was originally performed using planar techniques, by mid-1990 tomographic imaging had become the standard method for imaging the myocardium. Stress single-photon emission computed tomography (SPECT) is now a well-established technique for the noninvasive evaluation of coronary artery disease and is the most widely utilized modality of stress testing in the United States (1). Nevertheless, the accuracy of the technique is limited by image artifacts (2). These artifacts cause false-positive perfusion defects and impact test specificity. Although the initial validation of SPECT reported a specificity of greater than 90% (3,4), widespread clinical application of the technique has been associated with specificity in the range of 60% or lower (5,6).

One of the most common image artifacts is caused by nonuniform reduction of photon activity from attenuation by soft tissue (7). These artifacts are often recognized or at least suspected because of their typical locations and shapes relative to the heart (8,9). Attenuation artifacts usually occur in the anterior wall of women with large breasts and in the inferior wall of obese men. An example of the effect of diaphragmatic attenuation on apparent myocardial activity in the short axis and vertical long axis slices of a normal myocardium is shown in Figure 1. The true prevalence of soft-tissue artifacts is unknown, but estimates have ranged from 20% to 50% (10,11).

Several approaches have been used to address this problem. These include simple awareness of the typical location of the artifacts and review of the raw imaging data (9,12), imaging the patient in the prone versus supine position (13), quantitation using sex-matched normal databases (12,14,15), use of the higher energy radioisotope ^{99m}Tc instead of [201]Thallium (16), and gated imaging of the myocardium to assess wall motion (16,17). Reliance on a "typical" attenuation pattern to distinguish attenuation artifacts from perfusion defects is only a partial solution to a complex problem, as both the artifact and the true defect may coexist in the same location and there is no assurance that a breast artifact will be exactly consistent from one

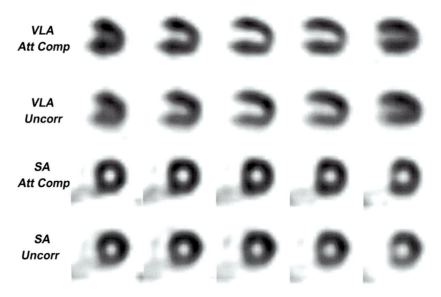

Figure 1 VLA and SA slices from a ^{99m}Tc myocardial perfusion study in a patient with low likelihood of coronary artery disease. Note the reduced uptake in the inferior wall of the myocardium in the conventional (uncorr) images, due to diaphragmatic attenuation, which is corrected on the attenuation-compensated (att comp) images. *Abbreviations*: VLA, vertical long axis; SA, short axis.

imaging session to another (18). Hence, while the above approaches improve artifact recognition, they all have limitations.

Determination of the true activity distribution in the myocardium requires correction of the root causes of the artifacts. Any factor that distorts the apparent distribution may lead to the incorrect interpretation of the image data with a consequential increase in the number of false-positive studies and reduced specificity. The key factor that alters the apparent distribution of activity is photon attenuation. Correction for this type of artifact is one of the main goals of the attenuation correction techniques that have been developed over the last five to eight years. It should be remembered, however, that attenuation is only one of a myriad of factors that influence apparent myocardial activity and simply correcting for one factor and ignoring the others can result in overcorrection of the image data and introduce new artifacts into the clinical studies (19). Before discussing methods for compensation for soft-tissue attenuation, the importance of some of these other factors, in particular scattered radiation and depth-dependent changes in resolution, need to be understood.

SCATTER CORRECTION

Apart from soft-tissue attenuation, the other major factor that degrades image quality is scattered radiation. The main effect of scatter is the loss of contrast in the image data. In cardiac studies, however, the primary need for scatter correction is to enable attenuation compensation techniques to be applied to the emission data. As described later, all current methods for generating the attenuation correction map produce measurements of the true attenuation coefficients for soft tissue, bone, and lung. These values cannot be applied to correct the emission data due to the presence of scatter, which effectively reduces the apparent attenuation present in the data. Application of the measured attenuation coefficients may therefore result in overcorrected emission data. Failure to apply scatter correction and depth-dependent resolution recovery are probably the primary reasons why early studies on cardiac attenuation correction yielded poor results (20,21). In addition, computer simulations by King et al. (22) have shown that after attenuation compensation using the true linear attenuation coefficients for ^{99m}Tc, the total change in counts due to scatter was higher than 30%, and the shape of the distortion was such that there was a slight increase in apparent activity as one went from the apex of the heart towards the base. Hence, in order to both solve this type of problem and to eliminate overcorrection by the attenuation correction maps, it is important to correct for scatter in the emission data.

Correction for scatter is a complex issue and numerous techniques have been developed over the years (23). The most common techniques in clinical use are the basic methods, such as the dual-energy or triple-energy window methods (24,25). The dual-energy window is known to be wrong in assuming that the fraction of the scatter to be subtracted from the photopeak window is spatially invariant. Likewise, the triple-energy window method may generate noisy estimates of the scatter fraction in low-count data. These methods have not yet proved to yield superior clinical results. A major disadvantage to these two scatter-correction techniques is the need to perform image subtraction of the scatter from the photopeak data for attenuation compensation studies. The attenuation compensation process amplifies the increased noise in the emission data. In low-count cardiac studies, this may degrade the quality of the attenuation-compensated data to the point where results are inferior to the uncorrected emission data. This point is illustrated in Figure 2,

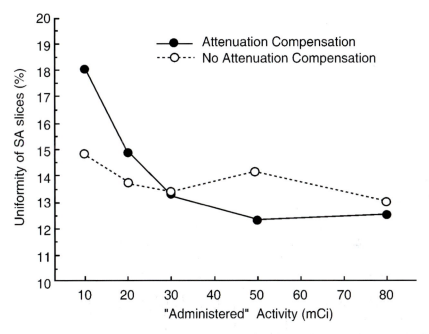

Figure 2 Results from a phantom study showing the uniformity of short axis slices from a normal myocardium with and without attenuation compensation, as a function of the "administered" ^{99m}Tc activity.

which shows the measured uniformity of activity in short axis slices from a cardiac phantom with a normal myocardium. Using different acquisition times to simulate different administered activities of ^{99m}Tc, the results show that at high administered activities, attenuation compensation improves the measured uniformity. At low administered activities, however, the uncorrected images are more uniform than the attenuation correction images. Hence, depending on the type of scatter correction employed in conjunction with an attenuation compensation algorithm, myocardial perfusion imaging protocols using one-day protocols (low-dose rest injections/ high-dose stress injections) may be compromised by the impact of the scatter correction on the low-dose study. A possible solution to this problem is the development of Monte Carlo-based scatter compensation techniques that generate low-noise scatter projections from the emission data (26), thereby avoiding the pitfalls of many of the current scatter-correction methods.

DEPTH-DEPENDENT RESOLUTION RECOVERY

In addition to scatter and attenuation, there are a number of other factors that influence the quality of myocardial perfusion studies. There are significant differences between myocardial perfusion SPECT imaging and conventional SPECT imaging of the torso or head. Myocardial perfusion studies have traditionally been acquired using a 180° orbit rather than the conventional 360° orbit. While the validity of this acquisition mode has been debated over the years (27), a close examination of the effect of various acquisition parameters on cardiac SPECT imaging shows that the optimum acquisition of cardiac data is a complex issue depending on the orbital range (180° vs. 360°), the type of orbit (circular vs. elliptical), and the location of the myocardium in the field of view (27). A recent study by Liu et al. (28) and work by Knesaurek in 1989 (29) suggested that a 360° rotation provides more accurate imaging results than a 180° rotation, particularly when the heart is off-center. Studies reporting significant nonuniformity and overestimation of defects in the images acquired with a 180° rotation have led many researchers to recommend a 360° acquisition rotation (30–32). The 360° rotation is theoretically favorable because complete spectral information is used to reconstruct the image and it averages opposing views, providing better image uniformity. It can be argued that a 180° rotation, however, provides increased image contrast and can allow the user to halve the acquisition time (33). Several studies have reported that using a 180° rotation over a 360° rotation has little or no effect on image quality (34,35). Hence, the de facto standard remains the 180° acquisition (36), primarily because the improved contrast and shorter imaging time appear to outweigh any small gain in quantitative accuracy and image fidelity with a 360° orbit.

An elliptical or body-contouring orbit is intended to minimize the distance between the camera head and the myocardium; however, the angular changes in system resolution resulting from these types of orbits may introduce artifacts in the reconstructed images. While a study by White et al. (37) in 50 patients showed improved image quality with a noncircular orbit, most studies have shown that an elliptical orbit introduces more severe image artifacts than a circular orbit (38–40). At present, the American Society of Nuclear Cardiology recommends that elliptical orbits only be used with the detector "backed-off" from the apex of the heart by about 2 cm to prevent apical artifacts (36). Maniawski et al. (40), in a series of phantom studies using 180° acquisitions, recommended use of a circular orbit and further recommended that the myocardium be positioned close to the center of rotation to minimize artifacts. While this last recommendation has been considered impractical on a routine clinical basis (41), it is one additional factor to be considered in the quest for optimal image quality in myocardial perfusion imaging.

The primary underlying factor that comes into play with changes in type of orbit, angular range, and myocardial location is the change in spatial resolution that occurs with distance from the collimator face. Failure to account for these factors can make it difficult to assess the efficacy of compensations for attenuation and scatter in clinical studies (42,43). Rather than attempting to eliminate changes in resolution with rotation, a more elegant solution to these factors is the use of depth-dependent resolution recovery algorithms that can compensate for the changes in system resolution during the SPECT acquisition. The purpose of depth-dependent resolution recovery algorithms is not necessary to improve resolution, but rather to improve the uniformity of resolution over the area of the myocardium. In a study of 100 patients, Narayanan et al. (44) showed that compensating for changes in resolution in addition to compensation for scatter and attenuation significantly improved the overall detection of coronary disease when compared with compensation for only attenuation and scatter.

Consequently, as attenuation compensation techniques have matured over the last five years, commercially available algorithms have generally included additional corrections for the effects of scattered radiation and depth-dependent resolution changes. Hence, the term "attenuation correction" that is used by vendors is a misnomer as these commercial hardware and software packages become more complete in the types of corrections they apply to the clinical data.

With compensation for attenuation, scatter, and changes in resolution, residual artifacts are generally due to patient or respiratory motion. While these cannot be ignored, the magnitude of errors caused by respiratory motion is generally less than those caused by the above factors. Current work is therefore heavily focused on the development of robust clinical software to address the above causes of image artifact.

ATTENUATION COMPENSATION

Attenuation compensation methods are not new to nuclear medicine. For many years, most reconstruction algorithms incorporated some type of simple attenuation compensation method. The most commonly used was the Chang method (45), where the contours of the body are estimated from the reconstructed transaxial slices. From the contour, an estimate of the attenuation path length is obtained for each pixel in the transaxial image. This method assumes a constant attenuation coefficient for the body, presuming that the body has an elliptical or circular cross-sectional area. Because this method uses a constant value for the linear attenuation coefficient for body tissue, it only works well for relatively uniform and symmetric body sections such as the head and abdomen. It yields incorrect results in the thorax because of the variable shapes and attenuation characteristics of the internal structures. In addition, the patterns of bone, lung, and soft-tissue attenuation in the thorax are complex and highly variable from patient to patient. No assumptions can be made about the attenuation pattern in a given subject. Consequently, for cardiac studies, it is necessary to develop an accurate, patient-specific attenuation map in order to be able to perform attenuation compensation, a fact that has been known for many years (46). Over the years, several techniques have been developed for generating patient-specific attenuation maps. The three primary methods are (i) estimation of the attenuation map from the emission data, (ii) utilizing anatomical data available from another modality such as computed tomography (CT) or magnetic resonance imaging (MRI) to compute the attenuation map, and (iii) generating an attenuation map using a transmission source at the time of the emission scan (47).

Several studies have described methods for the estimation of the attenuation map using the emission data. Investigators have used the photopeak and scatter images to estimate the body outline and the lung region. Segmentation can then be used to generate an attenuation map. Alternatively, a second radiopharmaceutical can be used to label lung tissue and utilize this information to segment out lung from soft-tissue. Neither of these methods has gained widespread use, and have not proved to be sufficiently reliable in the routine clinical setting (48–50).

High-resolution images from another modality can be imported and registered with the patient's emission scan (51). The principal limitation of this method is the difficulty in correctly matching the anatomical and functional datasets. Since the images are acquired on different systems at different time points, the shape of the imaging table (e.g., flat vs. curved), patient positioning (e.g., arms up or down), and scan conditions (e.g., CT breath-hold) may make accurate matching of the two datasets difficult (52).

The most commonly used method for generation of the patient-specific attenuation map is by means of a transmission source matched to the gamma camera. A variety of different transmission geometries have been developed over the last 10 years. These include a sheet transmission source (53,54), multiple lines sources (55), scanning line sources (56), a fixed line source positioned at the focal line of a fan-beam collimator (57–59), and a moving point source. Each of these methods has its advantages and limitations. King et al. has presented a full review of their inherent characteristics (47). The primary disadvantage of the sheet source was its cumbersome application, and the fact that it required source collimation in order that "good geometry" attenuation coefficients were estimated (55). In addition, ^{99m}Tc was usually employed as the source, which required that the sheet source be prepared every day. With the multiple line source method, there is high cross-talk between emission/transmission images. Correction for this increases the noise in the data. This method does have the advantage, however, of allowing the strongest sources to be located in the center of the field of view where body thickness is greatest. This yields high-quality attenuation maps and minimizes problems with decay of the transmission sources (55). The fixed-line source method was developed as an alternative to be used on triple-head systems where the scanning line source technique was not applicable. Disadvantages with this method include a small field of view, which can increase truncation over the multiple or scanning line source methods. A number of techniques have been developed to minimize the impact of truncation on the final attenuation correction applied to the emission data, though with limited effect (57,59,60).

The scanning line source is probably the most widely used transmission method. This technique has become the standard method applied in most commercial systems for acquiring attenuation maps in clinical studies. While it has been shown to work well, it has a number of inherent limitations. Commercial scanning line source systems have predominantly employed ^{153}Gadolinium (100 keV), although some laboratories have developed their own transmission sources using ^{241}Americium, ^{133}Barium, ^{157}Cobalt, or ^{99m}Technetium. ^{153}Gd has a 100-day half-life and current commercial suppliers can only provide a maximum activity of 400 to 500

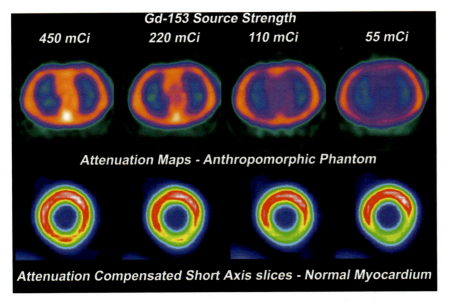

Figure 3 Attenuation maps of an anthropomorphic phantom generated using a ^{153}Gd scanning line source of varying strength, and the corresponding effects of these maps on the attenuations compensated short axis slices of a normal myocardium. *Abbreviation*: Gd, gadolinium.

mCi/source. As the source decays, the reduced photon flux can lead to degradation of the attenuation map. Examples of the attenuation maps from a cardiac phantom with a ^{153}Gd scanning line source system are shown in Figure 3. As the activity drops, the apparent attenuation map changes with an underestimation of the true attenuation coefficients in the center of the thorax. This reduction in the attenuation coefficients then corrupts the attenuation-compensated short axis slices of the heart. The gradual degradation in the attenuation coefficients (Fig. 4) with source strength leads to an insidious decrease in image quality and requires that the user frequently alter the acquisition parameters to ensure adequate counting statistics in the transmission data. An increase in the acquisition time of the transmission map can partly alleviate the problem of decreased ^{153}Gd source strength. However, with ^{99m}Tc radiopharmaceuticals, the software must also correct for downscatter from ^{99m}Tc into the ^{153}Gd energy window, by subtracting an image of ^{99m}Tc events in the ^{153}Gd energy window from the transmission data. Extending the transmission scan time increases the amount of downscatter with a consequential increase in the noise in the downscatter-corrected transmission data. Despite these limitations, once an adequate ^{153}Gd source strength is maintained, this method provides an accurate estimate of the patient-specific attenuation coefficients.

SPECT/CT

Lang et al. first described a prototype SPECT/CT in 1991 (61,62). This device used the same detector for both SPECT and CT, thereby providing perfect registration between the CT and emission data. Using this system, it has been possible to perform absolute quantitation of myocardial activity in animal models (63). This group has further developed a combined SPECT/CT system utilizing a conventional

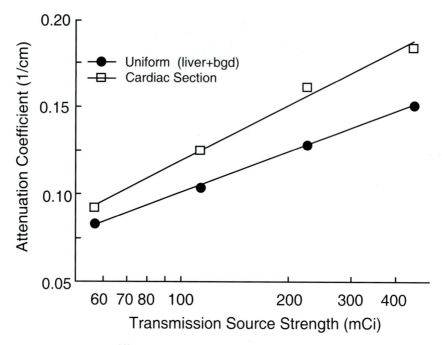

Figure 4 Effect of ^{153}Gd activity in the transmission source on the measured attenuation coefficients in an anthropomorphic phantom. Results are shown for regions over the liver and heart. *Abbreviations*: bgd, background; Gd, gadolinium.

single-slice CT scanner and a large field of view SPECT system (64) and has demonstrated that with corrections for attenuation and partial volumes effects, absolute regional activity concentration can be measured accurately in the porcine myocardium. On the commercial side, until recently, only one vendor had developed a combined SPECT/CT system. The Hawkeye^{TM} (GE Medical Systems) integrates a CT device into the gantry of the gamma camera. In this system, both the X-ray tube and CT detector assembly are offset from the gamma camera, requiring that the patient be moved axially between the emission and CT scans. X-ray transmission imaging can be performed before or after the emission scan. The CT device acquires information over an arc of 213° (180° + fan angle), at a rotation speed of 22 seconds/orbit. Each slice has an axial thickness of 5 or 10 mm and is acquired over approximately 13 seconds. Examples of attenuation maps generated with a variety of scanning or fixed line sources, relative to one generated using a CT device, are shown in Figure 5. Clearly, the image noise in the attenuation maps from a CT-based system is very low, and the in-plane resolution is high compared to maps generated using the more conventional transmission methods. While the CT system generates a higher resolution map, the improvement in resolution is not necessarily a factor in the accuracy of the attenuation compensation since we are primarily concerned with obtaining an accurate estimate of the attenuation path length for each pixel in the transaxial image. Examples of attenuation maps of a cardiac phantom generated using this CT device and a ^{153}Gd transmission source on two similar gamma cameras are shown in Figure 6. The phantom contained a "normal" myocardium and attenuation compensation was performed using identical software algorithms on the two systems, thereby allowing one to determine the impact of the differences

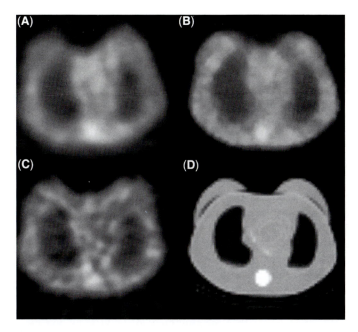

Figure 5 Attenuation maps of an anthropomorphic phantom with breast attachments generated by (**A**) multiple line sources, (**B**) scanning point source, (**C**) scanning line sources, and (**D**) CT system. *Abbreviation*: CT, computed tomography.

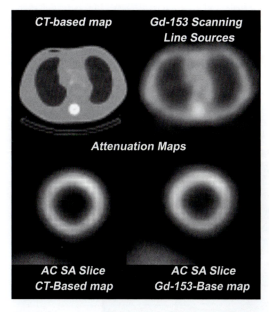

Figure 6 Attenuation maps of an anthropomorphic phantom generated using a CT-system and a ^{153}Gd scanning line source system (*top row*). The corresponding effect of these maps on the attenuation-compensated short axis slices of a normal myocardium are shown (*bottom row*). *Abbreviations*: CT, computed tomography; Gd, gadolinium.

in the quality of the attenuation map on image quality. The short axis slices of the myocardium indicate that both maps produce equivalent results.

CT-based transmission systems can produce optimal attenuation maps under ideal conditions, but have, however, certain limitations that are different from those of radioisotope-based attenuation correction. With CT-based systems, there is no need to correct for source decay or crosstalk. Because of constraints in where the CT device can be located on the gantry, the fan-beam of radiation has a limited field of view and hence the system is unable to adequately image patients with a chest circumference of greater than 55 cm. Truncation of parts of the torso results in underestimation of the attenuation and leads to corruption of the attenuation-corrected emission studies. Examples of the effects of truncation of the CT attenuation map on reconstructed image quality are shown in Figure 7. Mild truncation at the shoulders does not have a major impact on image quality, as it does not seriously corrupt the estimate of the attenuation path length for pixels in the myocardium. With increasing truncation, however, the attenuation path lengths are underestimated, resulting in undercorrection of the emission images. A second major issue for a CT-type system is misregistration between the emission and transmission data (65). This may occur for a number of reasons, including sagging of the emission table or patient motion. The impact of misregistration on the emission scans is highly patient-dependent. Artifacts can occur with a misalignment of as little as one pixel (~7 mm) in the y-direction (ventral to dorsal direction), which can superimpose parts of the myocardium over lung tissue (65). This can result in underestimation of attenuation path length in this part of the myocardium and an undercorrected emission scan that can lead to the creation of apical or anterior wall defects in the attenuation-compensated images (Fig. 8). This significant problem affects approximately 20% of clinical studies performed on this type of system (65). Recent upgrades to the support arm and rollers of the imaging table have helped minimize sagging and poor tracking of the imaging table. However, this will probably continue to be an issue as this first-generation SPECT/CT system was not designed as an integrated unit, but rather represents an existing SPECT system to which a CT device was added. It is expected that the next generation of SPECT/CT devices will overcome some of the limitations described earlier.

While no extensive large-scale clinical studies have been reported on the value of SPECT/CT devices in attenuation correction, preliminary reports have reported

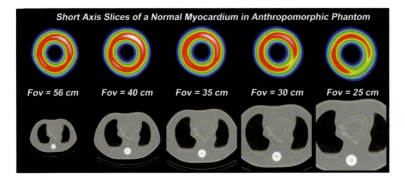

Figure 7   Effect of truncation of the attenuation map on the appearance of short axis slices from a normal myocardium located in an anthropomorphic phantom. *Abbreviation*: Fov, field of view.

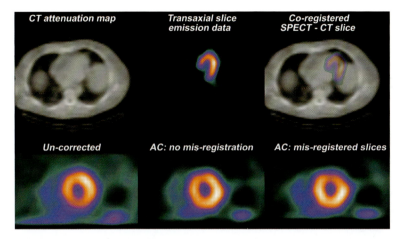

Figure 8 Misregistration between the CT and ^{99m}Tc sestamibi transaxial images in a patient study, with part of the myocardium overlapping the lung tissue. Bottom line shows the non-attenuation-compensated short axis slice and the attenuation-compensated slices, both with and without correction for the misregistration. *Abbreviation*: CT, computed tomography.

some discrepancies between attenuation-compensated myocardial perfusion studies using CT and ^{153}Gd transmission sources, and have highlighted the problems of noise and misregistration between the CT and SPECT images that are discussed earlier (66–68).

THE FUTURE OF SPECT/CT IN ATTENUATION COMPENSATION

Several vendors are now pursuing the development of integrated SPECT/CT devices. These systems integrate multislice CT scanners with large field-of-view gamma cameras and offer a number of advantages over the current SPECT/CT design. The second-generation systems utilize a dual-table configuration that eliminates table-sagging between the emission and CT scans. Gantry aperture is typically 70 cm, permitting larger patients to be imaged. This larger field-of-view should minimize the occurrence of truncation problems encountered with the first-generation SPECT/CT system. With the fully integrated diagnostic multislice-CT, these systems can provide an accurate attenuation map in less than 30 seconds of additional scan time. These new systems may provide the possibility of quantifying coronary artery calcium, evaluating the patency of vascular and coronary arteries, and assessing myocardial perfusion and viability in one clinical setting. What additional clinical benefit will accrue from such a procedure has yet to be determined (69).

The next generation SPECT/CTs, however, do come with some problems of their own. Since the CT component is now a multislice-CT unit capable of high-speed rotation, it can acquire anywhere from 2 to 16 slices per step. Acquisition time/slice is expected to be of the order of 200 to 300 milliseconds. It is assumed that the SPECT and CT images are perfectly registered, since the system has a common bed and the images are acquired sequentially without the patient moving from the table. Apart from involuntary patient motion, however, cardiac and respiratory motion can cause misalignment between SPECT and CT. For a cardiac study, the CT component is acquired in 30 seconds and represents a snapshot at certain points of the respiratory and cardiac cycles. The SPECT component, on the other hand, typically takes 15 to 20 minutes and the images represent an average of the respiratory and cardiac cycles. These two independent motions can be a major source of error and will need to be addressed before the full benefits of these systems for cardiac applications can be evaluated.

These problems are similar to those encountered in PET/CT systems. Current studies indicate that in PET cardiac studies, the use of radioisotope-based transmission imaging may be preferable to CT. Chin et al. (70) found that in clinical PET/CT studies, diaphragm misregistration between the CT and PET data is associated with relative decreased emission activity in inferior, inferoseptal, and inferolateral walls, and recommended further studies to determine if the frequency of these findings warrants the use of [68]Germanium transmission attenuation correction in myocardial Fluorodeoxyglucose positron emission tomography.

The future of SPECT/CT devices is certain, particularly in the area of oncology. For cardiac applications, the ability to rapidly generate a high-quality map and the potential for calcium-scoring are attractive features of this technology. Work still needs to be performed, however, to ensure that the effects of cardiac and respiratory motion can be adequately corrected in clinical studies.

REFERENCES

1. Gibbons RJ, Chatterjee K, Daley J, et al. ACC/AHA/ACP-ASIM guidelines for the management of patients with chronic stable angina: a report of the American College of Cardiology/American Heart Association Task Force on Practice Guidelines (Committee on Management of Patients with Chronic Stable Angina). J Am Coll Cardiol 1999; 33:2092–2197.
2. Garcia EV, Votaw JR, Cullom SJ, Galt JR, Cooke CD. Principles and Instrumentation. In: Radionuclide Imaging. 2d ed. Philadelphia: W.B. Saunders Company, 1996.
3. Tamaki N, Yonekura Y, Mukai T, et al. Stress thallium-201 transaxial emission computed tomography quantitative versus qualitative analysis for evaluation of coronary artery disease. J Am Coll Cardiol 1984; 4:1213–1221.
4. Mahmarian JJ, Pratt CM, Cocanougher MK, et al. Altered myocardial perfusion in patients with angina pectoris or silent ischemia during exercise as assessed by quantitative thallium-201 single-photon emission computed tomography. Circulation 1990; 82: 1305–1315.
5. Van Train KF, Garcia EV, Maddahi J, et al. Multi-center trial validation for quantitative analysis of same-day rest-stress technetium-99m-sestamibi myocardial tomograms. J Nucl Med 1994; 35:609–618.
6. Fleischmann KE, Hunink MGM, Kuntz KM, et al. Exercise echocardiography or exercise SPECT imaging? A meta-analysis of diagnostic test performance. JAMA 1998; 280:913–920.
7. Corbett JR, Ficaro EP. Clinical review of attenuation-corrected cardiac SPECT. J Nucl Cardiol 1999; 6:54–68.
8. Friedman TD, Greene AC, Iskandrian AS, et al. Exercise thallium-201 myocardial scintigraphy in women: correlation with coronary arteriography. Am J Cardiol 1982; 49: 1632–1637.
9. Goodgold HM, Rehder JG, Samuels LD, et al. Improved interpretation of exercise Tl-201 myocardial perfusion scintigraphy in women: characterization of breast attenuation artifacts. Radiology 1987; 165:361–366.
10. Desmarias R, Kaul S, Watson D, et al. Do false positive thallium-201 scans lead to unnecessary catheterization? Outcome of patients with perfusion defects on quantitative planar thallium scintigraphy. J Am Coll Cardiol 1993; 21:1058–1063.

11. Holly TA, Parker MA, Hendel RC. The prevalence of non-uniform soft tissue attenuation in myocardial SPECT perfusion imaging and the impact of gated SPECT. J Nucl Cardiol 1997; 4:S103.

12. Eisner RL, Tamas MJ, Clononger K, et al. Normal SPECT thallium-201 bull's-eye display: gender differences. J Nucl Med 1988; 29:1901–1909.

13. Esquerre J-P, Coca FJ, Martinez SJ, et al. Prone decubitus: a solution to inferior wall attenuation in thallium-201 myocardial tomography. J Nucl Med 1989; 30:398–401.

14. DePasquale EE, Nody AC, DePuey EG, et al. Quantitative rotational thallium-201 tomography for identifying and localizing coronary artery disease. Circulation 1988; 77:316–327.

15. Van Train KF, Areeda J, Garcia EV, et al. Quantitative same-day rest-stress technetium-99-m-sestamibi SPECT: definition and validation of stress normal limits and criteria for abnormality. J Nucl Med 1993; 34:1494–1502.

16. Taillefer R, DePuey E, Udelson J, et al. Comparative diagnostic accuracy of Tl-201 and Tc-99m sestamibi SPECT imaging (perfusion and ECG-gated SPECT) in detecting coronary artery disease in women. J Am Coll Cardiol 1997; 29:69–77.

17. DePuey EG, Rozanski A. Using gated technetium-99-m-sestamibi SPECT to characterize fixed myocardial defects as infarct or artifact. J Nucl Med 1995; 36:952–955.

18. Ficaro EP, Fessler JA, Shreve PD, et al. Simultaneous transmission/emission myocardial perfusion tomography: diagnostic accuracy of attenuation-corrected 99mTc-sestamibi single-photon emission computed tomography. Circulation 1996; 93:463–473.

19. Hutton BF. Cardiac single-photon emission tomography: is attenuation correction enough? Eur J Nucl Med 1997; 24:713–715.

20. Lee DS, So Y, Cheon GJ, et al. Limited incremental diagnostic value of attenuation-non-corrected gating and ungated attenuation-correction to rest/stress myocardial perfusion SPECT in patients with an intermediate likelihood of coronary artery disease. J Nucl Med 2000; 41: 852–859.

21. Cullom SJ, Case JA, Bateman TM. Attenuation correction for cardiac SPECT: clinical and developmental challenges. J Nucl Med 2000; 41:860–862.

22. King MA, Tsui BM, Pan TS, et al. Attenuation compensation for cardiac single-photon emission computed tomographic imaging. Part 2. Attenuation compensation algorithms. J Nucl Cardiol 1996; 3:55–64.

23. Zaidi H, Koral KF. Scatter modeling and compensation in emission tomography. Eur J Nucl Med Mol Imag 2004; 31:761–782.

24. Jaszczak RJ, Greer KL, Floyd CE, et al. Improved SPECT quantification using compensation for scattered photons. J Nucl Med 1984; 25:893–900.

25. Ogawa K, Harata Y, Ichihara T, et al. A practical method for position-dependent Compton-scatter correction in single-photon emission CT. IEEE Trans Med Imag 1991;10:408–412.

26. Beekman FJ, de Jong HW, van Geloven S. Efficient fully 3-D iterative SPECT reconstruction with Monte Carlo-based scatter compensation. IEEE Trans Med Imag 2002; 21:867–877.

27. O'Connor MK, Hruska CB. Effect of tomographic orbit and type of rotation on apparent myocardial activity. Nucl Med Commun 2005; 26:25–30.

28. Liu Y, Lam PT, Sinusas AJ, et al. Differential effect of 180 and 360 degree acquisition orbits on the accuracy of SPECT imaging: quantitative evaluation in phantoms. J Nucl Med 2002; 43:1115–1124.

29. Knesaurek K, King MA, Glick SJ, et al. Investigation of cause of geometric distortion in 180° and 360° angular sampling in SPECT. J Nucl Med 1989; 30:1666–1675.

30. Coleman RE, Jaszczak RJ, Cobb F. Comparison of 180 and 360 data collection in thallium-201 imaging using single photon-emission computerized tomography (SPECT): Concise communication. J Nucl Med 1982; 23:661–666.

31. Go RT, MacIntyre WJ, Houser TS, et al. Clinical evaluation of 360 and 180 data sampling technique for transaxial SPECT thallium-201 myocardial perfusion imaging. J Nucl Med 1985; 26:695–706.

32. Hoffman, EJ. 180 compared with 360 sampling in SPECT (editorial). J Nucl Med 1982; 23:745–747.

33. Eisner RL, Nowak DJ, Pettigrew R, et al. Fundamentals of 180° acquisition and reconstruction in SPECT imaging. J Nucl Med 1986; 27:1717–1728.

34. Freeman MR, Konstantinou C, Barr A, et al. Clinical comparison of 180° and 360° data collection of technetium 99m sestamibi SPECT for detection of coronary artery disease. J Nucl Cardiol 1998; 5:14–18.

35. LaCroix KJ, Tsui BMW, Hasegawa B. A comparison of 180 and 360 acquisition for attenuation-compensated thallium-201 SPECT images. J Nucl Med 1998; 39:562–574.

36. DePuey EG, Garcia EV. Updated imaging guidelines for nuclear cardiology procedures. Part 1. J Nucl Cardiol 2001; 8:G1–G58.

37. White MP, Russell A, Mascitelli VA, et al. Clinical comparison of circular versus noncircular acquisition using technetium-99m myocardial perfusion SPECT imaging. J Nucl Med Technol 1997; 25:37–40.

38. Abufadel A, Eisner RL, Schafer RW. Differences due to collimator blurring in cardiac images with use of circular and elliptical camera orbits. J Nucl Cardiol 2001; 8: 458–465.

39. Folks R, Van Train K, Wong C, et al. Evaluation of Tc-MIBI SPECT acquisition parameters: circular vs. elliptical and 180° vs. 360° orbits (abstract). J Nucl Med 1989; 30: 795–796.

40. Maniawski PJ, Morgan HT, Wackers FHT. Orbit-related variation in spatial resolution as a source of artifactual defects in Thallium-201 SPECT. J Nucl Med 1991; 32:871–875.

41. Keyes JW. SPECT and artifacts—In search of the imaginary lesion. J Nucl Med 1991; 32:875–877.

42. Corbett JR, Ficaro EP. Attenuation corrected cardiac perfusion SPECT. Curr Opin Cardiol 2000; 15:320–336.

43. Hendel RC, Corbett JR, Cullom SJ, et al. The value and practice of attenuation correction for myocardial perfusion SPECT imaging: a joint position statement from the American Society of Nuclear Cardiology and the Society of Nuclear Medicine. J Nucl Med 2002; 43:273–280.

44. Narayanan MV, King MA, Pretorius PH, et al. Human-observer receiver-operating characteristic evaluation of attenuation, scatter, and resolution compensation strategies for 99mTc myocardial perfusion imaging. J Nucl Med 2003; 44:1725–1734.

45. Chang LT. A method for attenuation correction in radionuclide computed tomography. IEEE Trans Nucl Sci 1978; 25:638–643.

46. Jaszczak RJ, Chang LT, Stein NA, et al. Whole-body single-photon emission computed tomography using dual, large-field-of-view scintillation cameras. Phys Med Biol 1979; 24:1123–1143.

47. King MA, Tsui BMW, Pan T-S. Attenuation compensation for cardiac single-photon emission computed tomographic imaging: Part 1. Impact of attenuation and methods of estimating attenuation maps. J Nucl Card 1995; 2:513–524.

48. Madsen MT, Kirchner PT, Edlin JP, et al. An emission-based technique for obtaining attenuation correction data for myocardial SPECT studies. Nucl Med Comm 1993; 14:689–695.

49. Wallis JW, Miller TR, Koppel P. Attenuation correction in cardiac SPECT without a transmission measurement. J Nucl Med 1995; 36:506–512.

50. Pan T-S, King MA, Luo D-S, et al. Estimation of attenuation maps from scatter and photopeak window single photon emission computed tomographic images of technetium-99m labeled sestamibi. J Nucl Card 1997; 4:42–51.

51. Fleming JS. A technique for using CT images in attenuation correction and quantification in SPECT. Nucl Med Commun 1989; 10:83–97.

52. Kashiwagi T, Yutani K, Fukuchi M, et al. Correction of nonuniform attenuation and image fusion in SPECT imaging by means of separate X-ray CT. Evaluation studies. Ann Nucl Med 2002; 16:255–261.

53. Bailey DL, Hutton BF, Walker PJ. Improved SPECT using simultaneous emission and transmission tomography. J Nucl Med 1987; 28:844–851.
54. Cao ZJ, Tsui BMW. Performance characteristics of transmission imaging using a uniform sheet source with a parallel-hole collimator. Med Phys 1992; 19:1205–1212.
55. Celler A, Sitek A, Stoub E, et al. Multiple line source array for SPECT transmission scans: simulation, phantom, and patient studies. J Nucl Med 1998; 39:2183–2189.
56. Tan P, Bailey DL, Meikle SR, et al. A scanning line source for simultaneous emission and transmission measurements in SPECT. J Nucl Med 1993; 34:1752–1760.
57. Tung C-H, Gullberg GT, Zeng GL, et al. Nonuniform attenuation correction using simultaneous transmission and emission converging tomography. IEEE Trans Nucl Sci 1992; 39:1134–1143.
58. Jaszczak RJ, Gilland DR, Hanson MW, et al. Fast transmission CT for determining attenuation maps using a collimated line source, rotatable air-copper-lead attenuators and fanbeam collimation. J Nucl Med 1993; 34:1577–1586.
59. Gullberg GT, Morgan HT, Zeng GL, et al. The design and performance of a simultaneous transmission and emission tomography system. IEEE Trans Nucl Sci 1998; 45:1676–1698.
60. Narayanan MV, King MA, Pan T-S, et al. Investigation of approaches to reduce truncation of attenuation maps with simultaneous transmission and emission SPECT imaging. IEEE Trans Nucl Sci 1998; 45:1200–1206.
61. Lang TF, Hasegawa BH, Liew SC, et al. A prototype emission-transmission CT system. Conf Rec. IEEE Nucl Sci Symp 1991; 3:1902–1906.
62. Lang TF, Hasegawa BH, Liew SC, et al. Description of a prototype emission-transmission computed tomography imaging system. J Nucl Med 1992; 33:1881–1887.
63. Kalki K, Blankespoor SC, Brown JK, et al. Myocardial perfusion imaging with a combined X-ray CT and SPECT system. J Nucl Med 1997; 38:1535–1540.
64. Da Silva AJ, Tang HR, Wong KH, et al. Absolute quantification of regional myocardial uptake of Tc-99m- sestamibi with SPECT: experimental validation in a porcine model. J Nucl Med 2001; 42:772–779.
65. Fricke H, Fricke E, Weise R, et al. A method to remove artifacts in attenuation-corrected myocardial perfusion SPECT introduced by misalignment between emission scan and CT-derived attenuation maps. J Nucl Med 2004; 45:1619–1625.
66. Kjaer A, Cortsen A, Federspiel M, et al. Comparison of attenuation correction in myocardial SPECT using gadolinium-153 line source or low-dose CT. Eur J Nucl Med 2004; 31(Suppl 2):S318.
67. Namdar M, Siegrist PT, Koepfli P, et al. CT-Attenuation correction in myocardial perfusion SPECT using a hybrid SPECT/CT. Eur J Nucl Med 2004; 31(Suppl 2):S261.
68. Halama JR, Henkin RE, Sajdak RA, et al. X-Ray CT registration requirements for attenuation correction in myocardial perfusion SPECT. J Nucl Med 2004; 45:417P.
69. Cerqueira M, Rumsey MP. Coronary calcium scoring: what does it really mean? J Nucl Cardiol 2003; 10:692–695.
70. Chin BB, Nakamoto Y, Kraitchman DL, et al. PET-CT evaluation of 2-deoxy-2-[18F]fluoro-D-glucose myocardial uptake: effect of respiratory motion. Mol Imag Biol 2003; 5:57–64.

12

SPECT/CT in the Evaluation of Infection

Zohar Keidar and Ora Israel
Rambam Health Care Campus, B. Rapaport School of Medicine, Technion—Israel Institute of Technology, Haifa, Israel

INTRODUCTION

Infections represent a major cause of illness and death worldwide. In the last few decades, important advances have been made in the development of potent and safe antimicrobial agents. The end of the "infection era" seemed to be closer than ever. Nevertheless, as new and powerful antibiotics are developed, microbes develop the ability to resist these drugs and to attack with new survival strategies. Therefore, instead of being eradicated, infectious processes become more varied and more difficult to diagnose and treat (1).

Timely diagnosis and localization of infection is a critical step in the appropriate clinical management of this group of diseases. Radiological imaging modalities such as computed tomography (CT), magnetic resonance imaging (MRI), and ultrasonography (US) are all being currently used for the evaluation of infection. CT and MRI can show bone destruction, periosteal reaction, and soft tissue changes in case of osteomyelitis and abscess collections related to infectious process in soft tissue. MRI has also been proven to be sensitive in the evaluation of osteomyelitis owing to its high resolution and contrast and its ability to detect bone marrow changes. US is most useful in the diagnosis of fluid collections, but not for evaluating the presence of osseous infection (2). These imaging techniques are, however, based on morphological changes and are therefore of limited value in the early stage of the infectious process, which may show only insignificant structural infection-related tissue modifications or none at all (3–5). For example, structural bone changes may appear on X-rays only one to two weeks after the onset of an infectious process. The time lag is somewhat shorter using CT (4,5). Diagnosis of osteomyelitis using X-ray or CT may be further complicated by the fact that the infection may coexist with other bone injuries such as fractures or osteoarthropathy (6). MRI, also highly sensitive in diagnosis of skeletal and associated soft tissue abnormalities, may be limited in its specificity by osteoarthropathy, biomechanical stress changes related to altered weight-bearing, and bone marrow signal changes following orthopedic surgery and trauma (5,7). In addition, assessment of infection using conventional imaging modalities may be hampered by anatomical changes induced by previous treatment and/or the healing process itself, such as bone remodeling and osteogenesis, postoperative edema, scar tissue formation, and fibrosis (3,8).

Nuclear medicine plays an important role in the evaluation of infection (9,10). Scintigraphic tests are excellent noninvasive modalities of whole-body scanning that allow diagnosis of infectious foci and assessment of the extent of disease in any part of the body. Nuclear medicine procedures have been the tool of choice for assessment of musculoskeletal infections such as osteomyelitis, infected prosthetic joint implants, and diabetic foot. These functional imaging modalities are also used for the evaluation of patients presenting with fever of unknown origin. Infection-seeking radiotracers are used for diagnosis and localization of suspected abdominal abscesses, vascular graft infection, and pulmonary infection (11,12).

Nuclear medicine provides unique information on pathophysiological and pathobiochemical processes—as opposed to other imaging procedures such as CT, MRI, and US, which supply high-resolution data on morphological changes that take place as a result of a specific disease (13). Because they provide a different type of information, scintigraphic and anatomical imaging modalities play a complementary role in many clinical settings, including the assessment and management of patients with infectious processes. For example, in patients with fever of unknown origin, nuclear medicine techniques determine the presence of a suspicious site of infection. Following the scintigraphic detection, CT-, US-, or MRI-guided biopsy can be then performed to further investigate the nature of the process by obtaining tissue samples for microbiologic and histopathological tests. In an inverse order, a suspicious but equivocal finding on CT, US, or MRI can be confirmed as having an infectious etiology by nuclear medicine procedures using infection-seeking radiotracers that concentrate in these types of diseases (13).

Precise localization of infectious foci is crucial especially when surgical intervention is considered as the treatment strategy for cure. But nuclear medicine techniques are generally limited in their ability to localize pathological findings because of the relatively poor image resolution and the lack of anatomical landmarks of tracer-related disease specificity (14). Consequently, side-by-side evaluation of scintigraphic and anatomical imaging data has been used, but visual fusion of scintigraphy with CT or MRI often proves difficult (14,15). To overcome these limitations, the information obtained from both functional and anatomical modalities can be fused into a single data set. The incremental value of data registration in the evaluation of cancer patients has been thoroughly investigated in recent years and was found to have a significant contribution both to diagnosis accuracy and to the clinical management of patients with malignancies (15,16).

The role of fusion imaging in the evaluation of infection is currently under investigation, with only limited data available in the literature. Preliminary reports indicate that combined imaging modalities in general and single-photon emission computed tomography (SPECT)/CT in particular may play an important role in the diagnosis and treatment planning of infection, as the merging of complementary data on structure, function, and localization of infectious diseases can overcome many of the limitations inherent to each modality.

The decision to use a specific infection-seeking radiolabeled tracer for scintigraphy is based on the specific clinical question at hand. Several radiopharmaceuticals are available for the routine clinical assessment of infection, including ^{67}gallium citrate, ^{99m}technetium (^{99m}Tc)- or ^{111}Indium (^{111}In)-labeled white blood cells (WBC), ^{99m}Tc-labeled antigranulocyte antibodies, and ^{99m}Tc- or ^{111}In-labeled human immunoglobulin. Recently, fluor-18-fluorodeoxyglucose (FDG), an indicator of increased intracellular glucose metabolism, has been found to accumulate in sites of infectious processes, and may also show promise in the evaluation of this type of disease with positron emission tomography (PET) (17–20).

^{67}GALLIUM CITRATE

^{67}Ga citrate (Ga) has been used for imaging infection for almost four decades, since its discovery in the early 1970s (21). Ga is currently used for the investigation of fever of unknown origin, acute or chronic osteomyelitis, and the diagnosis of infections in immunocompromised patients. In assessment of musculoskeletal infection, Ga scintigraphy is usually performed in conjunction with ^{99m}Tc methylene diphosphonate (MDP) bone scintigraphy. Ga can also be used for the detection of intra-abdominal abscesses, although ^{111}In-labeled WBC scintigraphy, when available, is considered the imaging modality of choice (22).

Although Ga scintigraphy shows high sensitivity in detection of both acute and chronic infections, it has several drawbacks that limit its clinical application. Its specificity is relatively low because of early physiologic excretion through the urinary system and delayed excretion through the bowel, and because of the normal biodistribution to the liver. Ga is a bone-seeking tracer, accumulating in areas of bone remodeling. It is also a tumor-seeking agent, accumulating in such tumors as lymphoma or hepatoma. Physiological hepatic, bowel, and renal uptake can either mimic or mask foci of infection localized inside or in the vicinity of these organs. Optimal Ga imaging may therefore require delayed imaging of up to a few days after injection, a disadvantage when diagnosing acute infectious processes that may require rapid therapeutic interventions (21).

Hybrid SPECT/CT imaging performed as soon as 24 hours after injection may overcome some of these drawbacks. By accurately localizing increased focal Ga uptake, fused images may help differentiate an infectious focus from physiological bowel uptake and therefore eliminate the delay in diagnosis due to the required three- to seven-day waiting period. Precise registration of Ga-SPECT and CT can facilitate the differential diagnosis of osteomyelitis from an adjacent soft-tissue abscess or cellulitis that spares the bone. In a study investigating the value of hybrid Ga-SPECT/CT in patients with suspected infection, Bar-Shalom et al. assessed 47 patients referred for evaluation of various infectious processes including a suspicion of osteomyelitis or soft-tissue infections, in addition to the assessment of patients with fever of unknown origin (23). Ga-SPECT/CT excluded the presence of an infectious process in four patients by localizing the increased tracer uptake to physiological bowel activity. In 13 patients, hybrid imaging led to a change in image interpretation by precise localization of the focus of abnormal Ga activity. The change in the diagnostic accuracy of Ga scintigraphy achieved by means of SPECT/CT may be a further impact on the clinical management of these patients by modifying the surgical approach, canceling planned surgery, and initiating or modifying conservative antibiotic therapy regimens. Figure 1 demonstrates the value of ^{67}Ga SPECT/CT in the evaluation of vertebral osteomyelitis. Figure 2 shows the value of ^{67}Ga SPECT/CT in assessing fever of unknown origin.

(A)

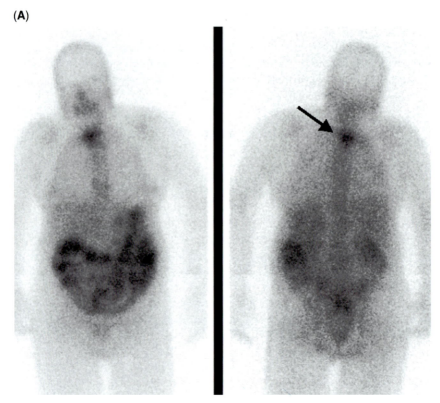

Figure 1   SPECT/CT-based diagnosis of osteomyelitis of the cervical spine in a 63-year-old woman after coronary bypass graft surgery presenting with prolonged fever, septicemia, and upper back pain. (A) Planar ^{67}Ga scintigraphy and (B) (*next page*) SPECT images, axial (*top row*) and coronal (*bottom rows*), show a pathological uptake of the tracer at the level of the lower neck (*arrows*). (C) (*next page*) ^{67}Ga SPECT/CT axial (*right column*), sagital (*center*), coronal (*left column*), CT (*upper row*), SPECT (*middle*), and fusion (*lower row*) precisely localize the site of uptake to the first thoracic vertebra (*arrows*). Osteomyelitis was further confirmed by high-resolution CT. *Abbreviation*: SPECT, single-photon emission computed tomography.

RADIOLABELED LEUKOCYTES

Labeled WBC imaging using either ^{99m}Tc- or ^{111}In has been shown to perform excellently in detecting acute and chronic infections. This test has become the nuclear medicine modality of choice in a variety of clinical settings when the suspicion of an infectious process exists. Clinical indications for labeled leukocytes scintigraphy include osteomyelitis, especially in cases where infected joint prosthesis or posttraumatic osteomyelitis is suspected. WBC scintigraphy has also been advocated for assessment of suspected vascular graft infections, diabetic foot, and abdominal infections (13,22).

Early diagnosis of vascular graft infection is important in view of its high morbidity and mortality, but is difficult to achieve. CT is the first choice imaging modality, but it is not specific and differential diagnosis with extragraft infection is difficult (24,25). Labeled leukocytes scintigraphy is a good diagnostic imaging modality and has become part of the diagnostic work-up of patients who present

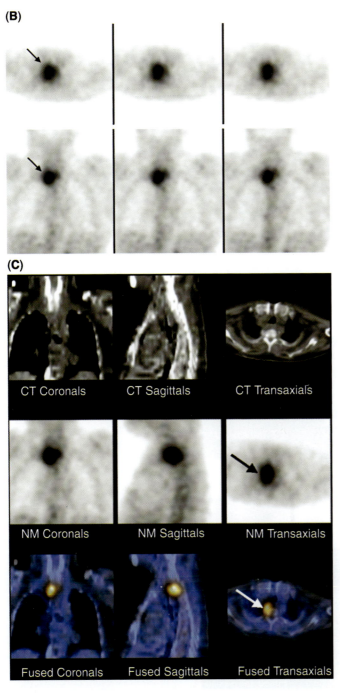

Figure 1 (*Caption on facing page*)

with a suspicion of graft infection (26–28). Scintigraphic assessment, however, lacks anatomical landmarks, rendering the topographic localization of sites of abnormal tracer uptake difficult. The close juxtaposition of the vascular graft and the surgical scar, which may often be infected, makes the diagnosis or exclusion of vascular graft

(A)

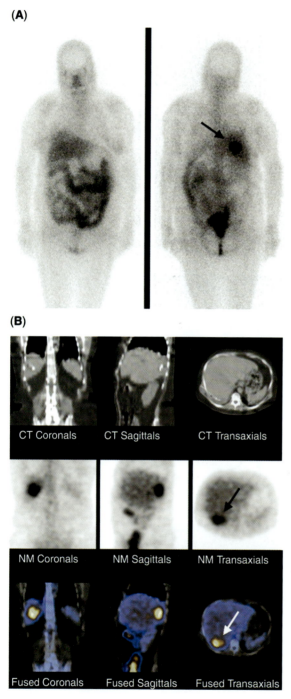

(B)

Figure 2   SPECT/CT-based diagnosis of a liver abscess in a 51-year-old woman presenting with fever of unknown origin. (**A**) Planar ^{67}Ga scintigraphy shows a focus of pathological uptake in the posterior aspect of the right upper abdomen (*arrow*). (**B**) ^{67}Ga SPECT/CT axial (*right column*), sagital (*center*), coronal (*left column*), CT (*upper row*), SPECT (*middle*), and fusion (*lower row*) precisely localize the focus of abnormal uptake to the right lobe of the liver (*arrows*). High-resolution, contrast-enhanced CT further confirmed the presence of a liver abscess. US-guided drainage was performed. The patient was further treated according to microbial culture sensitivity with good clinical response. *Abbreviation*: SPECT, single-photon emission computed tomography.

involvement more difficult. WBC-SPECT/CT addresses this diagnostic challenge in 16 of 24 patients assessed for suspected vascular graft infection (23). Hybrid imaging enabled the localization of WBC uptake to the vascular graft in nine patients and to soft tissues adjacent to the graft only, with no further evidence of graft involvement, in seven patients.

Osteomyelitis and septic arthritis complicate up to one-third of diabetic foot infections that require hospitalization and are, as a rule, the result of direct spread from contaminated soft tissue. Early diagnosis is difficult to achieve by noninvasive imaging studies (29). Nuclear medicine techniques and radiological imaging modalities are both used in the evaluation of the diabetic foot with only limited accuracy (30). Combined bone ^{99m}Tc MDP and WBC scintigraphy (in conjunction with bone marrow scans using ^{99m}Tc sulfur-colloid) is a highly sensitive procedure, performed usually as the first and most relevant diagnostic test. However, abnormal findings in coexisting pathological processes such as neuroarthropathy, trauma, or cellulitis may decrease the accuracy of this combination of nuclear medicine procedures. Although bone scintigraphy can indicate the presence of bone involvement, distinguishing between soft-tissue infection, with or without osteomyelitis of the adjacent bone, may be difficult. In this clinical setting, the precise localization achieved with SPECT/CT in a single imaging step and with greater diagnostic accuracy may provide a tool for guiding invasive tissue sampling procedures and further treatment planning.

WBC scintigraphy is currently the modality of choice for investigating osteomyelitis, especially when infected joint prosthesis or posttraumatic osteomyelitis is suspected. Despite its high sensitivity, its ability to define the precise localization of WBC uptake may be limited because of distorted anatomy in sites of previous fractures or surgery. SPECT has improved the image contrast of nuclear medicine studies and to some extent has increased lesion detectability and localization capabilities (15,16), but does not provide sufficient anatomical landmarks. A study aimed at investigating the potential advantage of ^{111}In-labeled WBC-SPECT/CT compared with SPECT alone evaluated 14 patients with various sites of suspected infection (31). SPECT/CT led to a change in the diagnosis of five of the patients and had an impact on the clinical management in four of them. In addition, in about one-third of the patients, SPECT/CT provided greater confidence in image interpretation although diagnosis was not changed. The authors concluded that SPECT/CT improved the accuracy of anatomic localization of foci of abnormal WBC uptake and led to modification in clinical management (31). Figure 3 demonstrates the value of labeled WBC-SPECT/CT in vascular graft infections.

LABELED ANTIGRANULOCYTE ANTIBODIES

The preparation of radiolabeled leukocytes is laborious, requires specialized staff and dedicated expensive equipment, and can be hazardous because of blood-product-handling. The advantages of radioimmunoscintigraphy over techniques involving labeled autologous leukocytes for imaging infection have to do mainly with the simplicity of its use due to the fact that there is no need for blood-product-handling. Indications for labeled antigranulocyte antibody scintigraphy include the suspicion of osteomyelitis, fever of unknown origin, abdominal infections, and vascular graft infections (13,32,33).

Correlation of results from this scintigraphic technique to anatomical imaging modalities is considered mandatory for accurate diagnosis. False-positive studies

(A)

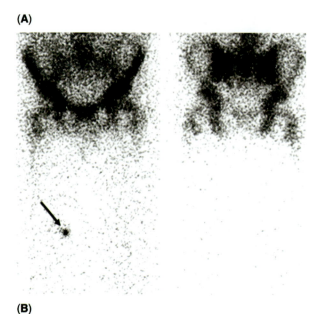

(B)

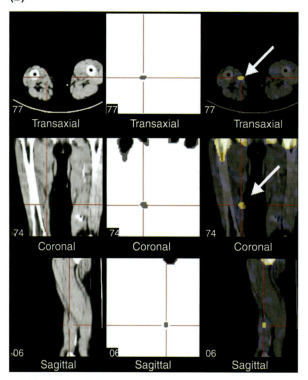

Figure 3 SPECT/CT-based diagnosis of vascular graft infection in a 59-year-old man after femoro-popliteal bypass, presenting with fever and an inflamed surgical wound at the medial aspect of the right thigh. (**A**) Planar ^{111}In labeled WBC scintigraphy shows a focal uptake of the tracer at the medial aspect of the right distal thigh (*arrow*). (**B**) [111]In labeled WBC SPECT/CT, axial (*upper row*), coronal (*middle*), and sagital (*lower row*), CT (*left column*), SPECT (*middle*), fusion (*right column*) precisely localizes the site of uptake to vascular graft (*arrows*). Vascular graft infection was further confirmed at surgery. *Abbreviations*: SPECT, single-photon emission computed tomography; WBC, white blood cells.

may be the result of increased vascular permeability. Increased uptake of labeled antibodies has been observed in perivascular hematomas and contusions, especially in the delayed phase images obtained 24 hours after injection. With the additional anatomical information it provides, SPECT/CT can be of value in excluding or confirming the presence of a hematoma as the cause of abnormal tracer uptake. Similar to the performance capabilities of other nuclear medicine procedures, the precise localization of the infected site may be difficult on scintigraphy alone. A study investigating the value of SPECT/CT in chronic osteomyelitis assessed 27 patients with 29 sites of suspected bone infection and compared planar and SPECT/CT imaging after injection of ^{99m}Tc-labeled antigranulocyte antibodies (14). SPECT/CT was able to correctly localize all positive foci detected on planar and SPECT images. SPECT/CT also allowed the differential diagnosis of soft-tissue infection, septic arthritis, and osteomyelitis. Furthermore, following the diagnosis of bone involvement, hybrid imaging was also able to differentiate between cortical, corticomedullar, and subperiostal foci of disease involvement. The authors concluded that combined SPECT/CT imaging improves the accuracy of radioimmunoscintigraphy for diagnosis of soft-tissue infection and osteomyelitis (14). The potential impact of SPECT/CT on patient management and treatment planning in particular appears to be significant.

LABELED HUMAN NONSPECIFIC IMMUNUGLOBULIN

Human nonspecific immunoglobulin (HIG) is a nonantigen specific immunoglobulin G (IgG) antibody. These antibodies can be labeled with both ^{111}In and ^{99m}Tc. The uptake mechanism of HIG in infectious lesions is not clearly understood. The main process of the antibody migration to the infection site is most likely related to increased blood vessel permeability. Additional potential pathways could be related to specific binding mechanisms to leukocytes or bacteria (32–34). Labeled HIG has been described for diagnosis of musculoskeletal and pulmonary infections, in particular in immunocompromised patients (34,35). The technique is limited by the very high blood pool uptake of the tracer up to 24 hours after the injection, which results in high background activity. These findings limit the value of the test for diagnosis of low-grade (13) and vascular infections (21). The use of SPECT/CT can minimize some of the biodistribution–related limitations of labeled HIG. By fusing anatomical information with the functional data obtained from HIG scintigraphy, blood vessels can be precisely delineated and major areas of increased blood pool in different body regions identified. These foci of increased HIG uptake can be further excluded as sites of infection, while suspicious but equivocal foci of increased HIG uptake can be precisely localized to sites of infection.

ADDITIONAL RADIOLABELED MOLECULES FOR INFECTION IMAGING

There have been many innovations in nuclear medicine imaging of infections since the first use of Ga, a tracer with a nonspecific mechanism of uptake in infection sites. Over the years, more specific agents (such as labeled WBC) and even receptor specific proteins (such as antigranulocyte and antibody fragments) have been introduced (9,21). The cumulative knowledge and better understanding of the biochemical processes of infection have led to the development of labeled receptor-specific small proteins and peptides. These agents are mediators of the inflammatory response that expresses a high affinity to their receptors, and therefore show avid and relatively high specific accumulation by infected tissue (3).

This group of agents includes chemotactic peptides such as formyl-methionyl-leucyl-phenylalanine, which bind to receptors on both polymorphonuclears (PMNs) and monocytes. Tuftsin, derived from the Fc portion of IgG, promotes chemotaxis and phagocytosis of neutrophils, monocytes, and macrophages by binding to receptors on these cells (9,36). Cytokines, including Interleukin-1 receptor antagonists, which bind to receptors expressed on various leukocytes, and Interleukin-2, which specifically binds to IL-2 receptors expressed on activated lymphocytes, can also play a role in the evaluation of chronic inflammatory process. Chemokines such as Interleukin-8 bind to receptors on neutrophils, and platelet factor 4 binds to PMNs and monocytes (9,36). Somatostatin analogs bind to receptors expressed on normal and activated lymphocytes and macrophages and may therefore be used for imaging chronic processes. Substance P, a neuropeptide, has also been found to accumulate in inflammatory processes (3).

All the above-mentioned peptides were investigated mainly in animal models and were labeled using ^{111}In and/or [99m]Tc. Infectious and inflammatory foci can be visualized with most radiolabeled peptides within several hours after injection (3,21,36). Their early and specific focal uptake, combined with rapid clearance from the background, make these labeled infection mediators compare favorably with conventional agents. However, all these agents also accumulate to a certain extent in several nontarget tissues. High background uptake has been described in the first few hours after injection (3). The major drawback of using small proteins and peptides for scintigraphic imaging is the toxicity due to their specific biological characteristics (9,21), which at the same time make radiolabeled peptides potentially highly specific and accurate agents in the evaluation of infection and inflammation.

For detection of acute processes, agents with high affinity to PMNs are preferable. Thus, formyl-methionyl-leucyl-phenylalanine, tuftsin antagonists, IL-1, IL-8, and platelet factor 4 appear to be the best choice. In contrast, in chronic infection and inflammation processes, IL-2, octreotide, and substance-P that bind specifically to mononuclear cells should be more suitable (3,9,21,36). In view of the promising role of labeled receptor-specific small proteins and peptides for the evaluation of infection and inflammation, the precise localization of tracer uptake may play a crucial role in the diagnostic accuracy of receptor-specific imaging. Labeled peptide SPECT/CT can take these agents a step forward in their ability to detect infection. Hybrid-imaging can overcome some of the limitations of labeled peptide-imaging, such as nontarget accumulation and high background activity. Potentially, SPECT/CT will successfully combine information provided by nuclear medicine tracers specifically targeted to the infectious process with structural data provided by CT for precisely characterizing the target tissue or organ.

CONCLUSIONS

Imaging plays a major role in the diagnosis and management of patients with suspected infection. There is a need for both anatomic and functional information provided noninvasively by various imaging modalities. By combining CT and scintigraphic studies, SPECT/CT merges this information and provides complementary data of incremental value on the structure, function, and localization of infectious processes in various regions of the body. The enhanced diagnostic ability of the

SPECT/CT modality makes this a promising technique in the investigation, treatment planning, and management of patients with suspected infectious process.

REFERENCES

1. Berkelman RL, Hughes JM. The conquest of infectious diseases: who are we kidding? Ann Intern Med 1993; 119:426–428.
2. Santiago RC, Gimenez CR, McCarthy K. Imaging of osteomyelitis and musculoskeletal soft tissue infections: current concepts. Rheum Dis Clin North Am 2003; 29:89–109.
3. Van der Laken CJ, Boerman OC, Oyen WJ, et al. Scintigraphic detection of infection and inflammation: new developments with special emphasis on receptor interaction. Eur J Nucl Med 1998; 25:535–546.
4. Greenspan A, Stadalnik RC. A musculoskeletal radiologist's view of nuclear medicine. Semin Nucl Med 1997; 27:372–385.
5. Schweitzer ME, Morrison WB. MR imaging of the diabetic foot. Radiol Clin North Am 2004; 42:61–71.
6. Tomas MB, Patel M, Marwin SE, et al. The diabetic foot. Br J Radiol 2000; 73:443–450.
7. Ledermann HP, Morrison WB, Schweitzer ME. MR image analysis of pedal osteomyelitis: distribution, patterns of spread, and frequency of associated ulceration and septic arthritis. Radiology 2002; 223:747–755.
8. Ledermann HP, Kaim A, Bongartz G, Steinbrich W. Pitfalls and limitations of magnetic resonance imaging in chronic posttraumatic osteomyelitis. Eur Radiol 2000; 10: 1815–1823.
9. Rennen HJ, Boerman OC, Oyen WJ, et al. Imaging infection/inflammation in the new millennium. Eur J Nucl Med 2001; 28:241–452.
10. Palestro CJ, Torres MA. Radionuclide imaging in orthopedic infections. Semin Nucl Med 1997; 27:334–345.
11. Peters AM. The choice of an appropriate agent for imaging inflammation. Nucl Med Commun 1996; 17:455–458.
12. Becker W. The contribution of nuclear medicine to the patient with infection. Eur J Nucl Med 1995; 22(10):1195–1211.
13. Becker W, Meller J. The role of nuclear medicine in infection and inflammation. Lancet Infect Dis 2001; 1:326–333.
14. Horger M, Eschmann SM, Pfannenberg C, et al. The value of SPET/CT in chronic osteomyelitis. Eur J Nucl Med Mol Imag 2003; 30:1665–1673.
15. Israel O, Keidar Z, Iosilevsky G, et al. The fusion of anatomic and physiologic imaging in the management of patients with cancer. Semin Nucl Med 2001; 31:191–205.
16. Keidar Z, Israel O, Krausz Y. SPECT/CT in tumor imaging: technical aspects and clinical applications. Semin Nucl Med 2003; 33:205–218.
17. Zhuang H, Duarte PS, Pourdehnad M, et al. The promising role of 18F-FDG PET in detecting infected lower limb prosthesis implants J Nucl Med 2001; 42:44–48.
18. Winter F, Vogelaers D, Gemmel F, et al. Promising role of 18-F-fluoro-D-deoxyglucose positron emission tomography in clinical infectious diseases. Eur J Clin Microbiol Infect Dis 2002; 21:247–257.
19. Bleeker-Rovers CP, de Kleijn EM, Corstens FH, et al. Clinical value of FDG PET in patients with fever of unknown origin and patients suspected of focal infection or inflammation. Eur J Nucl Med Mol Imag 2004; 31:29–37.
20. Sugawara Y, Braun DK, Kison PV, et al. Rapid detection of human infections with fluorine-18 fluorodeoxyglucose and positron emission tomography: preliminary results. Eur J Nucl Med 1998; 25:1238–1243.
21. Boerman OC, Rennen H, Oyen WJ, et al. Radiopharmaceuticals to image infection and inflammation. Semin Nucl Med 2001; 31:286–295.

22. Love C, Palestro CJ. Radionuclide imaging of infection. J Nucl Med Technol 2004; 32:47–57.

23. Bar-Shalom R, Yefremov N, Guralnik L, et al. SPECT/CT using Gallium-67 and In-111 labeled leukocyte scintigraphy for diagnosis of infection. J Nucl Med 2006. In press.

24. Ramo OJ, Vorne M, Lantto E, et al. Postoperative graft incorporation after aortic reconstruction–comparison between computerized tomography and Tc-99m-HMPAO labeled leukocyte imaging. Eur J Vasc Surg 1993; 7:122–128.

25. Liberatore M, Iurilli AP, Ponzo F, et al. Clinical usefulness of technetium-99m-HMPAO-labeled leukocyte scan in prosthetic vascular graft infection. J Nucl Med 1998; 39: 875–879.

26. Orton DF, LeVeen RF, Saigh JA, et al. Aortic prosthetic graft infections: radiologic manifestations and implications for management. Radiographics 2000; 20:977–993.

27. Williamson MR, Boyd CM, Read RC, et al. 111 In-labeled leukocytes in the detection of prosthetic vascular graft infections. Am J Roentgenol 1986; 147:173–176.

28. Samuel A, Paganelli G, Chiesa R, et al. Detection of prosthetic vascular graft infection using avidin/indium-111-biotin scintigraphy. J Nucl Med 1996; 37:55–61.

29. Marcus CD, Ladam-Marcus VJ, Leone J, et al. MR imaging of osteomyelitis and neuropathic osteoarthropathy in the feet of diabetics. Radiographics 1996; 16:1337–1348.

30. Stumpe KD, Dazzi H, Schaffner A, et al. Infection imaging using whole-body FDG-PET. Eur J Nucl Med 2000; 27:822–832.

31. Mirtcheva RM, Kostakoglu SJ, Goldsmith SJ. SPECT+CT fusion imaging in 111In WBC scintigraphy [abstr]. J Nucl Med 2003; 44:341P.

32. Hakki S, Harwood SJ, Morrissey MA, et al. Comparative study of monoclonal antibody scan in diagnosing orthopaedic infection. Clin Orthop 1997; 335:275–285.

33. Palestro CJ, Kipper SL, Weiland FL, et al. Osteomyelitis: diagnosis with (99m)Tc-labeled antigranulocyte antibodies compared with diagnosis with (111)In-labeled leukocytes—initial experience. Radiology 2002; 223:758–764.

34. Oyen WJ, Claessens RA, van der Meer JW, et al. Indium-111-labeled human nonspecific immunoglobulin G: a new radiopharmaceutical for imaging infectious and inflammatory foci. Clin Infect Dis 1992; 14:1110–1118.

35. Nijhof MW, Oyen WJ, van Kampen A, et al. Evaluation of infections of the locomotor system with indium-111-labeled human IgG scintigraphy. J Nucl Med 1997; 38:1300–1305.

36. Rennen HJ, Corstens FH, Oyen WJ, Boerman OC. New concepts in infection/inflammation imaging. Q J Nucl Med 2001; 45:167–173.

13

SPECT/CT Imaging of Small Animals

Bruce H. Hasegawa
*UCSF Physics Research Laboratory, Department of Radiology,
University of California, San Francisco, California, U.S.A.*

ROLE OF SMALL ANIMALS IN BIOMEDICAL RESEARCH

The modern use of animals in biological research and medical discovery arguably began over 200 years ago with Jenner's use of cowpox as a vaccine for smallpox in 1796. In the ensuing two centuries, animal research has affected virtually all aspects of biological discovery and medical technology. Progress in the biomedical sciences accelerated with the dawn of molecular biology 50 years ago (1), including the development of monoclonal antibodies for diagnosis and treatment of disease, organ transplantation, laparoscopic surgery, gene therapy, and molecularly targeted drugs for cancer and other diseases. These developments have largely relied on studies that involve animal experimentation and now are having a profound and widespread effect on the health and well-being of the human species.

While the use of larger mammalian species (e.g., dogs, cats, rabbits, pigs, and primates) remains an important element in biomedical research, most animal research has been and continues to be performed with rodents, especially mice. Mice have obvious advantages that facilitate their role in biomedical research. Mice are mammals and share similar anatomical and physiological characteristics with humans. Mice are small, easy-to-house, multiply rapidly, and are relatively inexpensive to maintain. The average lifespan of a wild mouse typically is one to two years; mice can live in captivity for three years, and as long as six years. Approximately 90% of all research involving vertebrate biology now uses mice as the predominant mammalian model (2). Advances in molecular biology and genetic engineering and the modern growth of the pharmaceutical and biotechnology industries have increased the need for biological studies that involve mice. Approximately 25 million mice were raised in 2001 for experimental studies, and this number is expected to increase at a rate of 10% to 20% annually over the next decade (2). Murine models are available for a wide variety of biological conditions and transgenic animals (3–6) now account for a sizable and growing fraction of animal models used in biological research.

The availability of small animal models representing specific biological conditions and disease states has motivated the development of laboratory protocols and methods designed to extract definitive biological and medical information from these studies. Transplantation methods for the heart, kidneys, and other organs have been developed for both mice and rats (7–12), and represent important model systems for evaluating organ preservation, graft rejection, and immunosuppression. In the arena of neurodegenerative disease, genetically manipulated mouse models for disorders such as Alzheimer's disease, Parkinson's disease, amyotrophic lateral sclerosis, and spinal muscular atrophy are now available and provide researchers with a better understanding of the mechanisms of the disease (13–16), which presents new targets for therapeutic intervention (16). Cancer researchers now have significant insights into the genetic mechanisms underlying carcinogenesis, and have established many important parallels between tumor development in mice and humans, thereby confirming that the mouse indeed is an excellent model for the study of human cancer (17).

Traditional techniques of biological research (e.g., microscopy, histology, organ sampling, and autoradiography) generally require that an animal must be sacrificed and tissue harvested for the measurement to be obtained. These methods eliminate the possibility of serial studies on an individual animal and do not allow the investigator to follow the progression of disease in an individual animal, or to follow the response of the animal to diagnostic procedures or therapeutic intervention. Surgical techniques now have been developed to assess physiological function with organ systems in intact animals in both mice (6) and rats (18), but these procedures ultimately culminate with euthanasia.

For these reasons, over the past decade, there has been an increasing interest in imaging systems designed specifically for biological research, with a special focus on high-resolution small animal imaging (19–21). These noninvasive measurement tools have the capability to improve the reliability and statistical power of the gathered data, while reducing experimental costs and the number of animals needed for the experiment. Clinical imaging techniques, i.e., those designed for imaging humans, obviously can and will continue to be used for noninvasive measurements involving larger animals, but generally have inadequate spatial resolution for studies involving small animals such as mice and rats. For example, the left heart of a mouse in short axis view is approximately 5 mm in diameter with the left ventricular wall having a thickness of approximately 1 mm. In addition, the mouse heart rate is 600 to 900 beats per minute, making it difficult to acquire image data even with cardiac gating (6,22). Similarly, tumors having a diameter of a few millimeters lie within the spatial resolution limit of clinical computed tomography (CT) and magnetic resonance imaging (MRI) systems, but cannot be resolved by conventional radionuclide imaging methods such as single-photon emission computed tomography (SPECT) or positron emission tomography (PET). For this reason, methods such as high-resolution MRI (19,23–28) and ultrasonography (29) provide detailed anatomical information at submillimeter length scales. There is also significant need for other methods such as fluorescence (30), bioluminescence (31), PET (32,33), SPECT (34–36), and magnetic resonance spectroscopy (37–39) that allow one to monitor functional, metabolic, and molecular changes associated with biological function and disease in the intact animal. These methods can bridge results extracted from in vitro studies at the cellular and subcellular levels using microscopy and other invasive techniques, as well as those obtained from noninvasive studies that can be performed by observing the behavior of the animal. Small animal imaging thereby provides unique insights into fundamental biological processes in the intact animal and is becoming critical for the development and evaluation of new methods of clinical diagnosis and therapy.

Specialized small animal imaging systems, with features that parallel those found in clinical imaging, are now available from commercial sources, and are utilized at a growing number of academic and corporate centers. The field of small animal imaging is expanding rapidly and has led to the development of dual-modality systems specifically for studies involving small animals. As is true for clinical dual-modality imaging, the development of small animal dual-modality imaging and specifically of small animal SPECT/CT has depended strongly on advances in instrumentation and techniques for both radionuclide imaging and X-ray imaging.

SMALL ANIMAL SPECT IMAGING

Technical Capabilities and Design of Systems

The design of small animal SPECT/CT systems is similar, but not identical, to clinical SPECT/CT systems, which obviously incorporate design elements from independent SPECT and CT systems. Ultimately, the design of a small animal SPECT system must be motivated by characteristics needed to perform the experimental study. It is obvious that the spatial resolution of the small animal radionuclide imaging system (i.e., microPET and microSPECT) must be sufficiently fine to resolve the target sites of interest (e.g., tumors, myocardium, and brain). Consequently, both microSPECT and microPET systems offer spatial resolution in the range of 1 to 2 mm. A higher resolving power obviously would be better, but generally must be obtained at the cost of decreased detection efficiency or increased system complexity. Detection efficiency is a more difficult characteristic to specify. The efficiency of the system must accommodate the range of radioactivity (i.e., generally in the approximate range of 5–50 MBq, roughly 100 µCi to 1 mCi) administered to the animal, as radiopharmaceutical levels in this range are compatible with imaging studies. These factors therefore require the detection efficiency to have a minimum level of around $\varepsilon = 10^{-5}$ for microSPECT studies; ideally the detection efficiency should be in the range of 2×10^{-3} to 1×10^{-4}, or better. Furthermore, one would like to complete the imaging study in one hour or less, so that the animal can be maintained under anesthesia during the imaging study and so the imaging system can be used efficiently for several studies in a single day. Finally, if one would like to visualize a comparable level of detail in animals as one obtains in clinical studies, it is logical that both studies would require the same quantity of radionuclide in each voxel (40). This introduces a significant challenge in terms of radionuclide concentration as well as the radiation dose delivered to the animal due to the large differences in body mass between small animals (i.e., 20–30 g for mice, 200–300 g for rats) versus humans (e.g., approximately 70,000 g). A large concentration of radiopharmaceutical can jeopardize the tracer principle and may lead to unwanted pharmacological effects (40). The associated radioactivity levels may be significantly high to alter genetic expression or physiological response, or ultimately to jeopardize life, although these questions about radiation dose in small animal imaging studies have not yet been analyzed in detail.

Other issues must be considered that are beyond the physical factors associated with radiation detection but nevertheless are important for performing the experiment and for maintaining the animal during the study. First, the animal must be given fluids intravenously to administer radiopharmaceuticals, radiographic contrast media, anesthesia, or fluids for hydration. The total blood volume (TBV) of a mouse is 6% to 8% of its body weight (41), corresponding to a volume in the range of 1200 to 1600 µL in a 20 g mouse. As a result, the volume of injectate should be limited to 10% to 15% of TBV (or approximately 100–200 µL) to avoid mechanical stress to the cardiovascular system. Correspondingly, the imager must be designed with appropriate plumbing to deliver fluids to the animal during the imaging session. Second, gaseous anesthesia (e.g., isoflurane) has several appealing characteristics in comparison to injectable agents in terms of ease of use, relatively inexpensive cost, and rapid recovery. The animal imaging system, therefore, must also be designed with apparatus to deliver and exhaust the gaseous anesthesia during the experiment. Third, physiological monitoring may be needed to monitor the respiration pattern, the electrocardiogram, and body temperature. In addition, the EKG and respiratory signals are useful for synchronizing or gating the image acquisition, while body temperature can be used for monitoring and maintaining body temperature of the animal in concert with heat lamps, electric heaters, or delivery of forced warm air during the study.

Radionuclide Detection and Pinhole Collimation

A small animal SPECT system obviously must include a radionuclide imaging detector that typically has a rectangular (or square) face, but that also can be configured as a ring. As in clinical imaging, the detector used for small animal SPECT must be operated with a collimator and shielding so that it records gamma rays only from a known predetermined direction without contamination from stray radiation. However, unlike clinical imaging, which generally is performed with a parallel hole collimator, small animal SPECT generally is performed with a pinhole collimator to achieve millimeter or submillimeter spatial resolution with geometric efficiencies suitable for the levels of radioactivity administered to the animal in these procedures. The pinhole collimator allows gamma rays from the radioactive object to pass through the aperture and strike the detector where they are recorded to form the image. The pinhole collimator uses a "convergent" or "cone-beam" imaging geometry in which the image is magnified on the detector to improve spatial resolution.

The basic performance of pinhole SPECT in terms of its spatial resolution and detection efficiency can be quantified using the mathematics of pinhole collimation. A typical pinhole collimator (Fig. 1) can be represented as an aperture having a physical diameter d located on the shielded collimator cone. We can assume that the aperture is located a distance t away from the detector surface, and views a radioactive source at distance b from the pinhole aperture. The "cone angle" (α) is measured between the opposite walls of the collimator cone with respect to a vertex at the aperture. The spatial resolution of the pinhole collimator is given by the size [typically represented by the "full-width-at-half-maximum" (FWHM)] of the spot (or "point-spread function") that is projected through the pinhole onto the detector surface from a point source at distance b from the aperture. The spatial resolution is affected not only by the physical size and shape of the pinhole, but also by transmission of gamma rays through the edge of the pinhole aperture—an effect known as "septal penetration." If the collimator is fabricated from an absorptive material (e.g., lead) having a linear attenuation coefficient μ, the "effective aperture diameter" (d_e) that includes both the physical size of the aperture as wells as the effects of septal penetration is given by (43)

$$d_e = \sqrt{d[d + 2\mu^{-1}\tan \alpha]} \tag{1}$$

with the spatial resolution R_c measured on the detector surface calculated as

$$R_c = \frac{(t + b)d_e}{b} \tag{2}$$

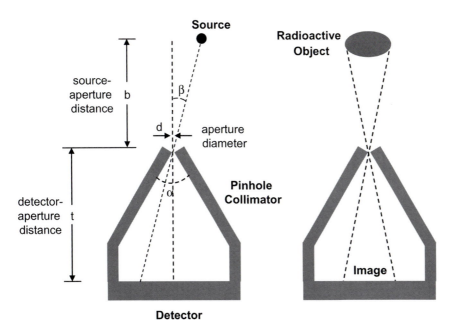

Figure 1 Schematic showing collimator with single pinhole. (*Left*) A pinhole collimator has an aperture diameter d, cone angle α, and aperture-detector distance t. A source is imaged at source-pinhole distance b and angle β from the central pinhole axis. (*Right*) The image of the radioactive object is projected onto the detector through the pinhole aperture.

The pinhole collimator magnifies the radionuclide distribution of the object onto the surface of the detector, where the spatial resolution of that projected image is represented by Eq. (2). Generally, it is more useful to represent the spatial resolution as measured not on the detector surface, but in the object space. This can be done by correcting for the magnification factor in Eq. (2), which produces the spatial resolution R_o of the collimator measured in the plane of the object

$$R_o = \frac{(t+b)d_e}{t}, \tag{3}$$

and which represents the spatial resolution with which we view the object in the final image.

The performance of the collimator can also be characterized by the geometric efficiency, i.e., the fraction of emitted photons that pass from the source through the pinhole aperture and strike the detector surface. The geometric efficiency can be evaluated by considering a point source located at a perpendicular distance b from the pinhole aperture. If d represents the aperture diameter and θ represents the angle between the central axis of the pinhole and the line from the source through the center of the pinhole aperture, then the geometric efficiency g of the pinhole can be calculated as (43)

$$g = \frac{d_e^2 \cos^3\theta}{16b^2}, \tag{4}$$

where d_e is the effective pinhole diameter [Eq. (1)]. As noted above, decreasing the source-pinhole distance b causes the spatial resolution and the geometric efficiency to improve [i.e., R_o becomes smaller in Eq. (2), and g becomes larger in Eq. (3)]. These improvements are accompanied by a reduction in the field of view by virtue of the geometry of the pinhole collimator (Fig. 1), which makes the pinhole collimator especially suitable for imaging small body parts (e.g., thyroid gland) or small animals.

The design of the pinhole collimator always represents a delicate balance in small animal SPECT, which requires a careful trade-off between spatial resolution and geometric efficiency, both of which are important factors in radionuclide imaging. That is, one would like to obtain the highest possible level of spatial resolution (ideally at submillimeter levels for small animal radionuclide imaging), but also with a high level of geometric efficiency that would allow the imaging study to be performed in short imaging times and minimal radiation dose. The need for spatial resolution and geometric efficiency unfortunately compete with one another since the former is obtained by decreasing the diameter of the pinhole while the latter is obtained by increasing the pinhole diameter. Methods to improve both spatial resolution and geometric efficiency simultaneously with pinhole imaging have been developed primarily through use of multiple detectors and collimators with multiple pinholes (44–50).

Small Animal SPECT with Conventional Scintillation Cameras

It is possible to perform either planar (40) or tomographic (51–55) imaging of small animals with conventional scintillation cameras. However, clinical radionuclide imaging is commonly performed with parallel hole collimators that offer spatial resolution in the range of 6 to 8 mm FWHM (or worse) that allow only a very rough visualization of radiopharmaceutical uptake within mice and rats. In comparison, a conventional scintillation camera with pinhole collimation can achieve excellent spatial resolution and adequate detection efficiency over small fields of view by magnifying the radionuclide distribution in the animal by factors of 8 to 10 using a pinhole collimator with an aperture diameter between 0.5 and 2 mm. The large magnification minimizes any degradation that otherwise would be introduced by the intrinsic spatial resolution of the camera. The spatial resolution is also improved by decreasing the size of the pinhole aperture at the cost of decreasing detection efficiency. However, to a certain degree, this can be offset by positioning the animal close to the pinhole to increase efficiency.

A practical small animal radionuclide imaging system therefore can be implemented with a standard scintillation camera operated with a pinhole collimator. Some of the earliest work of this type was performed by Jaszczak and Jianying at Duke University (51) and by Weber and Ivanovic at Brookhaven National Laboratory (52) to image radionuclide uptake in murine models of cancer. Similarly, Ishizu et al. (53) added specially designed pinhole collimators to a clinical four-headed SPECT system for cardiovascular studies in rats for static imaging of the rat brain with ^{99m}Tc-HMPAO and of the rat heart with ^{99m}Tc-MIBI. In addition, these authors used the same system to perform dynamic SPECT imaging of the rat brain with ^{123}I-iomazenil (54). Small animal pinhole SPECT also was developed by Wu et al. (55) for cardiovascular imaging of myocardial perfusion and gated blood pool studies of an anesthetized mouse. Weber and Ivanovic (52) and Wu et al. (55) rotated the animal in a vertical axis (nose up, tail down) in front of a stationary scintillation camera with a pinhole collimator. This approach was taken to support the large size and weight of the clinical scintillation camera while maintaining the geometric alignment of the camera and pinhole collimator with respect to the axis of rotation while the camera was rotated around the animal to acquire the SPECT data. This is especially important for imaging at the high magnification needed for small animal SPECT. For example, if one wishes to achieve a spatial resolution in the range of

1 mm FWHM, the alignment of the camera and the center of rotation must be maintained within an accuracy of 100 μm or better at a magnification of 10 to prevent blurring and the introduction of misregistration artifact into the tomographic reconstruction. However, it is generally believed that imaging the animal in a horizontal position reduces physiological stress in comparison to maintaining the animal in a vertical position for imaging, especially when the animal is anesthetized. Jaszczak et al. demonstrated that a conventional scintillation camera and gantry could provide sufficient mechanical stability and alignment for small animal radionuclide imaging (51). In addition, most small animal SPECT systems now are configured to rotate the camera around a stationary horizontal animal rather than rotating a vertical animal in front of a stationary pinhole camera, the preferred configuration for these studies.

Small Animal SPECT with Compact Scintillation Cameras

While these studies have demonstrated the feasibility of using conventional scintillation cameras for small animal SPECT, other investigators attempted to design cameras specifically for high-resolution radionuclide imaging (56). This work is motivated by the investigators' desire to develop camera of smaller size and less weight than conventional cameras that could be incorporated into compact systems with the stability and alignment needed for submillimeter spatial resolution. In addition, the camera could be designed for radiopharmaceuticals labeled with ^{125}I-iodine, which are useful for small animal imaging but emit photons having energies in the range of 25 to 27 keV that otherwise are impractical for clinical imaging. Finally, compact radionuclide cameras have been included in SPECT/CT systems designed specifically for imaging small animals.

Several approaches have been pursued in designing compact scintillation cameras for small animal imaging. The first detector type is an analog of the clinical scintillation camera that couples a sodium iodide or cesium iodide scintillator to an array of photomultiplier tubes (57). However, in comparison to conventional cameras that use continuous scintillation crystals, these small format cameras generally are designed to preserve spatial resolution with segmented crystals that channel the scintillation light from the site of photon interaction to the underlying photomultiplier tube array. In addition, the segmented scintillator material typically is read out using special multianode or position-sensitive photomultiplier tubes (PSPMTs) that, unlike conventional photomultiplier tubes, provide position information in a single device, again with the goal of achieving a higher level of spatial resolution than can be achieved using conventional designs. A compact scintillation camera with a segmented crystal and PSPMT can achieve spatial resolution in the range of 2 to 3 mm FWHM (58) in comparison to conventional scintillation cameras which typically offer spatial resolutions in the range of 3.5 to 4 mm FWHM.

Improved imaging performance can also be obtained by replacing the photomultiplier tube with solid-state detectors such as high-resolution pixellated silicon photodiodes (59,60) that have intrinsic spatial resolution of 2 mm or better. By replacing the photomultiplier tube with a semiconductor device, the overall imaging system can be reduced in size and weight in comparison to cameras with either conventional or PSPMTs. However, at present, the silicon photodiodes do not yet offer the same energy resolution performance as conventional scintillation cameras. This may not be a major limitation in small animal imaging where the small size of the animal torso minimizes the amount of scatter radiation that is produced in the studies, but may reduce performance when energy discrimination is needed for dual-isotope studies or for imaging low-energy radionuclides such as [125]I.

Specific examples of compact cameras designed for high-resolution imaging include the work of Dilmanian, who, in 1990, configured a scintillation camera having a PSPMT, a NaI(Tl) scintillator, and a parallel-hole collimator (61). The system produced a spatial resolution of 0.7 mm with a 26-mm diameter field of view, but had impractically low detection efficiency for in vivo studies. Other investigators developed compact gamma-ray imagers that used a NaI(Tl) scintillator coupled to a PSPMT with both parallel hole collimators (62–65) and pinhole collimators (51,66,67) for SPECT and planar imaging of small objects or body regions. Goertzen et al. performed pinhole SPECT with a CeraSPECT^{TM} scanner (Digital Scintigraphics, Inc., Watham, Massachusetts, U.S.A.) using a rotating 8-pinhole collimator that projected multiple images onto a stationary annular scintillation camera originally designed for brain imaging (56). The system produced SPECT images of a mouse with a spatial resolution of 1.7 mm and a sensitivity of 13.8 cps/μCi at the center of the field of view. The annular design of the camera is also conducive to fast dynamic imaging of the radiopharmaceutical distribution if the administered radioactivity level is sufficiently high.

Semiconductor Detectors for Small Animal Radionuclide Imaging

The use of a scintillator with an optical readout device such as an array of photomultiplier tubes or photodiodes represents a form of "indirect" detection of the gamma ray from the radionuclide. That is, in indirect detection, the gamma ray is absorbed in the scintillator and is converted to multiple optical photons (i.e., visible light). The optical photons are then collected and converted by the photomultiplier tube or photodiode into an electrical signal that is proportional in amplitude to the energy of the absorbed photon. This sequence of events occurs for each photon absorbed in the detector, with each step introducing a slight degradation in spatial and energy resolution.

Other radionuclide imaging detectors use direct conversion of the recorded gamma rays, which produce an electronic signal proportional to the energy of the absorbed photon without an intermediate step. Semiconductor detectors that offer direct conversion produce the highest level of energy resolution available for nuclear medicine and medical imaging. One such detector material is high-purity germanium, which can offer energy resolutions better than 1 keV FWHM, and intrinsic spatial resolution of better than 1 mm FWHM, but must be operated at liquid nitrogen temperatures (77 K) to maintain their performance (68). For this reason, virtually all investigators using semiconductor detectors for medical imaging rely on solid-state detectors that can be operated at room temperature rather than at cryogenic temperatures. Such candidate materials are cadmium zinc telluride (CZT) and cadmium telluride (CdTe) that can be pixelated with intrinsic spatial resolutions as low as 400 μm (58). In addition, CZT and CdTe can offer energy resolution performance in the range of 5% to 8% FWHM at the ^{99m}Tc photon energy of 140 keV (69,70). As radiation detectors, CdTe and CZT unfortunately are not ideal in that they can exhibit partial charge trapping that can reduce both energy resolution and detection efficiency. These effects are, however, minimized with imaging detectors having small pixels in which the electric field needed to collect electric charge from each detector pixel is localized to provide some immunity against charge carriers trapped deep in the material.

Pioneering work in small animal radionuclide imaging is being performed at the Center for Gamma Ray Imaging at the University of Arizona, where high-resolution SPECT and SPECT/CT systems for small animal imaging have been configured (71). Several detector technologies and system designs have been developed by this group, and are driven by Barrett's concept of maximizing the space-bandwidth product (72) in which high-resolution radionuclide detectors are used to acquire minified images at high-spatial resolution. The innovative strategy behind this approach is that individual images can be acquired in small detector areas when recorded with high spatial resolution. This allows multiple images to be recorded using multipinhole collimators or coded aperatures, or multiple detector modules to be configured around the object to maximize detection efficiency. High-resolution radionuclide imaging is performed with a custom modular semiconductor detector (36,73–75) that consists of a $2.5 \times 2.5 \times 0.2\,cm^3$ slab of cadmium zinc telluride (CdZnTe) operated with a continuous gold electrode to apply bias on one side, and a 64×64 array of pixelated gold electrodes on the opposite side connected to an application-specific integrated circuit (ASIC) for read-out of the individual pixel signals. The detector has a 380 μm pixel pitch, and 330 μm wide pixels, coupled with a 7 mm thick parallel-hole collimator for radionuclide imaging. The Arizona group also has developed a prototype SPECT/CT system (71) by combining radionuclide imaging subsystems mounted with their image axes perpendicular to one another with the animal rotated vertically within the common field of view. The X-ray and radionuclide projection data are acquired sequentially, and corrected for distortions and nonuniformities introduced by each of the detectors, then reconstructed with an OSEM algorithm for the X-ray tomogram and an MLEM algorithm for the radionuclide data.

Small Animal Multipinhole SPECT

An appealing approach to small animal SPECT uses collimators with multiple apertures rather than a single pinhole aperture for imaging small objects. Beekman et al. have developed and tested one system (76) that uses a cylindrical collimator insert that is positioned along the common rotational axis of a stationary triple-headed scintillation camera arrangement. The collimator insert has 75 gold micropinholes, each having a diameter of 0.6 mm that provides a peak sensitivity of 0.22% and a sensitivity of $>0.12\%$ for the central 12 mm diameter of the central plane with a 15% wide energy window. Phantom measurements with this configuration have resolved radioactivity-filled capillaries having a diameter of 0.5 mm with a separation of 0.5 mm, to provide a volumetric resolution of approximately 0.1 μL. The collimator geometry from Schramm et al. views the radioactive object (e.g., animal) through a multipinhole collimator, which thereby projects multiple images (Fig. 2) onto the detector surface (45). The tomographic reconstruction algorithm can be designed to account for the multipinhole geometry and is used to reconstruct the three-dimensional distribution of radioactivity in the object. Of course, there is the potential for imaging the object in a way that allows the projections data through the individual pinholes to overlap on the detector surface. This presents a complication since the projection data through one pinhole can overlap and contaminate the projection data acquired through an adjacent pinhole. Schramm et al. have found that a small degree of overlap (i.e., $<30\%$ of the projected size of the object) can be tolerated and allows the tomographic data to be reconstructed (45). Ploux and Mastrippolito have proposed a novel system for in vivo brain imaging of small animals that directly measures the radioactive concentration in a small tissue volume with a large focused multipinhole collimator coupled to an array of radiation detectors (78). Tomographic imaging is enhanced using coincidence imaging of radionuclides that emit two photons

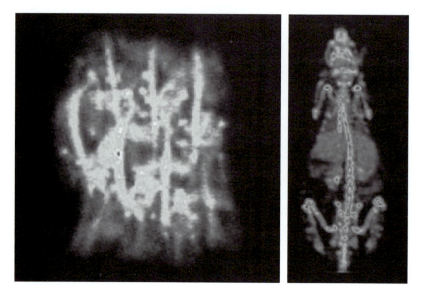

Figure 2   Skeletal image of mouse acquired with multipinhole SPECT. (*Left*) Projection data acquired with 10-pinhole collimator and clinical triple-head scintillation camera. (*Right*) Whole-body coronal section showing SPECT reconstruction of multipinhole data using iterative reconstruction algorithm. The multipinhole collimator affords an increase of 5.2-fold in detection sensitivity in comparison to a single-pinhole system. *Abbreviation*: SPECT, single-photon emission computed tomography. *Source*: Courtesy of Nils Schramm, Research Center Juelich, Germany. From Ref. 50.

of different energies (e.g., 171 and 245 keV for [111]indium) (79). The use of these multi-pinhole techniques allows radionuclide images of mice to be reconstructed with levels of spatial resolution and detection efficiency that exceed those that can be obtained with single-pinhole imaging systems (45,76).

SMALL ANIMAL CT IMAGING

The implementation of small animal dual-modality imaging has relied strongly on the development of compact high-resolution CT systems designed specifically for in vivo visualization of anatomical structure in mice and other small animals (19,27,28,80). These instruments, often called "microCT systems," offer spatial reso-lutions in the range of 50 μm or better with radiation dose levels in the range that allow the animal to survive serial imaging studies. The microCT system must be designed so that the imaging study can be completed in a few minutes, with some studies now performed with cardiac or respiratory gating (81) to minimize the effects of blurring due to physiological motion. Most microCT systems share operational elements analogous to those found in a clinical CT scanner. Radiation is provided by a microfocus X-ray tube operating with an anode potential in the range of 50 to 90 kVp and with a cathode current of approximately 1mA. The radiation from the X-ray tube passes through the anesthetized animal and is captured using a high-resolution X-ray detector. Common X-ray detectors in microCT (82,83) include a charge-coupled device (CCD), complementary metal oxide semiconductor (CMOS) cameras, amorphous silicon detectors that are used for indirect read-out of the optical signal produced when the X-ray signal is captured by a thallium-activated cesium iodide [CsI(Tl)] or gadolinium oxysulfide (GOS) scintillators, or direct X-ray detectors such as amorphous selenium (82).

Clinical CT systems used in medical practice generally have fan-beam geometries that irradiate a relatively small section of the patient at one time, with the resulting X-ray projection data reconstructed using filtered backprojection reconstruction algorithms. Most small animal CT scanners use a cone-beam X-ray geometry to expose a proportionally larger region of the animal (4–5 cm) and use an area X-ray detector to record the X-ray transmission data as a two-dimensional image from which 256 or 512 slices can be reconstructed simultaneously (84). The use of the cone beam geometry complicates the process of tomographic reconstruction and requires reconstruction algorithms that are specifically designed for this configuration. The most common algorithm for reconstructing cone-beam data from X-ray CT is an analytic technique called the Feldkamp algorithm, which uses approximations to account for the divergent geometry of volumetric CT imaging in the reconstruction method (85). In addition, the cone-beam X-ray geometry radiates a volume that is widest at the middle and tapered at the ends, and that produces a CT reconstruction volume in which the central slices have the largest diameter, while those on the ends have successively smaller diameters (84). Nevertheless, this technique has demonstrated its usefulness for recording and reconstructing volumetric CT data, especially for small animal imaging applications where the reconstruction volume is relatively small. Small animal CT scanners are now available from several manufacturers, and provide spatial resolutions in the range of 33 to 50 μm with scan times of 1 to 30 minutes (27).

Cone-beam CT scanners are sensitive to perturbations due to scattered radiation, and also are subject to the "cone-beam artifacts" mentioned above, which arise due to the divergence of the X-ray beam. For these reasons, other small animal CT scanners are designed with a "fan-beam" geometry (similar to those used in clinical CT scanners) that minimizes the effect of X-ray scatter but provides only a few image sections in a single rotation of the detector around these objects (86). These "fan beam" systems typically adopt linear silicon diode arrays coupled with a cadmium tungstate (CdWO$_4$) or other phosphor material (27) similar to X-ray detectors used in clinical CT scanners. The object to be imaged (i.e., the animal) generally is supported on a small platform between the source and detector, both of which are rotated around the animal to record the X-ray projection data. Typical acquisition geometries rotate the detector from 180° to 360° around the animal while acquiring several hundred projection data sets. The X-ray projection data are then transferred to a host computer for processing and tomographic reconstruction.

One significant challenge of small animal CT imaging is the need to maintain the radiation dose at a level where it has minimal effect on the biology or life of the animal (27). Changes in gene expression are seen with whole-body irradiation levels in the range of 5 to 6 cGy; the lethal dose for small animals is in the range of 5 to 6 Gray (27). This is a significant challenge for small animal scanners. At present the best method to reduce radiation dose is to use highly efficient X-ray detectors, but this benefit is offset by the increased radiation dose needed to visualize structures at higher spatial resolutions. The effect of increasing spatial resolution can be fairly dramatic since the radiation dose is approximately proportional to the third power of the spatial resolution level (27). For example, by increasing the spatial resolution of the system from 50 to 25 μm, the radiation dose needed to form the image would be increased by a factor of 2^3 or eightfold (87).

Finally, CT can reveal structural changes in normal anatomy, but otherwise does not intrinsically offer sufficient soft-tissue contrast to visualize functional or molecular changes in diseased or ischemic tissue. The vascular contrast media used in clinical studies demonstrate rapid clearance from the circulatory system over times (from seconds to a few minutes) comparable with the acquisition speed of modern volumetric CT studies. However, the longer acquisition time of small animal CT requires contrast agents with slower clearance to maintain opacification over the times (5–20 minutes) needed to complete the microCT study. One approach uses an intraperitoneal (i.p.) injection of a nonionic water-soluble iodinated contrast medium (e.g., Nycomed Omnipaque 300), which opacifies the peritoneal cavity to improve visualization of the liver, kidneys, intestines, and other abdominal structures (27). Other investigators have tested and used lipid-based contrast media that have slower clearance and longer retention in the rodent over times that are compatible for microCT imaging (88–91). Studies in rats used iodine concentrations of 25 to 50 mg/kg and demonstrated maximum opacification at two hours postinjection, and that remained elevated for an additional one to two hours, with contrast returning to baseline levels after 24 hours (89). Finally, new higher-speed microCT systems are now becoming available for use in small animal research studies. Some microCT manufacturers now offer upgrades to their small animal SPECT/CT system in which a full set of volumetric microCT data can be acquired with a 60-second acquisition, sufficiently fast to capture the opacification introduced by vascular iodinated contrast media. GE Healthcare, Inc., has modified a clinical multislice spiral scanning with a high-resolution detector and X-ray source to produce volumetric microCT with an acquisition of a few seconds, but has not yet incorporated this high-end capability into a small animal dual-modality imaging system.

IMPLEMENTATION OF SMALL ANIMAL SPECT/CT SYSTEMS

Design of Small Animal SPECT/CT Systems

Just as conventional imaging techniques are being adapted as single modalities (i.e., SPECT, PET, CT, MRI, and ultrasound) for experimental studies involving small animals, dual-modality imaging methods (i.e., PET/CT, SPECT/CT) have been developed to facilitate biological research involving small animals such as mice and rats. The clinical implementation and use of dual-modality imaging systems began with the commercial introduction of SPECT/CT in 1999 and of PET/CT in 2000. Historically, the development of dual-modality imaging also coincided with the development of high-resolution nuclear imaging and CT for small animal studies. It is therefore not surprising that researchers recognized the potential role of dual-modality methods for biological research and soon began to design dual-modality systems for imaging small animals. These developments have occurred both at academic laboratories and by corporate research and development groups (20,33,35,71,92–99). Both SPECT/CT and PET/CT are now available for small animal imaging, and provide several potential advantages over radionuclide or X-ray imaging performed alone (93). It is also not surprising that small animal dual-modality systems share many features with those designed for clinical use. A small animal dual-modality system includes a radionuclide imaging subsystem, a CT subsystem, a mechanical gantry to support and rotate the imaging components around the animal, and a platform to support the animal during imaging. The system also includes one or more computers for system control and for acquisition, reconstruction, display, processing, and analysis of the

Digital x-ray image   **^{99m}Tc-MDP image**   **Fused image**

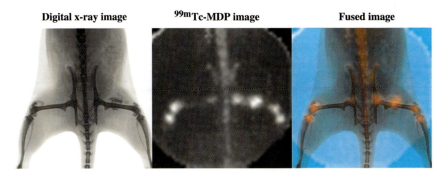

Figure 3 Superposition of digital X-ray radiography (*left*) and radionuclide ^{99m}Tc-MDP bone scan (*middle*) resulting in a fused image of the rear portion of a rat (*right*). *Source*: Courtesy of Mark B. Williams, University of Virginia. From Ref. 77.

dual-modality image data. While these elements are similar to those in clinical imaging systems, the mechanical subsystems obviously must be reduced in size to accommodate the geometries needed for imaging small animals rather than humans.

Perhaps the first small combined X-ray and radionuclide imaging system was developed by a collaboration among the Thomas Jefferson National Accelerator Facility, the University of Virginia, and the College of William and Mary (94–96). This system used a compact scintillation camera developed at Jefferson Laboratory configured with multiple Hamamatsu R3292 PSPMTs coupled to a pixelated array of cesium iodide [CsI(Tl)] crystals and was operated using both pinhole and parallel-hole collimators. The X-ray data are acquired using a small fluoroscopic X-ray system (Lixi, Inc., Downers Grove, Illinois, U.S.A.). Representative planar dual-modality images acquired with this dual-modality system from a 250-g rat injected with ^{99m}Tc-MDP (Fig. 3) demonstrated radiopharmaceutical uptake in the bone superimposed with an anatomical X-ray image of the rat (95,100).

Several other academic investigators also have or are developing SPECT/CT systems designed specifically for small animal imaging. Kastis and Furenlid at the University of Arizona (71) have developed a SPECT/CT system that rotates the animal in front of a high-performance 64×64 pixelated CZT detector (380 μm detector pitch) with a parallel hole tungsten collimator and a CCD X-ray detector with a gadolinium oxysulfide (GOS) phosphor. Other groups, including those at Johns Hopkins University (101), the University of California, San Francisco (35,97,98,102), and the above-mentioned consortium of the University of Virginia, the Research Center Jülich (103), Thomas Jefferson Laboratory, and College of William and Mary (94–96), are developing small animal SPECT/CT systems in which compact scintillation cameras with pinhole collimators and a microCT system are rotated around a stationary animal. These systems mimic the geometry of clinical SPECT/CT systems in which the object is maintained along a stationary horizontal axis with separate SPECT and CT subsystems to acquire the radionuclide and X-ray data, respectively.

Commercial Small Animal SPECT/CT Systems

Significant development in small animal dual-modality imaging has occurred over recent years and small animal SPECT/CT systems are now available from three companies (Fig. 4). It is likely that other companies are also developing small animal SPECT/CT systems which will be introduced soon to the marketplace.

The X-SPECT™ system (Gamma Medica Ideas Inc., Northridge, California, U.S.A.) is a small animal SPECT/CT system with two compact scintillation cameras and a high-resolution CT subsystem for dual-modality imaging of mice, rats, and other small animals (98,99,104,105). The scintillation camera can be operated with pinhole collimators that provide submillimeter spatial resolution in the reconstructed images or with parallel-hole collimators when higher detection sensitivity or whole-body imaging is desired. The system includes a high-resolution microCT subsystem implemented with a complementary metal oxide semiconductor (CMOS) X-ray

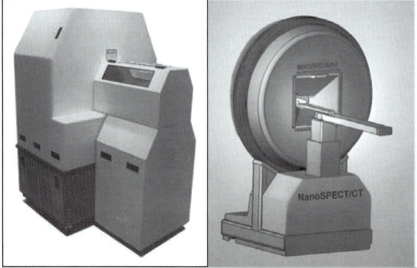

Figure 4 Small animal SPECT/CT imaging systems. (*Upper*) Gamma Medical X-SPECT™ system has a dual-headed compact SPECT system. (*Lower left*) "MicroCAT®+SPECT" system with integrated dual-headed SPECT detectors. (*Lower right*) NanoSPECT™ system is equipped with up to four SPECT detectors. *Abbreviations*: CT, computed tomography; SPECT, single-photon emission computed tomography. *Source*: From Gamma Medica-IDEAS, Inc. (Northridge, California, U.S.A.), Siemens Preclinical Solutions, Inc. (Knoxville, Tennessee) and Bioscan, Inc. (Washington, D.C., U.S.A.).

detector coupled to a GOS scintillator and a low-power X-ray tube (99). The microCT system provides anatomical imaging with a spatial resolution of approximately 50 µm; the resulting X-ray data can be used both for attenuation correction and for anatomical localization of the radionuclide data. In addition, the radionuclide data can be acquired with ECG-gating for cardiovascular imaging applications when visualization of wall-motion abnormalities, ejection fraction calculations, or other assessments of ventricular function are necessary. A coincidence detector ring can be incorporated into the X-SPECT gantry to perform high-resolution PET of small animals. The system is available with microCT system only, microSPECT only, microPET only, or a combination of any two or all three systems in a single gantry.

A small animal SPECT/CT system has been developed by Imtek, Incorporated, now part of Siemens Preclinical Solutions, Inc. (Knoxville, Tennessee, U.S.A.). The system incorporates the Siemens microCAT® II high-resolution X-ray CT system for laboratory animal imaging. The microCAT II performs volumetric CT imaging with an adjustable field of view suitable for whole-animal studies, or with smaller fields of view with resolution as high as 15 µm. The microCAT II can be integrated with up to two 150×150 mm^2 SPECT detectors configured with PSPMTs and coupled to a 100×100 array of NaI(Tl) crystals, each having a pixel spacing of 1.5×1.5 mm^2. The system has an animal bed that can be interchanged with the microPET system of the same manufacturer, and has fiducial markers that are visible in CT, PET, MRI, and SPECT. These fiducial markers facilitate the use of image registration and analysis software suitable for spatial registration and analysis of multimodality image data.

A combined SPECT/CT system called "NanoSPECT^{TM}" (Bioscan Nano-SPECT^{TM} System, Washington, D.C., U.S.A.) is also available for small animal imaging (103). The microSPECT system is based on the multipinhole imaging technique developed by Schramm et al. (45,58). The system can include up to four rectangular SPECT cameras with multipinhole collimators to record tomographic data that are reconstructed with an iterative OSEM reconstruction algorithm. The radionuclide detectors are configured as a scintillation camera with NaI(Tl) scintillation crystal, each having a field of view of 230×215 mm^2. The cameras can be operated with multipinhole collimators having 7 to 10 pinholes per detector, offering spatial resolutions in the range of 0.8 to 1.2 mm FWHM and sensitivities in the range of 400 to 2000 cps/MBq. The system can also be integrated with a cone-beam CT system with a rectangular detector having an area of 204.8×102.4 mm^2 that offers a spatial resolution better than 200 µm FWHM. Both the SPECT detectors as well as the CT subsystem are integrated on a common gantry with an animal bed (available in two sizes for mice and rats) mounted on a horizontal translation stage to enable helical scanning of the animal.

PRACTICAL ISSUES ASSOCIATED WITH SMALL ANIMAL SPECT/CT IMAGING

For a laboratory interested in small animal imaging, acquiring the imaging system is only one of several steps needed to develop the environment and resources for small animal imaging studies. There are significant regulatory issues associated with the handling of small animals, and additional requirements that must be met for handling radionuclides. In the United States, all institutions that use animals for research must establish an "Institutional Animal Care and Use Committee" (IACUC) to oversee all aspects of the institution's animal care and use program;

similar programs exist in other countries. There are also practical factors related to small animal imaging, many of which extend beyond the requirements needed for clinical imaging studies. The special requirements for contrast media used in small animal versus clinical imaging studies have been discussed earlier in this chapter. The animal must also be anesthetized for virtually all imaging studies. In addition, even though experiments involving small animals can include serial imaging procedures over days or weeks, the animal eventually is almost always euthanized. This requires that the researcher must adopt procedures that allow the animal to be euthanized humanely without causing pain (106). Finally, the researcher must be concerned about issues involving disposal of biological tissues; this is especially a concern when the tissues contain long-lived radionuclides (e.g., ^{125}I having a physical half-life of 60 days).

Small animal imaging almost always must be performed after the animal has undergone general anesthesia to prevent motion during the study, as well as to promote ethical and humane handling of the animal (40). The anesthetic agent and protocol must be selected so that anesthesia is maintained during the entire imaging study but does not perturb the animal's physiological status so that the resultant data truly represent the biological process or disease state of interest (107). The choice of anesthesia must take into account issues related to limited animal access, small sample volumes, anesthetic complications, cardiovascular status, strain and gender variability, and the possible introduction of image artifacts (108). Because each imaging study presents unique challenges, a thorough understanding of the imaging modality used, the animal's health status, and the research data desired are required. Anesthesia that is not carefully administered and monitored can jeopardize both the validity of the imaging and the health and survival of the animal. Anesthesia can also affect temperature regulation, and the anesthetized animal can rapidly lose heat and suffer hypothermia unless an external heat source from a heat lamp or heating pad is applied (40).

The proper use of anesthesia eliminates the possibility of voluntary motion and motion-induced blurring of the image data during the imaging procedure. For studies that take longer than a few seconds, respiration must be maintained either naturally or by external mechanical ventilation. Furthermore, cardiac motion also naturally occurs as a function of the cardiovascular system. For these reasons, motion-induced blurring of the heart, of pulmonary structures or of structures near the diaphragm may occur when the animal is anesthetized. In some cases, excellent images can be acquired even when respiratory and cardiac motion occurs; for example, myocardial perfusion data can be obtained with SPECT studies that extend over several minutes from a mouse having a heart rate of several hundred beats per minute. When the highest degree of spatial resolution is required to achieve the goals of the experiment, the imaging study can be performed with physiological gating to account for respiratory or cardiac motion. For example, Vanhove et al. (109) have used pinhole SPECT to acquire gated myocardial perfusion images of normal male rats administered with 439 ± 52 MBq of ^{99m}Tc-sestamibi. Similarly, blood pool SPECT images were obtained with 520 ± 49 MBq in 300 mL of ^{99m}Tc-labeled red blood cells. To acquire the radionuclide data with maximum sensitivity, both the myocardial perfusion and blood pool images were acquired with a 3 mm diameter pinhole with a single-headed SPECT camera and with heart rates ranging from 260 to 420 beats per minute (46).

Similarly, other investigators have performed studies using microCT with respiratory gating (81,110). One possible approach uses a high-power X-ray tube with a large focal spot that produces a photon fluence rate measured at the detector approximately 250 times higher than that produced by a conventional microCT system (81). Data can be synchronized with respiratory and cardiac gating using a 10 millisecond frame to minimize motion blurring to compensate for both cardiac and respiratory motion. Respiratory-gated microCT images have been acquired by Cody et al. (111) at 100 millisecond per frame, 5 frames per view, with 360 views over 360° at 80 kVp and 450 μA. A 20-minute acquisition provided contiguous slices with voxel dimensions of $(91 \, \mu m)^3$.

RADIONUCLIDE QUANTIFICATION WITH SMALL ANIMAL SPECT/CT

An area that still is undergoing development is the use of quantitative, rather than qualitative, methods of interpreting small animal imaging data (78). The importance of quantification of radionuclide data from these experimental studies arises from several important motivations. First, animal research studies are performed by biologists and other scientists who prefer quantitative methods of recording and analyzing data in testing hypotheses and establishing correlations. This is in comparison to clinical studies where visual or qualitative interpretation often provides sufficient information to reach a diagnostic decision and a course of action for patient management. Second, quantitative measurements are important because small animal radionuclide imaging is affected by physical effects similar to those that affect image quality and quantitative accuracy in radionuclide studies involving humans. Several investigators are exploring methods to improve the quantitative accuracy of radionuclide data acquired with small animal SPECT.

Correction for photon attenuation is often the first process that draws attention from investigators interested in quantitative radionuclide measurements. Dual-modality imaging with SPECT/CT facilitates this process by providing a correlated X-ray CT image, which represents the spatial distribution of linear attenuation coefficients at the effective photon energy of the X-ray beam. In most cases, the effective photon energy of the X-ray beam is different than that of the radionuclide imaged in the SPECT study. However, the CT data can be converted to a corresponding map of linear attenuation coefficients calibrated for the photon energy of the radionuclide. Hwang et al. implemented a calibration technique by developing a phantom that contains small cylindrical rods having known material properties from which the linear attenuation coefficient could be derived (102). By imaging the calibration phantom with microCT at the same X-ray energy used to acquire the small animal CT data, the authors produced a calibration curve that related the measured CT number for each calibration material versus its linear attenuation coefficient calculated for the photon energy of the radionuclide used to acquire the SPECT image. This calibration curve then is available to convert the pixel values in the CT image to corresponding values of the linear attenuation coefficient for each point in the reconstruction volume. The resulting attenuation map can be incorporated into an iterative algorithm to reconstruct the cone-beam SPECT data with corrections for photon attenuation (102).

It can be argued that perturbations arising from photon attenuation will be relatively inconsequential in small animal imaging due to the limited size of the objects encountered in these studies. For example, the 140 keV gamma rays from ^{99m}technetium suffer 26.5% attenuation across 2 cm of tissue (the approximate diameter of a mouse) compared with 95.4% attenuation across 20 cm of tissue (the approximate anteroposterior thickness of the human torso). However, small animal imaging can be performed with low-energy radionuclides, which are impractical for human imaging. For example, small animal imaging studies can be acquired with low-energy isotopes such as ^{125}I-iodine, which emit photons having energies in the range of 27 to 35 keV. Correspondingly, the 27 keV X-rays (113% emission) from ^{125}I-iodine suffer 57.7% attenuation across 2 cm of tissue. (In comparison, the 27 keV X-rays from ^{125}I-iodine undergo 99.98% attenuation across 20 cm of tissue, indicating that they are completely absorbed and worthless in human imaging.) Hwang et al. performed phantom studies with a 2.5-cm diameter cylindrical container filled with a uniform aqueous solution of ^{125}I-iodine, and demonstrated a 30% "cupping" nonuniformity artifact in the SPECT image reconstructed without attenuation correction (102). They also demonstrated that this nonuniformity artifact can be resolved when X-ray-based attenuation correction for ^{125}I-iodine is implemented with small animal SPECT/CT using the approach described earlier. These results indicate that small animal SPECT/CT can improve the quantitative accuracy and potentially the visual quality of the image data acquired with the small animal imaging system in comparison to small animal radionuclide imaging alone.

Another major source of quantification error arises from partial volume effects contributed by the limited spatial resolution of the radionuclide imaging system. Partial volume errors occur when the radionuclide content of a target region is blurred by the imaging process into the background (an effect known as "spill-out") and when the radionuclide content in the background is blurred into the target region (an effect known as "spill-in"). Partial volume errors may occur for structures that are smaller than twice the size of the spatial resolution of the imaging system (112) measured in terms of its FWHM. Correction techniques have not been thoroughly investigated or developed for microSPECT or microSPECT/CT. For simple geometries such as spheres and cylinders, the partial volume effect can be estimated mathematically to derive recovery coefficients to correct the radionuclide data for these errors. The error-correction process must also account for the size and shape of more complex target and background structures to improve the quantitative accuracy of the radionuclide data, which in concept can be derived from the correlated CT data.

APPLICATIONS OF SMALL ANIMAL SPECT AND SPECT/CT IMAGING

As noted earlier, experimental research involving laboratory animals traditionally relies on invasive assessments in which the animal is sacrificed to obtain sample tissues needed to complete the experiment. Common methods of assessing tissue from animals include techniques such as autoradiography, organ sampling, tissue counting, and histology. Small animal imaging offers significant advantages as it allows measurements to be performed noninvasively without sacrificing the animal (113). Radionuclide imaging can be used for biodistribution studies to assess the temporal characteristics of biological uptake of radiolabeled molecules (21). For example, a typical biodistribution study using autoradiography or tissue counting might involve five time-points with four to six animals each, for a total of 20 to 30 animals. In comparison, radionuclide imaging would allow the biodistribution measurement to be performed at multiple time-points in individual animals. The availability of serial imaging on an individual animal also is important for monitoring the effects of therapeutic intervention or following the progression of disease.

Pharmaceutical and Biotechnology Research

There is significant need for improving the methodology and approach used to test and develop new drugs and pharmacological agents. Before clinical trials can be initiated, the agent must be tested and evaluated in animal models. Much of this initial work now is pursued with mice as a necessary step for testing safety and toxicity and to evaluate efficacy in a mammalian species. In addition, studies with small animals allow genetic assays at a relatively low cost (114). Furthermore, transgenic technologies now allow animal models to be developed with specific receptors and targets that, for example, may indicate that the target is rate-limiting for disease development but not essential for normal physiological function (115). Transgenic technologies can also produce animal models that contain the human target or that are better in mimicking specific human pathologies to demonstrate that a potential therapy has a desired effect in vivo, while allowing studies that investigate whether the potential drug has adverse or toxicological effects on the organism's metabolism or function.

Cardiovascular Research

Over the past 40 years, nuclear medicine has served an essential role in research investigations of the cardiovascular system, and especially for assessment of functional parameters such as myocardial metabolism, myocardial perfusion, ejection fraction, cardiac output, adrenergic innervation, and lipid metabolism (109,116–118). Traditionally, these studies have been performed with dogs, pigs, and other large animals, which represent excellent models of human anatomy and function, and where the size of the heart and other cardiovascular structures make these species especially accessible for conventional radionuclide imaging. However, the use of large mammalian species in research is expensive, and has increasing regulatory burden. For these reasons, cardiovascular research increasingly is being performed with small animal imaging (microSPECT, microPET, high-resolution ultrasound, and high-resolution magnetic resonance imaging) of mice and rats, including animal lines that have undergone genetic manipulations to express specific phenotypes.

One important area where microSPECT has established a key role is in the development of new tracers to assess cardiovascular and myocardial function. For example, Liu et al. have used a custom high-resolution microSPECT system (FASTSPECT) at the University of Arizona to image ^{99m}Tc-glucarate in a rat model of varying levels of myocardial ischemia with reperfusion (47). The superb detection efficiency and spatial resolution of this system allows dynamic in vivo imaging of the ^{99m}Tc-glucarate (an infarct avid agent) in a way that allows detection and quantification of acute necrotic myocardium. Boschi et al. used small animal imaging for in vivo assessments of a novel class of monocationic ^{99m}Tc-nitrile heterocomplexes for myocardial imaging (119). Biodistribution studies on rats were performed with in vivo high-resolution SPECT and demonstrated rapid clearance from the blood and lung, virtually negligible retention in the liver, and both fast extraction and high retention in the myocardium with heart-to-lung and heart-to-liver ratios at levels much higher than those measured for ^{99m}Tc-sestamibi and ^{99m}Tc-tetrofosmin.

MicroSPECT also is becoming an important tool to discern the molecular processes that dictate biological processes in vascular and myocardial disease. For example, cell proliferation and DNA synthesis can be followed with radiolabeled thymidine analogs such as 5-iodo-2'-deoxy-uridine that can be labeled with ^{125}I, ^{123}I, ^{124}I, or ^{131}I

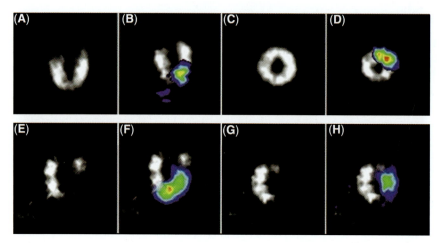

Figure 5 SPECT study of rat myocardium showing retention of ^{111}In-labeled stem cells in simultaneous dual-isotope SPECT study of myocardial perfusion with ^{99m}Tc-sestamibi. Cardiac long- and short-axis SPECT images of normal (**A** and **C**), and infracted (**E** and **G**) rat myocardium using perfusion tracer ^{99m}Tc-sestamibi. [111]In signal (color) was overlaid on gray-scale ^{99m}Tc-sestamibi images for normal (**B** and **D**) and infracted (**F** and **H**) myocardium. *Abbreviation*: SPECT, single-photon emission computed tomography. *Source*: From Ref. 128.

and 5-bromo-2′-deoxy-uridine that can be labeled with [76]Br (120). Other investigators have used radiolabeled Annexin-V to image apoptosis (programmed cell death) in vivo (121), which may be useful to track rejection of cardiac transplants as well as to evaluate and delineate the extent of infarction for patients undergoing reperfusion therapy (122,123). Work by Schäfers et al. (124) and Kopka et al. (125) has shown that in vivo scintigraphic imaging of metalloproteinases can be performed using a radiolabeled MMP-inhibitor for assessment of vascular remodeling, inflammation, or atherosclerosis in the myocardium and other tissues, and may eventually lead to strategies to assess vascular cell proliferation associated with postangioplasty restenosis or pulmonary hypertension (120). Other investigators are starting to image myocardial angiogenesis using ^{111}In-labeled $VEGF_{121}$ in skeletal muscle (126) and ^{111}In-labeled quinolone for targeting the $\alpha_v\beta_3$ integrin in the myocardium (127). These studies suggest that it may be possible to identify ischemic tissues or tissues where there are sites of angiogenesis that would augment clinical evaluations of flow and perfusion, and for evaluation of angiogenic response after injury or applied therapies that stimulate angiogenesis (120). MicroSPECT can also assess functional and molecular interactions such as perfusion or angiogenesis while following myocardial repair mediated by contemporary therapeutic approaches (Fig. 5) such as stem cell engraftment (128). It is virtually certain that both the basic biological mechanisms as well as the preclinical evaluations of diagnostic and therapeutic approaches will occur first in small animal models before they are advanced to clinical application.

Cancer Research

To date, cancer research is perhaps the one area of biomedical research where small animal imaging has had the biggest impact. In this context, small animal imaging is being used to answer questions of tumorigenesis, tumor progression, and metastasis, to advance our understanding of cancer biology, and to develop and evaluate new diagnostic and therapeutic approaches (Fig. 6) (130). Advances in molecular biology and genetic engineering have produced a wide spectrum of transgenic and knock-out mice which recapitulate a more natural expression of human cancer in an intact murine model. The noninvasive nature of in vivo imaging allows the researcher to measure these parameters at multiple time-points in a single animal, and in a way that allows disease progression or therapeutic response to be measured without sacrificing the animal. Many of the same methods described above for cardiovascular research also can be used for cancer. For example, radiolabeled Annexin V can be used to detect apoptosis (131) as a means of following the response of a tumor to chemotherapy, while radiolabeled peptides having an affinity for the $\alpha_v\beta_3$ integrin can be used to assess tumoral angiogenesis (132,133) or for the vascular endothelial growth factor (VEGF) using SPECT or PET (134). The response of the tumor to anticancer therapeutics also can, in principle, be monitored by imaging tumor-specific markers such as the HER2 antibody Herceptin (135). Molecular imaging strategies are also being

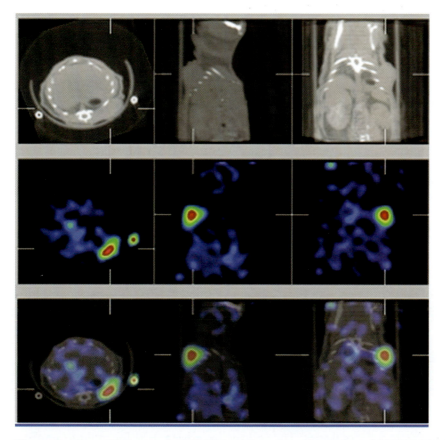

Figure 6 X-ray CT (*top row*), SPECT (*middle row*), and coregistered SPECT/CT (*bottom row*) obtained with small animal dual-modality system (X-SPECT, Gamma Medica, Northridge, California, U.S.A.) of colon carcinoma xenograft in right shoulder of mouse. SPECT images recorded 23 hours after administration of ^{123}I-labeled anti-CEA minibody that localizes in colon carcinoma. Images shown in transverse (*left column*), sagital (*middle column*), and coronal (*right column*) tomograms. *Abbreviations*: CT, computed tomography; SPECT, single-photon emission computed tomography. *Source*: Courtesy of AM Wu, UCLA. From Ref. 129.

developed to assess tumor growth and the tumor microenvironment using radiolabeled probes that target endogenous tumor cell receptors (136).

Small animal radionuclide imaging, including both microPET and microSPECT, can be used as a noninvasive investigational tool for studying gene expression in living organisms (137). This can be performed by radiolabeling probes that can be used in combination with the HSV1-tk reporter gene; noninvasive and quantitative imaging of gene expression can facilitate human gene therapy trials and assess animal modules of molecular and cellular therapy. Other investigators are assessing the mechanisms involved in multidrug resistance of chemotherapeutic drugs, and monitoring the effect of potential blocking agents in this mechanism (138). In oncology, antiangiogenic therapies are now being developed and tested in transgenic animals in a way that demonstrates their effect at different stages of multistage carcinogenesis (139).

Finally, small animal CT can be used to visualize the location and measure the size of tumors as small as 100 μm without respiratory gating (28). Serial CT measurements are also useful to follow the growth of tumors and to monitor tumor regression in response to therapeutic interventions (28). These studies illustrate the appeal of small animal imaging for noninvasive in vivo measurements of the molecular and genetic basis of disease in a way that could have immediate impact on our ability to follow disease and therapeutic efficacy in a wide area of cancer research.

Neurological Research

In vivo imaging of small animal models of neurological disease is complicated by the small size (<1 cm) of the mouse brain (140). For these studies, the spatial resolution performance of microSPECT must be honed to the submillimeter level needed to study murine models of several neurological disorders (141). Measurements of cerebral blood flow analogous to those used clinically (142) may be adapted for use in experimental studies using small animals. High-resolution imaging of the neurological systems of small animals is also being used to study the underlying mechanisms of neurological disorders such as depression, epilepsy (143), and schizophrenia, and to explore potential new treatments for these diseases (144). These therapeutic approaches could be used in combination with diagnostic radiolabeled agents for beta-amyloid plaques and neurofibrillary tangles that are being developed for detection and diagnosis of Alzheimer's disease (145). Saji et al. have used microSPECT imaging of the marmoset brain with ^{123}I-beta-CIT and ^{123}I-IBZM in marmosets treated with MPTP to induce symptoms mimicking Parkinson's disease (146).

FUTURE OF SMALL ANIMAL SPECT/CT IMAGING

At the time when this chapter was being prepared (i.e., October 2005), there were roughly 40 small animal SPECT/CT systems installed worldwide with approximately half installed at academic settings and half at corporate pharmaceutical and biotechnology laboratories. The systems have been used for multiple applications, including those discussed in the previous section. As is common for medical imaging technologies, small animal SPECT/CT systems experienced an initial rapid phase of development that is now being followed by a prolonged period of system refinement. In this second phase, work is being focused on improving the spatial resolution and detection efficiency of SPECT using novel collimator techniques including those having multipinhole geometries. These refinements are needed so that the imaging study can be performed with small amounts of the radiopharmaceutical, to reduce both the study time and the radiation dose delivered to the animal. A second important direction is focused on the development of in vivo methods to quantify radiopharmaceutical uptake using SPECT/CT. The primary focus of these techniques is on partial volume corrections to fix quantification errors introduced by the limited spatial resolution of the collimator and radionuclide imaging system. Investigators are also pursuing methods to correct photon attenuation and scattered radiation, which are significant sources of error in clinical imaging but may be of secondary importance for small animal radionuclide imaging.

Need for Combined Imaging Systems

An important question is raised by investigators who ask whether integrated SPECT/CT or dual-modality imaging systems are needed for experimental measurements of small animals. These investigators note that while dual-modality imaging systems in the form of PET/CT and SPECT/CT are being adopted rapidly in clinical imaging, the adoption of dual-modality systems specifically for small animal imaging is progressing at a slower pace. This is likely due to two important reasons: (i) small animal imaging still is a young field and (ii) this field currently represents a significantly smaller capital equipment market than clinical applications of medical imaging. Furthermore, the need for dual-modality imaging may be relaxed since the animal is anesthetized during the imaging study, unlike clinical imaging where the patient is awake and alert. Therefore, it can be argued that dual-modality imaging can be performed simply by moving the anesthetized animal on its bed between imaging systems followed by registration of the dual-modality image data in software. However, it is likely that small animal dual-modality imaging systems will continue to be developed and utilized in both the academic and corporate settings. This is especially true for small animal imaging studies that use highly specific radiopharmaceuticals, which need a correlated anatomical image to provide information about localization and uptake. In addition, the anatomical information is important in small animal imaging to correct the radionuclide data both for photon attenuation and partial volume errors. A third important direction is represented by the development of protocols and techniques beyond those that are available in current systems to facilitate biological measurements using small animal imaging. These include the development and routine use of improved physiological gating methods to reduce the effects of physiological motion in the image data. Overall, it is likely that the featured SPECT/CT systems will continue to be refined and advanced with the goal of obtaining higher spatial resolution and better detection efficiency so that studies can be performed with lower administered radioactivity and reduced radiation dose, as well as improved capability for quantitative measurements derived from the SPECT data.

Relationship of Small Animal SPECT/CT with Respect to Small Animal PET/CT Imaging

In parallel with the development of SPECT/CT, other technologies have been developed and are being refined for small animal imaging. These include single-modality systems for high-resolution ultrasound, microPET, microSPECT, microCT, and optical imaging specifically designed for small animal studies. Combined PET/CT systems designed specifically for small animal imaging (93) share attributes with small animal SPECT/CT in that they offer structural–functional correlation and improved quantification in comparison to radionuclide imaging alone.

There are important differences between SPECT/CT and PET/CT. Specifically, PET/CT is performed with positron emitting radiopharmaceuticals such as ^{18}F-fluorodeoxyglucose, and with other positron emitters such as ^{13}N,^{15}O, and ^{11}C. However, other than ^{18}F, most positron-emitting radionuclides have relatively short half-lives, which require them to be synthesized on-site using a medical cyclotron. PET has a significant advantage of improved detection efficiency, especially when compared with small animal SPECT performed with a single pinhole collimator. The spatial resolution of microPET is limited fundamentally by the finite range of the positron following emission and the noncolinearity of the emitted gamma rays following positron annihilation. The spatial resolution of the microPET is limited to the range of 1 to 2 mm, although submillimeter spatial resolution may be possible by optimizing the microPET system (147). In comparison, microSPECT does not share the same limitations in spatial resolution and may be capable of achieving spatial resolutions in the range of a few hundred microns (76). However, as discussed earlier, the superb spatial resolution of microSPECT is achieved at the cost of decreased detection efficiency, and the ultimate level of spatial resolution with a practical level of detection efficiency is still unknown. MicroSPECT also is compatible with a wide range of radiopharmaceuticals including those that are available commercially from central radiopharmacies for clinical studies. Much of clinical nuclear medicine is performed with radiopharmaceuticals labeled with ^{99m}Tc, which are very suitable for small animal imaging studies. Other single-photon radionuclides available for small animal SPECT include ^{201}Tl,^{123}I,^{111}In, and ^{131}I, which also are used for clinical imaging. The small size of mice and rats is compatible for imaging with ^{125}I, a common radionuclide for in vitro measurements that serves as an experimental analog for ^{123}I or ^{131}I-labeled radiopharmaceuticals being assessed for clinical applications in humans. SPECT is compatible with long-lived radionuclides such as ^{111}In for labeling antibodies being developed for diagnostic and therapeutic targeting of tumors. The long effective half-life of ^{111}In is suitable for labeling antibodies and other pharmaceuticals that can take from hours to days to disperse from background structures or to accumulate in the target. It is therefore possible to image the ^{111}In-labeled radiopharmaceutical several days after administration, when the target contrast is maximized. Finally, microSPECT can perform dual-isotope imaging, for example, as shown in Figure 5 (128), where ^{99m}Tc-sestamibi is used to assess myocardial perfusion that is imaged simultaneously with stem cells labeled with ^{111}In to monitor stem cell engraftment. Dual-isotope imaging is possible in both small animal studies with microSPECT, as well as clinical studies on humans with SPECT. For these reasons, small animal SPECT is likely to continue to be an important tool for biological studies involving small animals, especially with the higher efficiency and improved spatial resolution afforded by multipinhole imaging techniques.

Potential for Simultaneous SPECT/CT

To date, all commercially available dual-modality systems (i.e., both PET/CT and SPECT/CT), including those designed for small animal imaging, use separate detectors for the radionuclide (i.e., PET or SPECT) and the X-ray (i.e., CT) measurements. A few investigators have explored the possibility of developing a single detector technology that can perform both X-ray and radionuclide imaging simultaneously with the goals of reducing scan time, facilitating image registration, and simplifying system design (75,148). However, there are several significant challenges that make this approach difficult to implement. First, radionuclide imaging is performed best with the detector operated in a photon counting mode, where individual photons are recorded with sufficient energy resolution to discriminate primary photons from scattered events or background radiation. A detector designed for simultaneous X-ray and radionuclide imaging would also have to discriminate the radionuclide events from those emitted by the X-ray source. However, an X-ray tube easily can produce photons at a rate several orders of magnitude higher than that from the radionuclide source, and therefore at rates that far exceed the capability of the detectors used for radionuclide imaging. In comparison, X-ray detectors are almost always operated in the current integration mode, which does not provide any energy resolution and therefore cannot discriminate simultaneously arriving X-ray and radionuclide photons. An alternative approach would use a single detector with electronics that could be switched between photon counting mode and current integration mode. Prototype electronics of this type were developed by Heanue et al. (149) and by Boles et al. (150), which in theory could be used to operate a single solid-state detector (e.g., high-purity germanium or cadmium zinc telluride) in the counting mode for radionuclide imaging and current mode for X-ray imaging. Since the event rate from an X-ray tube is so much higher than that from the radionuclide source, the detector would be operated most of the time in the photon counting mode for radionuclide imaging and be switched quickly for X-ray imaging. This could be repeated for each angle of a tomographic acquisition as the detector and X-ray source are rotated around the animal. A third challenge is raised by differences in collimation used for X-ray and radionuclide imaging, since small animal SPECT is typically performed with pinhole collimation that is incompatible with small animal CT. This means that the radionuclide collimator would have to be inserted into the data path for acquiring the radionuclide data but removed from the imaging path for X-ray imaging. It is likely that these technical challenges can be overcome by careful design and clever engineering. However, they also are likely to compromise other aspects of imaging system performance, and may in toto offer marginal or negligible enhancement of system performance. For this reason, although a simultaneous SPECT/CT system with a single detector has conceptual appeal, it is the humble opinion of this author that this approach will remain challenging to implement or realize in a practical system in the foreseeable future.

CONCLUDING REMARKS

Dual-modality systems combine imaging modalities with the goal of producing information that is not available or is difficult to obtain from a single-imaging modality alone. Currently available dual-modality systems for small animal imaging include SPECT/CT and PET/CT, with which the radionuclide imaging technology (SPECT or PET) provides functional or metabolic information that is complementary to anatomical information provided by CT. For spatial and temporal correlation, the dual-modality data can be presented as a fused image in which the radionuclide data are displayed in color and are superposed on the gray-scale CT image. The resulting correlated data improves differentiation of foci of radionuclide uptake that can indicate disease from those benign foci representing normal physiological uptake. Dual-modality imaging has been available since the year 1999, and as such is a relatively recent development for both small animal studies and for clinical applications. However, the commercial emergence of both PET/CT and SPECT/CT has been rapid and has benefited significantly from recent technological advances in conventional SPECT, PET, and CT.

Newer high-resolution SPECT/CT and PET/CT systems are also becoming available for small animal imaging and are needed for molecular imaging, biological research, and pharmaceutical development in small animal models of human biology and disease. The technology of small animal dual-modality imaging will continue to advance in terms of its performance and capabilities. In the area of small animal SPECT/CT, these systems offer submillimeter spatial resolution but suffer reduced detection efficiency. Newer multipinhole SPECT systems are under development and offer both improved geometric efficiency and spatial resolution in comparison to conventional radionuclide imaging approaches. MicroCT systems with spatial resolution in the range of 50 mm and that allow cardiac and respiratory gating also are available. In addition, advances in computing power will enable the development and implementation of new anatomically guided statistical reconstruction algorithms and data processing techniques for both clinical and small animal dual-modality imaging. At present, dual-modality imaging is used primarily for structural–functional correlations, but likely will have an increasingly important role in radionuclide quantification, especially for biological studies involving small animals where the use and understanding of dual-modality imaging is just starting to emerge. The importance of these developments only will be understood and manifest over the ensuing years as PET/CT, SPECT/CT, and other forms of dual-modality imaging become available and are utilized for biological investigations involving animal models of biology and disease, as well as for clinical studies in humans.

ACKNOWLEDGMENTS

The author appreciates the encouragement and guidance of Drs. Ora Israel and Stanley Goldsmith in the preparation of this manuscript, and of Dr. Israel and her staff in editing the manuscript and preparing the bibliography for this chapter.

REFERENCES

1. Watson JD, Crick FH. Molecular structure of nuclei acids; a structure for deoxyribose nucleic acid. Nature 1953; 25:737–738.
2. Weissleder R, Mahmood U. Molecular imaging. Radiology 2001; 219:316–333.
3. Hanahan D. Transgenic mice as probes into complex systems. Science 1989; 246: 1265–1275.
4. Sigmund CD. Major approaches for generating and analyzing transgenic mice. Hypertension 1993; 22:599–607.
5. Wight DC, Wagner TE. Transgenic mice: a decade of progress in technology and research. Mutat Res 1994; 307:429–440.
6. James JF, Hewett TE, Robbins J. Cardiac physiology in transgenic mice. Circ Res 1998; 82:407–415.
7. Zhong R. Organ transplantation in mice: current status and future prospects. Microsurgery 1999; 19:52–55.
8. Schumacher M, Van Vliet BN, Ferrari P. Kidney transplantation in rats: an appraisal of surgical techniques and outcome. Microsurgery 2003; 23:387–394.
9. Corry RJ, Winn HJ, Russell PS. Heart transplantation in congenic strains of mice. Transplant Proc 1973; 5:733–735.

10. Skoskiewicz M, Chase C, Winn HJ, et al. Kidney transplants between mice of graded immunogenetic diversity. Transplant Proc 1973; 5:721–725.
11. Han WR, Murray-Segal LJ, Mottram PL. Modified technique for kidney transplantation in mice. Microsurgery 1999; 19(6):272–274.
12. Fisher B, Lee S. Microvascular surgical techniques in research, with special emphasis to rental transplantation in rat. Surgery 1965; 58:904–914.
13. Price DL, Tauzi RE. Alzheimer's disease: genetic studies and transgenic models. Annu Rev Genet 1998; 32:461–493.
14. Sedelis M, Schwarting MK, Huston JP. Behavioral phenotyping of the MPTP mouse model of Parkinson's disease. Behav Brain Res 2001; 125:109–125.
15. Hsieh-Li HM, Chang JH, Yong YJ, et al. A mouse model for spinal muscular atrophy. Nat Genet 2000; 24:66–70.
16. Wong PC, Cai H, Borchelt DR, et al. Genetically engineered mouse models of neurodegenerative diseases. Nat Neurosci 2002; 5:633–639.
17. Balmain A, Harris CC. Carcinogenesis in mouse and human cells: parallels and paradoxes. Carcinogenesis 2000; 21:371–377.
18. Chung WS, Cho C, Kim S, et al. Review of significant microvascular surgical breakthroughs involving the heart and lungs in rats. Microsurgery 1999; 19:71–77.
19. Paulus MJ, Gleason SS, Fasterly ME, et al. A review of high-resolution X-ray computed tomography and other imaging modalities for small animal research. Lab Animal (NY) 2001; 30:36–45.
20. Phelps ME. PET: the merging of biology and imaging into molecular imaging. J Nucl Med 2000; 41:661–681.
21. Wirrwar A, Schramm N, Vosberg H, et al. High resolution SPECT in small animal research. Rev Neurosci 2001; 12:187–193.
22. Wu MC, Gao DW, Sievers RE, et al. Pinhole single-photon emission computed tomography for myocardial perfusion imaging of mice. J Am Coll Cardiol 2003; 42:376–582
23. Jacobs RE, Ahreus ET, Meade TJ, et al. Looking deeper into vertebrate development. Trends Cell Biol 1999; 9:73–76.
24. Johnson GA, Benveniste H, Engelhardt RT, et al. Magnetic resonance microscopy in basic studies of brain structure and function. Ann NY Acad Sci 1997; 820:139–147.
25. Johnson GA, Cofer GP, Gewalt SL, et al. Morphologic phenotyping with MR microscopy the visible mouse. Radiology 2002; 222:789–793.
26. McDaniel B, Sheng H, Warner DS, et al. Tracking brain volume changes in C57BL/6J and ApoE-deficient mice in a model of neurodegeneration: a 5-week longitudinal micro-MRI study. Neuroimage 2001; 14:1244–1255.
27. Paulus MJ, Gleason SS, Kennel SJ, et al. High resolution X-ray computed tomography: an emerging tool for small animal cancer research. Neoplasia 2000, 2.62–70.
28. Kennel SJ, David IA. High resolution computed tomography and MRI for monitoring lung tumor growth in mice undergoing radioimmunotherapy: correlation with histology. Med Phys 2000; 27:1101–1107.
29. Foster FS, Pavlin CJ, Harasiewicz KA, et al. Advances in ultrasound biomicroscopy. Ultrasound Med Biol 2000; 26:1–27.
30. Weissleder R, Tung CH, Mahmood U, et al. In vivo imaging of tumors with protease-activated near-infrared florescent probes. Nat Biotechnol 1999; 17:375–378.
31. Contag CH, Jenkins D, Contag PR, et al. Use of reporter genes for optical measurements of neoplastic disease in vivo. Neoplasia 2000; 2:41–52.
32. Cherry SR, Shao Y, Silverman RW, et al. MicroPET: a high resolution PET scanner for imaging small animals. IEEE Trans Nucl Sci 1997; 44:1161–1166.
33. Cherry SR. In vivo molecular and genomic imaging: new challenges for imaging physics. Phys Med Biol 2004; 49:R13–R48.
34. Wu MC, Tang HR, O'Connell JW, et al. An ultra-high resolution ECG-gated myocardial imaging system for small animals. IEEE Trans Nucl Sci 1999; 46:1199–1202.

35. Hasegawa BH, Wu MC, Iwata K, et al. Applications of penetrating radiation for small animal imaging. Proc SPIE 2002; 4786:80–90.
36. Peterson TF, Kim H, et al. Semi-SPECT: a small-animal imaging system based on eight CdZnTe pixel detectors. IEEE Nucl Sci Symp Med Imag Conf Record 2003; 3:1844–1847.
37. Lee WT, Chang C. Magnetic resonance imaging and spectroscopy in assessing 3-nitropropionic acid-induced brain lesions: an animal model of Huntington's disease. Prog Neurobiol 2004; 72:87–110.
38. Chatham JC, Blackband SJ. Nuclear magnetic resonance spectroscopy and imaging in animal research. Ilar J 2001; 42:189–208.
39. Beckmann N, Hof RP, Rudin M. The role of magnetic resonance imaging and spectroscopy in transplantation: from animal models to man. NMR Biomed 2000; 13:329–348.
40. Green MV, Seidel J, Vaquero JJ, et al. High resolution PET, SPECT and projection imaging in small animals. Comput Med Imag Graph 2001; 25:79–86.
41. Hoff J. Methods of blood collection in the mouse. Lab Animal 2000; 29:47–53.
42. Meikle SR, Kench S, Kasslou M, et al. Small animal SPECT and its place in the matrix of molecular imaging technologies. Phys Med Biol 2005; 50:R45–R61.
43. Cherry SR, Sorenson J, Phelps M. Physics in Nuclear Medicine. 3rd ed. Philadelphia: WB Saunders 2003.
44. Smith MF, Meikle SR, Majewski SR, et al. Design of multipinhole collimators for small animal SPECT. IEEE Nucl Sci Symp Conf Record 2003; 4:2291–2295.
45. Schramm NU, Ebel G, Engeland U, et al. High-resolution SPECT using multipinhole collimation. IEEE Trans Nucl Sci 2003; 50:315–320.
46. Furenlid LR, Wilson DW, Chen YC, et al. FastSPECT II: a second-generation high-resolution dynamic SPECT imager. IEEE Trans Nucl Sci 2004; 51:631–635.
47. Liu Z, Barrett HH, Stevenson GD, et al. High-resolution imaging with (99m)Tc-glucarate for assessing myocardial injury in rat heart models exposed to different durations of ischemia with reperfusion. J Nucl Med 2004; 45:1251–1259.
48. Beekman FJ, Vastenhouw B. Design and simulation of a high-resolution stationary SPECT system for small animals. Phys Med Biol 2004; 49:4579–4592.
49. Beekman FJ, Vastenhouw B. Design and simulation of U-SPECT, an ultra-high resolution molecular imaging system. IEEE Nucl Sci Symp Conf Record 2002; 2:792–796.
50. Schramm NU, Schipper M, Schurrat T, et al. Performance of a multi-pinhole animal SPECT. IEEE Nucl Sci Symp Conf Record 2003; 2:2077–2079.
51. Jaszczak RJ, Li J, Wang H, et al. Pinhole collimation for ultra-high-resolution, small-field-of-view SPECT. Phys Med Biol 1994; 39:425–437.
52. Weber DA, Ivanovic M. Pinhole SPECT: ultra-high resolution imaging for small animal studies. J Nucl Med 1995; 36:2287–2289.
53. Ishizu K, Mukai T, Yonekura Y, et al. Ultra-high resolution SPECT system using four pinhole collimators for small animal studies. J Nucl Med 1995; 36:2282–2286.
54. Ogawa K, Kawade T, Nakamura K, et al. Ultra high resolution pinhole SPECT for small animal study. IEEE Trans Nucl Sci 1998; 45:3122–3126.
55. Wu MC, Hasegawa BH, Dae MW. Performance evaluation of a pinhole SPECT system for myocardial perfusion imaging of mice. Med Phys 2002; 29:2830–2839.
56. Goertzen AL, Jones DW, Seidel J, et al. First results from the high-resolution mouse-SPECT annular scintillation camera. IEE Trans Med Imag 2005; 24:863–867.
57. Mc Elroy DP, Hoffman EJ, MacDonald L, et al. Evaluation of performance of dedicated compact scintillation cameras. Nucl Med Biol 2003; 30:889–895.
58. Schramm N, Wirrwar A, Sonnenberg F, et al. Compact high resolution detector for small animal SPECT. IEEE Trans Nucl Sci 2000; 47:1163–1167.
59. Gruber GJ, Moses WW, Derenzo SE, et al. A discrete scintillation camera using silicon photodiode readout of CsI(Tl) crystals for breast cancer imaging. IEEE Trans Nucl Sci 1998; 45:1063–1068.

60. Choong WS, Moses WW, Tindall CS, et al. Design for a high-resolution small-animal SPECT system using pixellated Si(Li) detectors for in vivo ^{125}I imaging. IEEE Trans Nucl Sci 2005; 52:174–180.

61. Dilmanian FA, Weber DA, Coderre JA, et al. A high-resolution SPECT system based on a microchannel-plate imager. IEEE Trans Nucl Sci 1990; 37:687–695.

62. Aarsvold JN, Mintzer RA, Yasillo NJ, et al. A miniature gamma camera. Ann NY Acad Sci 1994; 720:192–205.

63. Bird AJ, Ramsden D. Images obtained with a compact gamma camera. Nucl Instr Meth A 1990; 299:480–483.

64. Saoudi A, Valda-Ochoa A, Mastrippolito R, et al. A miniaturized high-resolution gamma camera to image in vivo dopamine receptors in rat brain. In Hermann HJ, Wolf DE, Poppel E, eds. Supercomputer in Brain Research: From Tomography to Neural Networks. Singapore: World Scientific, 1995:99–103.

65. Yasillo NJ, Beck RN, Cooper M. Design considerations for a single-tube gamma camera. IEEE Trans Nucl Sci 1990; 37:609–615.

66. Moore RH, Ohtani H, Shaw BA. High resolution pinhole sequence of small laboratory animals. J Nucl Med 1991; 32:987–992.

67. Palmer J, Wollmer P. Pinhole emission computed tomography. Method and experimental evaluation. Phys Med Biol 1990; 35:339–350.

68. Hasegawa BH, Stebler B, Rutt BK, et al. A prototype high-purity germanium detector system with fast photon-counting circuitry for medical imaging. Med Phys 1991; 18:900–909.

69. Heanue JA, Brown JK, Hasegawa BH. The use of CdTe or CdZnTe for pulse-counting and current-mode medical imaging applications: a feasibility study. Med Phys 1996; 23:169–173.

70. Heanue JA, Hasegawa BH. Two-dimensional modeling of Cd(Zn)Te strip detectors. IEEE Trans Nucl Sci 1997; 44:701–707.

71. Kastis GA, Furenlid LR, Wilson DW, et al. Compact CT/SPECT small-animal imaging system. IEEE Trans Nucl Sci 2004; 51:63–67.

72. Barrett HH, Hunter WCJ. Detectors for small-animal SPECT. In: Barrett HH, Kupinski MA, eds. Small-Animal SPECT Imaging. New York: Springer Science & Business Media Inc, 2005:9–48.

73. Barber HB, Apotovsky BA, Augustine FL, et al. Semiconductor pixel detectors for gamma-ray imaging in nuclear medicine. Nucl Instrum Meth Phys Res 1997; 395:421–428.

74. Barber HB, Barrett HH, Augustine FL, et al. Development of a 64×64 CdZnTe array and associated readout integrated circuit for use in nuclear medicine. J Electron Mater 1997; 26:765–772.

75. Matherson KJ, Barber HB, Barrett HH, et al. Progress in the development of large-area modular 64×64 CdZnTe imaging arrays for nuclear medicine. IEEE Trans Nucl Sci 1998; 45:354–358.

76. Beekman FJ, van der Have F, Vastenhouw B, et al. U–SPECT–I: a novel system for submillimeter-resolution tomography with radiolabeled molecules in mice. J Nucl Med 2005; 46:1194–1200.

77. Hasegawa BH, Zaidi H. Dual-modality imaging: more than the sum of its parts. In: Zaidi H, ed. Quantitative Analysis in Nuclear Medicine imaging. New York: Spinger, 2006:29–64.

78. Ploux L, Mastrippolito R. In vivo radiolabel quantification in small-animal models. Nucl Med Biol 1998; 25:737–742.

79. Liang Z, Jaszczak RJ. Comparison of multiple photon coincidence imaging techniques. IEEE Trans Nucl Sci 1990; 37:1282–1292.

80. Paulus MJ, Sari-Sarraf H, Gleason SS, et al. A new X-ray computed tomography system for laboratory mouse imaging. IEEE Trans Nucl Sci 1999; 46:558–564.

81. Badea C, Hedlund LW, Johnson GA. Micro-CT with respiratory and cardiac gating. Med Phys 2004; 31:3324–3329.
82. Goertzen AL, Nagarkar V, Street RA, et al. A comparison of X-ray detectors for mouse CT imaging. Phys Med Biol 2004; 49:5251–5265.
83. Lee SC, Kim HK, Chun IK, et al. A flat-panel detector based micro-CT system:performance evaluation for small-animal imaging. Phys Med Biol 2003; 48:4173–4185.
84. Smith BD. Image reconstruction from cone-beam projections. Necessary and sufficient conditions and reconstruction methods. IEEE Trans Med Imag 1985; MI–4:14–25.
85. Feldkamp LA, Davis LC, Kress JW. Practical cone beam algorithm. J Opt Soc Am A 1984; 1:612–619.
86. Crawford CR, Gullberg GT, Tsui BMW. Reconstruction for fan beam with an angular-dependent displaced center-of-rotation. Med Phys 1988; 15:67–71.
87. Boone JM, Velazquez O, Cherry SR. Small-animal X-ray dose from micro-CT. Mol Imag 2004; 3:149–158.
88. Weichert JP, Longino MA, Bakan DA, et al. Polyiodinated triglyceride analogs as potential computed tomography imaging agents for the liver. J Med Chem 1995; 38:636–646.
89. Bakan DA, Lee FT, Weichert JP, et al. Hepatobiliary imaging using a novel hepatocyte-selective CT contrast agent. Acad Radiol 2002; 9(suppl 1):S194–S199.
90. Weber SM, Peterson KA, Durkee B, et al. Imaging of murine liver tumor using microCT with a hepatocyte-selective contrast agent: accuracy is dependent on adequate contrast enhancement. J Surg Res 2004; 119:41–45.
91. Weichert JP, Lee FT Jr, Longino MA, et al. Lipid-based blood-pool CT imaging of the liver. Acad Radiol, 1998; 5(Suppl 1):S16–S19; discussion S28–S30.
92. Shao Y, Cherry SR, Farahani K, et al. Simultaneous PET and MR imaging. Phys Med Biol 1997; 42:1965–1970.
93. Goertzen AL, Meadors AK, Silverman RW, et al. Simultaneous molecular and anatomical imaging of the mouse in vivo. Phys Med Biol 2002; 21:4315–4328.
94. Weisenberger AG, Wojcik R, Bradley EL, et al. SPECT-CT system for small animal imaging. IEEE Trans Nucl Sci 2003; 50:74–79.
95. Williams MB, Zhang G, More MJ, et al. Integrated CT-SPECT system for small animal imaging. Proc SPIE 2000; 4142:265–274.
96. Welsh RE, Brewer P, Bradley EL, et al. An economical dual-modality small animal imaging system with application to studies of diabetes. IEEE Nucl Sci Symp Med Imag Conf Record 2002; 3:1845–1848.
97. Iwata K, Wu MC, Hasegawa BH. Design of combined X-ray CT and SPECT systems for small animals. IEEE Nucl Sci Symp Conf Record 2002; 3:1849–1852.
98. Hwang AB, Hasegawa BH. Attenuation correction for small animal SPECT imaging using X-ray CT data. Med Phys 2005; 32:2799–2804.
99. MacDonald LR, Iwata K, et al. Evaluation of X-ray detectors for dual-modality CT-SPECT animal imaging. Proc SPIE 2002; 4786:91–102.
100. Varady P, Li JZ, Alden TD, et al. CT and radionuclide study of BMP-2 gene therapy-induced bone formation. Acad Radiol 2002; 9:632–637.
101. Song X, Frey EC, Tsui BMW. Development and evaluation of a microCT system for small animal imaging. IEEE Nucl Sci Symp Med Imag Conf Record 2002; 3:1600–1604.
102. Hwang AB, Iwata K, et al. Gantry specifications for a dual modality imaging system for small animals. IEEE Nucl Sci Symp Med Imag Conf Record 2003; 2:1303–1307.
103. Lackas C, Schramm NU, Hoppin JW, et al. On the development of a high-resolution multi-pinhole SPECT/CT. IEEE Nucl Sci Symp Conf Record 2004; 6:2893–3895.
104. MacDonald LR, Patt BE, Iwanczyk JS, et al. Pinhole SPECT of mice using the Luma GEM gamma camera. IEEE Trans Nucl Sci 2001; 48:830–836.
105. McElroy DP, MacDonald LR, Beekman FJ, et al. Performance evaluation of A-SPECT: a high resolution desktop pinhole SPECT system for imaging small animals. IEEE Trans Nucl Sci 2002; 49:2139–2147.

106. Euthanasia, AVAP: 2000 Report on the AVMA Panel on Euthanasia. JAMA 2001; 218:669–696.

107. Colby LA, Morenko BJ. Clinical considerations in rodent bioimaging. Comp Med 2004; 54:623–630.

108. Cantwell SL. Ferrett rabbit, and rodent anesthesia. Vet Clin North Am Exot Anim Pract 2001; 4:169–191.

109. Vanhove C, Lahoutte T, Defrise M, et al. Reproducibility of left ventricular volume andejection fraction measurements in rat using pinhole gated SPECT. Eur J Nucl Med Mol Imag 2005; 32:211–220.

110. Cavanaugh D, Johnson E, Price RE, et al. In vivo respiratory-gated micro-CT imaging in small-animal oncology models. Mol Imag 2004; 3:55–62.

111. Cody DD, Nelson CL, Bradley WM, et al. Murine lung tumor measurement using respiratory-gated micro-computed tomography. Invest Radiol 2005:40:263–269.

112. Kessler RM, Ellis JR, Eden M. Analysis of emission tomographic scan data: limitations imposed by resolution and background. J Comput Assist Tomogr 1984; 8:514–522.

113. Myers R, Hume S, Bloomfield P, et al. Radio-imaging in small animals. J Psychopharm 1999; 13:352–357.

114. Paigen K, Eppig JT. A mouse phenome project. Mamm Genome 2000; 11:715–717.

115. Turnell J, Snaith M. Transgenic systems in drug discovery. from target identification to humanized mice. Drug Discover Today 2002; 7:461–470.

116. Constantinesco A, Choquet P, Monassier L, et al. Assessment of left ventricular perfusion, volumes, and motion in mice using pinhole gated SPECT. J Nucl Med 2005; 46:1005–1011.

117. Hosokawa R, Nohara R, Fujibayashi Y, et al. Myocardial metabolism of 123I-BMIPP in a canine model with ischemia: implications of perfusion-metabolism mismatch on SPECT images in patients with ischemic heart disease. J Nucl Med 1999; 40: 471–478.

118. Fujibayashi Y, Nohara R, Hosokawa R, et al. Metabolism and kinetics of iodine-123-BMIPP in canine myocardium. J Nucl Med 1996; 37:757–761.

119. Boschi A, Uccelli L, Bolzati C, et al. Synthesis and biologic evaluation of monocationic asymmetric 99mTc-nitride heterocomplexes showing high heart uptake and improved imaging properties. J Nucl Med 2003; 44:806–814.

120. Dobrucki LW, Sinusas AJ. Molecular cardiovascular imaging. Curr Cardiol Rep 2005; 7:130–135.

121. Strauss HW, Narula J, Blankenberg FG. Radioimaging to identify myocardial cell death and probably injury. Lancet 2000; 356:180–181.

122. Blankenberg FG, Katsikis PD, Tait JF, et al. In vivo detection and imaging of phosphatidylserine expression during programmed cell death. Proc Natl Acad Sci USA 1998; 26:6349 6354.

123. Narula J, Acio ER, Narula N, et al. Annexin-V imaging for noninvasive detection of cardiac allograft rejection. Nat Med 2001; 7:1347–1352.

124. Schafers M, Riemann B, Kopka K, et al. Scintigraphic imaging of matrix metalloproteinase activity in the arterial wall in vivo. Circulation 2004; 109:2554–2559.

125. Kopka K, Breyholz HJ, Wagner S, et al. Synthesis and preliminary biological evaluation of new radioiodinated MMP inhibitors for imaging MMP activity in vivo. Nucl Med Biol 2004; 31:257–267.

126. Lu E, Wagner WR, Schellenberger U, et al. Targeted in vivo labeling of receptors for vascular endothelial growth factor. approach to identification of ischemic tissue. Circulation 2003; 108:97–103.

127. Meoli DF, Sadeghi MM, Krassilnikova S, et al. Noninvasive imaging of myocardial angiogenesis following experimental myocardial infarction. J Clin Invest 2004; 113:1684–1691.

128. Zhou R, Thomas DH, Qiao H, et al. In vivo detection of stem cells grafted in infarcted rat myocardium. J Nucl Med 2005; 46:816–822.

129. Hasegawa BH, Barber WC, Funk T, et al. Implementation and applications of dual-modality imaging. Nucl Instr Meth Phys Res 2004; 525:236–241.

130. Lyons SK. Advances in imaging mouse tumour models in vivo. J Pathol 2005; 205: 194–205.

131. Mandl SJ, Mari C, Edinger M, et al. Multi-modality imaging identifies key times for annexin V imaging as an early predictor of therapeutic outcome. Mol Imag 2004; 3:1–8.

132. Janssen BJ, Smiths JF. Autonomic control of blood pressure in mice: basic physiology and effects of genetic modification. Am J Physiol Regul Integr Comp Physiol 2002; 282:R1545–R1564.

133. Chen X, Liu S, Hou Y, et al. MicroPET imaging of breast cancer alphav-integrin expression with 64Cu-labeled dimeric RGD peptides. Mol Imag Biol 2004; 6:350–359.

134. Collingridge DR, Carroll VA, Glaser M, et al. The development of [(124)I]iodinated-VG76e: a novel tracer for imaging vascular endothelial growth factor in vivo using positron emission tomography. Cancer Res 2002; 15:5912–5919.

135. Robinson MK, Doss M, Shaller C, et al. Quantitative immuno-positron emission tomography imaging of HER2-positive tumor xenografts with an iodine-124 labeled anti-HER2 diabody. Cancer Res 2005; 15:1471–1478.

136. Winnard P Jr, Raman V. Real time non-invasive imaging of receptor–ligand interactions in vivo. J Cell Biochem 2003; 90:454–463.

137. Gambhir SS, Herschman HR, Cherry SR, et al. Imaging transgene expression with radionuclide imaging technologies. Neoplasia 2000; 2:118–138.

138. Hendrikse NH, Franssen EJ, van der Graaf WT, et al. Visualization of multidrug resistance in vivo. Eur J Nucl Med 1999; 26:283–293.

139. Bergers G, Javaherian K, Lo KM, et al. Effects of angiogenesis inhibitors on multistage carcinogenesis in mice. Science 1999; 284:808–812.

140. Acton PD, Kung HF. Small animal imaging with high resolution single photon emission tomography. Nucl Med Biol 2003; 30:889–895.

141. Andringa G, Drukarch B, Bol JG, et al. Pinhole SPECT imaging of dopamine transporters correlates with dopamine transporter immunohistochemical analysis in the MPTP mouse model of Parkinson's disease. Neuroimage 2005; 26:1150–1158.

142. Devous MD. Single-photon emission computed tomography in neurotherapeutics. NeuroRx 2005; 2:237–249.

143. Hashizume K, Tsuda H, Hodozuka A, et al. Clinical and experimental studies of epilepsy associated with focal cortical dysplasia. Psychiatr Clin Neurosci 2004; 58:S26–S29.

144. Garcia-Alloza M, Bacskai BJ. Techniques for brain imaging in vivo. Neuromol Med 2004; 6:65–78.

145. Mathis CA, Wang Y, Klunk WE. Imaging beta-amyloid plaques and neurofibrillary tangles in the aging human brain. Curr Pharm Des 2004; 10:1469–1492.

146. Saji H, Iida Y, Kawashima H, et al. In vivo imaging of brain dopaminergic neurotransmission system in small animals with high–resolution single photon emission computed tomography. Anal Sci 2003; 19:67–71.

147. Ziegler SI, Development of a small-animal PET system. Proc IEEE 2005; 93:763–770.

148. Hasegawa BH, Gingold EL. Description of a simultaneous emission–transmission CT system. Proc SPIE 1990; 1231:50–60.

149. Heanue JA, Boser BE, Hasegawa BH. CMOS detector readout electronics for an emission–transmission imaging system. IEEE Trans Nucl Sci 1995; 42:1133–1138.

150. Boles CD, Boser BE, Hasegawa BH, et al. A multimode digital detector readout for solid-state medical iamging detectors. IEEE J Solid-State Circuits 1998; 33:733–742.

Index